SEDIMENTARY PETROLOGY

by

W. v. ENGELHARDT, H. FÜCHTBAUER, G. MÜLLER

PART III

THE ORIGIN OF SEDIMENTS AND SEDIMENTARY ROCKS

by

Prof. Dr. WOLF v. ENGELHARDT

Mineralogical-Petrographic Institute of the University of Tübingen

translated by

Prof. WILLIAM D. JOHNS

Dept. of Geology, University of Missouri-Columbia

Second revised edition

With 134 figures and 55 tables in the text

1977

E. Schweizerbart'sche Verlagsbuchhandlung
(Nägele u. Obermiller) Stuttgart

A Halsted Press Book
John Wiley & Sons, New York - Toronto - Sydney

Library of Congress Cataloging in Publication Data (Revised)

Engelhardt, Wolf, Freiherr von.
    Sedimentary petrology.
    Translation of Sediment-Petrologie.
    Pts. published in U.S. by Halsted Press, New York.
    Vol. 1 has pt. of illustrative matter in pocket.
    Includes bibliographies.
    CONTENTS: pt. 1 Methods in sedimentary petrology, by G. Müller. — pt. 2. 1. Sediments and sedimentary rocks, by H. Füchtbauer, with a contribution by H.-U. Schmincke. 2d rev. and enl. ed. 1974. [etc.]
    1. Rocks, Sedimentary. I. Füchtbauer, H., joint author. II. Müller, German, joint author. III. Title.

Library of Congress Cataloging in Publication Data
QE471.E523   552'.5   67-28575
ISBN 0-470-99142-9

First edition (in German) © E. Schweizerbart'sche Verlagsbuchhandlung (Nägele u. Obermiller) Stuttgart 1973
Second edition © E. Schweizerbart'sche Verlagsbuchhandlung (Nägele u. Obermiller) Stuttgart 1977
All rights reserved, including translations into foreign languages.
This book, or parts thereof, may not be reproduced in any form without permission from the publishers.
Published in the U.S.A., Canada, Australia by Halsted Press a Division of
John Wiley & Sons, Inc., New York
In all other countries: E. Schweizerbart'sche Verlagsbuchhandlung (Nägele u. Obermiller) Stuttgart.
ISBN 3-510-65077-8

Design of the jacket: Wolfgang Karrasch
Printer: Gebrüder Ranz, Dietenheim
Printed in Germany

Dedicated to

CARL W. CORRENS

FOREWORD TO PART III

In the second part of *Sedimentary Petrology,* published in 1970, H. Füchtbauer and G. Müller described the mineralogical compositions, the textures and the conditions for the formation of sediments. The latter were based on petrographic evidence, geologic association, and observations from recent sedimentary environments. They considered also the diagenetic transformation of different deposits into solid rocks which form the uppermost part of the earth's crust. After this systematic presentation of sediments and sedimentary rocks, an attempt is made in the third part of this series to describe and explain comprehensively the most important processes by which sediments are formed and sedimentary rocks derived.

While the presentation in the second part starts from the descriptive picture of sediments and sedimentary rocks and their natural occurrence, the aspect under which the present volume deals with sedimentary petrology, presupposes this petrographic knowledge. This presentation has the objective in mind to explain what happens in the sedimentary domain at the earth's surface and in the uppermost parts of the earth's crust by physical and chemical principles. The purpose is to relate, as far as possible, the extremely complex natural processes to simple models and to show the resulting possibilities for quantitative treatment of sedimentary petrological problems.

Limitation to the general principles of sediment- and rock-forming processes determines the relation of this volume to the second part of this series. On the one hand the presentation, even where not expressly indicated, is based on the systematic and detailed descriptions in the second part. On the other hand some phenomena and processes already considered there, are repeated and discussed from the viewpoint of physical and chemical interpretation and quantitative treatment. Since sedimentary petrology is not a closed system, but still a fully evolving science, this leads in some cases to other definitions of concepts, occasionally also to other proposed interpretations.

The analysis of processes follows the natural course of events. Solid rocks experience chemical alteration by the influence of water and atmosphere at the earth's surface and are broken down into loose masses. The products of this weathering are transported by wind and water. Solid particles are separated from the solutions and are sorted according to shape and size with continuing change in shape and composition. In the continental and marine depositional environments these clastic masses accumulate. Partially through the influence of organisms, other constituents precipitate out of solution. As a result of geologic processes, these sediments subside and are buried at greater depths. Thereby they experience textural changes as well as changes in their chemical and mineralogical composition.

Accordingly the presentation is organized into five main chapters which deal with parent materials, weathering, transport and deposition of clastic constituents by waters and wind, the formation of chemical sediments, and diagenesis.

The book is directed to advanced students of all disciplines of geoscience, and as well to those who work on or carry on research dealing with problems whose solutions depend on the knowledge of how sediments originate and become solid rocks. It was attempted to present a picture which corresponds to the present state of inquiry. If this was not accomplished always to the extent desirable, may the reader excuse the author in not being able in all areas, in writing the manuscript, to keep abreast of the rapid developments in observational data and theory.

The author thanks H. FÜCHTBAUER for his careful review of the manuscript and for numerous fruitful suggestions.

My thanks go to Mr. J. NÄGELE and Dr. E. NÄGELE of the E. Schweizerbart'sche Verlagsbuchhandlung in Stuttgart, for suggesting this book, for their generous patience during the preparation of the manuscript and their cooperation in composing the volume.

The first edition of this book appeared in German in the fall of 1973. A number of minor additions were made to this present second edition.

I am grateful to Professor WILLIAM D. JOHNS for his careful translation.

Tübingen, February 1977   W. v. ENGELHARDT

TABLE OF CONTENTS

1. Parent Materials 1

2. Subaerial Weathering 8
   - 2.1 General 8
   - 2.2 Breakdown of primary rocks and minerals into loose materials 9
   - 2.3 Chemical decomposition 16
     - 2.31 General 16
     - 2.32 Solution processes 21
     - 2.33 Formation of new minerals 33
       - 2.331 Oxides and hydrated oxides 33
       - 2.332 Clay minerals 39
         - 2.332.1 Classification of clay minerals 39
         - 2.332.2 Alteration of primary layer silicates to clay minerals . . . 41
         - 2.332.3 Neoformation of clay minerals 44
   - 2.4 Soils 46
     - 2.41 General 46
     - 2.42 Factors in soil formation 47
     - 2.43 Soil types 48

3. Transport and Deposition of Clastic Constituents 52
   - 3.1 Transport and deposition in water 52
     - 3.11 General 52
     - 3.12 Laminar and turbulent flow 53
     - 3.13 Transport in suspension (suspended load) 58
     - 3.14 Bottom transport (bed load) 70
     - 3.15 Transport capacity of streams 77
     - 3.16 Marine transport processes 81
       - 3.161 General 81
       - 3.162 Transport by ocean currents 81
       - 3.163 Wave transport 82
       - 3.164 Submarine landslides and suspension currents 92
     - 3.17 Changes during transport and deposition 99
       - 3.171 Mechanical effects of water transport 99
         - 3.171.1 General 99
         - 3.171.2 Definition of form and roundness 99
         - 3.171.3 Experiments on the mechanical effects of water transport . 103
         - 3.171.4 Observations of the mechanical effects of water transport . 109
         - 3.171.5 Summary 113

                3.172 Chemical changes during transport and deposition:
                      subaquatic weathering 114
                      3.172.1 General 114
                      3.172.2 Subfluvial weathering 115
                      3.172.3 Submarine weathering 116
           3.18 Deposition in water 124
                3.181 Grain size distribution 124
                3.182 Streams 129
                3.183 Deltas 136
                3.184 Sea coasts 137
                3.185 Lakes and sea 141
      3.2 Transport and deposition in air 145
           3.21 General 145
           3.22 Transport in suspension (suspended load) 146
           3.23 Bottom transport (bed load) 149
           3.24 Transport capacity of winds 155
           3.25 Mechanical effects of aeolian transport 157
           3.26 Deposition in air 158
      3.3 Distinguishing aeolian and aquatic clastic sediments 161
      3.4 Transport and deposition by ice 164

4. Formation of Chemical Sediments 166

      4.1 General 166
      4.2 Limestone 168
      4.3 Dolomite 182
      4.4 Gypsum and anhydrite 185

5. Diagenesis 188

      5.1 General 188
      5.2 Pore space in sediments and sedimentary rocks 189
           5.21 General 189
           5.22 Porosity 189
           5.23 Homogeneous flow processes 193
           5.24 Heterogeneous flow processes and equilibria in pore space 203
      5.3 Formation waters 208
           5.31 General 208
           5.32 Composition of formation waters 210
           5.33 Diagenesis of formation waters 231
      5.4 Diagenesis of sands 239
           5.41 Mechanical diagenesis 239
           5.42 Chemical diagenesis 241
      5.5 Diagenesis of clays 262
           5.51 Mechanical diagenesis (compaction) 262
           5.52 Chemical diagenesis 293
      5.6 Diagenesis of carbonates 304
           5.61 Mechanical diagenesis 304
           5.62 Isochemical diagenesis 306

           5.621 General 306
           5.622 Exogenic isochemical diagenesis 308
           5.623 Endogenic isochemical diagenesis 312
           5.624 Stylolites 313
           5.625 Transformation of Mg-calcite and aragonite into calcite 315
           5.626 Recrystallization of calcite 317
   5.63 Allochemical diagenesis 318
           5.631 Dolomitization and dedolomitization 318
           5.632 Formation of SiO_2-concentrations (chert) 324
           5.633 Neoformation of silicates 327

References and Authors' Index 329

Subject Index 353

1. Parent Materials

The parent materials which are involved in sediment-forming processes are, on the one hand, igneous and metamorphic rocks (primary rocks), and on the other, older sedimentary rocks whose constituents have already been subject one or more times to sedimentary processes, without having been transformed by high temperatures and pressures.

The abundances of the igneous rock-forming minerals are given in table 1-1, from data gathered by BARTH. The first column relates to igneous rocks of the Oslo region, where primarily syenitic-granitic intrusives and their extrusive equivalents occur. The values given were calculated, based on the mineralogical composition of the individual rock types and taking into consideration their areal distributions. The second column is based on lavas from the Pacific islands, assuming proportions of 95% basalt and 5% phonolite-trachyte. The data in the third column have been calculated from the average composition of all igneous rocks as estimated by CLARKE & WASHINGTON (1924).

Table 1-1. Abundance of igneous rock-forming minerals (after BARTH, 1949): In weight percent.

	Plutonic and volcanic rocks of the Oslo region	Volcanic rocks of the Pacific Islands	Average of all igneous rocks
Quartz	8	3.5	12.4
Alkali feldspar	62	18	31.0
Plagioclase	16	30	29.2
Pyroxene	5	27	12.0
Hornblende	2	—	1.7
Biotite	2	—	3.8
Muscovite	0.2	—	1.4
Olivine	0.3	9.5	2.6
Nepheline	0.3	2.2	0.3
Ores	2	7.4	4.1
Others	2.2	2.4	1.5

Similar compilations for the average mineral content of metamorphic rocks have not been made. They would be most difficult to determine, since the mineral composition of a metamorphic rock depends not only upon its chemical composition, but also upon the pressure and temperature conditions of metamorphism. Qualitatively it can be said, however, that metamorphic rocks contain less feldspars and pyroxenes than igneous rocks, and thus more amphiboles and micas. In addition they contain a variety of minerals, such as chlorites, epidote, carbonates, garnets, and others which are absent or rare in igneous rocks.

Estimates are available of the relative proportions of each of the major sedimentary rock types, claystones, sandstones, and carbonates, based on geochemical calculations. A study by WICKMAN (1954) led to the following values, which compare favorably with the estimates of other authors:

        Claystones and clays . . . . 80—82%
        Sandstones . . . . 8—12%
        Carbonates . . . . 8—10%

Sandstones contain appreciable quartz as well as lesser amounts of feldspars, micas, and rock fragments. The claystones consist predominantly of phyllosilicates: in addition to mica they contain different alkali-free minerals of the chlorite, kaolinite, and montmorillonite families. Carbonate rocks consist predominantly of calcite and/or dolomite.

Simple qualitative comparison of the mineralogy of igneous, metamorphic, and sedimentary rocks reveals the fundamental changes which the former experience as a result of sediment-forming processes. The feldspars, which are so abundant in igneous and in many metamorphic rocks, are largely lost in sediments, as are the pyroxenes and amphiboles. Accordingly, the phyllosilicates play a governing role in sediments. Many of these phyllosilicates may originate from metamorphites. However, it should be borne in mind that sedimentary micas belong almost exclusively to the dioctahedral types, whereas in metamorphic, as in igneous rocks, trioctahedral micas are by far more common. Also the dioctahedral micas in sediments belong for the most part to the group of deficient micas, the illites, which contain less potassium and have a higher Si:Al-ratio than normal muscovite. When it has not formed anew during the process of sediment formation, illite can only have been formed as a result of chemical conversion of primary mica. For the most part kaolin and montmorillonite minerals and a portion of the chlorites and carbonates are formed anew during the sediment-forming process.

In order to understand weathering and transport processes, it is interesting to inquire about the grain size distribution of the most important minerals in the primary source rocks. In the case of metamorphic rocks this question cannot be answered realistically, since the grain sizes depend on the nature of the parent rock as well as the kind and grade of metamorphism. Therefore, metamorphic mineral grain sizes vary over wide ranges. Likewise extrusive rocks are not suitable for such comparisons, since glassy and very fine-grained groundmasses predominate. For the more coarsely crystalline intrusive rocks a study by FENIAK (1944) is informative. In this investigation 92 acid (granites, syenites), 55 intermediate (nepheline syenites, monzonites, diorites) and 53 basic rocks (gabbros, ultrabasites) were studied. By means of microscopic measurements from thin sections, maximum and minimum diameters were determined for each individual mineral grain. For each mineral species values for the arithmetic mean diameters were calculated. The results are summarized in table 1—2.

Table 1-2. Average grain sizes of intrusive rock minerals (after Feniak 1944).

Mineral	Acid rocks					Intermediate rocks					Basic rocks				
	(n)	d_1	d_2	F	$\bar{d}$	(n)	d_1	d_2	F	$\bar{d}$	(n)	d_1	d_2	F	$\bar{d}$
Quartz	(88)	0.85	0.55	0.468	0.67	(26)	0.49	0.34	0.166	0.41	(4)	0.39	0.26	0.101	0.32
Orthoclase	(76)	1.25	0.80	1.00	1.00	(21)	1.4	0.75	1.05	1.03	(5)	1.6	0.90	1.44	1.20
Microcline	(49)	1.15	0.75	0.87	0.94	(4)	2.05	1.15	2.36	1.54	—	—	—	—	—
Microperthite	(7)	1.8	1.05	1.89	1.38	(2)	1.95	0.75	1.46	1.21	—	—	—	—	—
Plagioclase	(87)	1.3	0.75	0.975	0.99	(48)	1.1	0.60	0.66	0.81	(43)	1.3	0.60	0.78	0.88
Nepheline	—	—	—	—	—	(10)	1.1	0.85	0.935	0.97	(2)	0.80	0.45	0.36	0.60
Muscovite	(11)	0.445	0.23	0.103	0.32	(2)	0.26	0.103	0.027	0.17	—	—	—	—	—
Biotite	(62)	0.70	0.31	0.217	0.47	(36)	0.70	0.375	0.262	0.51	(16)	0.70	0.38	0.266	0.52
Hornblende	(38)	0.80	0.48	0.384	0.62	(27)	0.85	0.50	0.425	0.65	(9)	0.95	0.55	0.525	0.73
Pyroxene	(3)	0.77	0.44	0.339	0.58	(27)	0.824	0.421	0.347	0.59	(35)	1.114	0.696	0.697	0.84
Olivine	—	—	—	—	—	(2)	0.50	0.37	0.185	0.43	(26)	0.9	0.6	0.54	0.74
Zircon	(43)	0.063	0.042	0.0026	0.053	(7)	0.066	0.046	0.003	0.050	(4)	0.057	0.035	0.002	0.045
Orthite	(6)	0.20	0.108	0.0216	0.15	(3)	0.43	0.29	0.125	0.35	(1)	0.335	0.195	0.065	0.26
Sphene	(20)	0.23	0.114	0.0262	0.16	(20)	0.31	0.154	0.048	0.22	(2)	0.20	0.132	0.0264	0.16
Apatite	(58)	0.12	0.05	0.006	0.089	(51)	0.17	0.07	0.012	0.11	(27)	0.23	0.10	0.023	0.15
Fluorite	(5)	0.35	0.22	0.077	0.28	(1)	0.17	0.085	0.0145	0.12	—	—	—	—	—
Magnetite	(66)	0.185	0.123	0.023	0.15	(42)	0.185	0.14	0.026	0.16	(47)	0.37	0.28	0.103	0.32
Pyrite	(3)	0.137	0.088	0.012	0.11	(6)	0.23	0.17	0.039	0.20	(8)	0.20	0.132	0.026	0.16

(n) Number of rocks studied; d_1 max. aver. diameter in mm; d_2 min. aver. diameter in mm; $F = d_1 \times d_2$ in mm^2; $\bar{d} = \sqrt{F}$ = average diameter in mm.

Since igneous extrusive and metamorphic rocks consist on the average of smaller grains, the data in table 1—2 can be considered representative of the maximum values that can be expected for the clastic mineral grain sizes in sedimentary rocks. For the most important clastic constituents, feldspar and quartz, these maximum values lie between 1.5 and 0.7 mm, that is, in the range of coarse sand. Actually the preponderance of sediments are finer-grained than coarse sand.

In addition to absolute grain size of primary minerals, the relative sizes are also important. It is concluded from table 1—2 that the feldspars form the largest primary grains. In quartz-rich rocks quartz follows in grain size. Among the dark minerals, hornblende and pyroxenes are the largest crystals. Grains of zircon are especially small.

The parent material for a definite sedimentation sequence can consist of quite varied proportions of igneous, metamorphic and sedimentary rocks. It has always been one of the goals of sedimentary petrology to deduce the nature of the contributing parent material from the mineral composition of a sediment. This can be done primarily with the help of characteristic guide minerals. They must be able to survive the chemical and mechanical stresses of sediment formation. An exact understanding of the effects of sedimentary processes on individual minerals is just as important for correctly reconstructing the parent rocks, as is knowledge of the occurrence and development of characteristic guide minerals in the primary rock. Generally, it can be said that the kind of evidence pertaining to parent materials which survives, depends upon the kind of sediment and its previous history. The more intensive the action of weathering, the longer the path of transport, the more intensive the diagenetic alteration, the more obliterated are the clues to the parent material. In coarse-grained clastic rocks, characteristic guide minerals can be found and recognized more readily than in fine-grained clays. The latter consist predominantly of newly formed minerals resulting from weathering. Guide minerals are the more useful, the better they can resist weathering and selective transport.

Without anticipating the detailed discussion in following sections on the effects of weathering, transport, and diagenesis on the primary minerals, some important information on guide minerals is summarized in the next section.

Q u a r t z, because of its high resistance to chemical agents as well as mechanical stresses, could be well suited as a guide mineral, provided it is possible to establish diagnostic criteria which are characteristic of specific parent rocks. Examples of such criteria which have been the subject of investigation are crystal morphology, inclusions, color, undulose extinction and the nature of so-called quartz overgrowths (see FÜCHTBAUER in part II/1 of this series).

The morphology of quartz grains in sediments is of little diagnostic value. Quartz grains in many sands and sandstones often exhibit elongate forms, in which the long axis of the grain coincides with the c-axis or lies in the plane of the unit rhombohedron (10$\bar{1}$1) (SCHUMANN 1941, BLATT & CHRISTIE 1963). It had been suggested that this was derived from the primary morphology of quartz in crystalline schists. It has been shown, however, that there are sediments derived from metamorphic rocks which contain more rounded quartz and also that some sediments with elongated quartz have to have been derived from plutonic rocks. Elongate morphology of sediments quartz is, therefore, an uncertain index of origin. Also in many cases it probably arises during transport as a result of preferential fracture along $\{10\bar{1}1\}$ and of the anisotropic resistance to abrasion.

It has been assumed repeatedly than one could infer the origin of sedimentary quartz from the presence of undulose extinction. It was assumed that quartz in igneous rocks should show little or no undulose extinction. Marked extinction should be a characteristic criteria of quartz from crystalline schists. As the result of systematic study by BLATT & CHRISTIE (1963), these assumptions were proved to be incorrect. They studied 51 plutonic, 18 volcanic, and 50 metamorphic rocks as well as 44 sediments of different ages. Quartz with undulose extinction occurs frequently and just as often in plutonic as in metamorphic rocks. No important distinction occurred. On the other hand such quartz is very rare in volcanic rocks. Among the quartz-bearing sediments studied the pure quartz orthoquartzites were almost lacking in undulose quartz, whereas in other sandstones, arkoses, and graywackes, undulose quartz was found with about the same frequency as in plutonic and metamorphic rocks. BLATT & CHRISTIE concluded from these observations that quartz with undulose extinction is mechanically less stable and is preferentially destroyed during weathering and transport. Thus highly abraded sediments such as orthoquartzites, whose constituents are especially thoroughly worked and reworked, contain very little undulose quartz.

Quartz which exhibits an overgrowth of new quartz deposited on a rounded quartz grain originates as a rule from older sediments, if it can be established that the phenomenon did not take place following deposition of the sediment being studied. This results during the course of diagenesis from the deposition of secondary quartz in the pore space of the sediment (see section 5.42). Such quartz with older overgrowths, therefore, indicates that a sediment consists of older sedimentary material.

Minerals of the f e l d s p a r g r o u p are useful as indicators of parent rocks only in special cases, since feldspars are strongly subjected to chemical and mechanical destruction during weathering and transport. Thus a deficiency of feldspar tells us nothing. High feldspar content always suggests a non-sedimentary source material. In these cases the identity of the parent rock can be deduced from the kind of feldspar.

As means of stratigraphic classification of unfossiliferous clastic sediments, as a means of correlation, and as guide minerals for the recognition of provenance of origin, accessory minerals of high specific weight, the so-called h e a v y m i n e r a l s have been called upon especially. To the extent that these minerals are sufficiently resistant to chemical and mechanical destruction and are characteristic of parent materials, their presence in a sediment permits us to conclude something about the origin of sedimentary materials. In the following section some of the more important heavy minerals are discussed briefly and are considered alphabetically.

A m p h i b o l e minerals succumb relatively easily to mechanical, as well as chemical destruction. However, they can be sufficiently abundant in some clastic sediments to serve as guide minerals. Common hornblende occurs in igneous rocks, especially in granites, syenites, and diorites and not in their extrusive equivalents. In regional metamorphic rocks of the amphibolite facies and in contact metamorphic rocks of the hornblende-hornfels facies it is very widely distributed. Basaltic hornblende is found in alkali-rich syenitic plutonic rocks, in hornblende-gabbros and in many low SiO_2-volcanites (andesites, basalts). Glaucophane occurs in crystalline schists of the glaucophane facies. Riebeckite occurs in alkali-rich plutonics (granite, syenite), corresponding volcanics, and also in contact metamorphic rocks and in gneisses. Arfvedsonite is common in nepheline syenites. Tremolite and minerals of the tremolite-actinolite series occur in crystalline schists of the greenschist facies.

Anatase is not a characteristic guide mineral, since it can originate from a variety of igneous and metamorphic rocks. It also often forms during weathering or diagenesis.

Andalusite occurs primarily in contact metamorphic rocks of the hornblende and pyroxene-hornfels facies, as well as in the low-pressure regional metamorphic rocks of the cordierite-amphibolite facies. It is also found in some granites.

Apatite can be destroyed easily by acid solution during the course of weathering and diagenesis. Still there are some clastic sediments which nevertheless contain apatite grains. As a primary mineral apatite is found in all igneous rocks and is likewise common in regional and contact metamorphic rocks.

Brookite, like anatase, is of little value as a guide mineral. As a primary mineral it is found especially in acid igneous rocks and in a variety of crystalline schists. Brookite also can form anew in sediments.

Cordierite occurs primarily in contact metamorphic rocks of the pyroxene and hornblende-hornfels facies, as well as the low-pressure regional metamorphic rocks of the cordierite-amphibolite facies. Cordierite is found also in silica-rich igneous rocks.

Kyanite is a typical mineral of regional metamorphic rocks. It is also found in many granulites and eclogites.

Of the minerals of the epidote-zoisite group, epidote especially occurs in Ca-rich crystalline schists of the greenschist facies, but is found also in many granites. Allanite occurs in SiO_2-rich plutonics, but also in crystalline schists. Clinozoisite and zoisite are found in crystalline schists which also carry epidote, especially metamorphosed basic rocks.

From the garnet group almandine and almandine-rich garnets, such as grossularite, spessartite, and andradite occur in regional metamorphic rocks of the amphibolite facies. Pyrope-rich garnets are found in rocks of the granulite facies; eclogites are particularly pyrope-rich. In contact metamorphic rocks of the hornblende-hornfels facies grossularite and spessartite occur. Melanite is typical of alkali-rich, low silica plutonic and eruptive rocks.

Ilmenite occurs as a primary mineral in all igneous rocks. It is widely distributed also in metamorphic rocks.

Likewise, magnetite occurs in all igneous and metamorphic rocks.

Pyrrhotite is found in all igneous rocks, especially those of low silica content, as well as in many metamorphites. It can also form anew in sediments.

Monazite occurs in SiO_2-rich igneous rocks, especially in granites, and also in quartz- and mica-rich crystalline schists.

Olivine is a typical mineral of low SiO_2-igneous rocks. Ultrabasic rocks (dunites, peridotites) are especially olivine-rich. Many basalts and gabbros contain olivine. Because of its chemical instability it rarely persists in sediments.

Orthite (allanite) occurs in granite pegmatites and is of pneumatolytic to high hydrothermal origin in different contact metamorphic rocks, but rarely in plutonic and eruptive rocks. It is found, however, in regional metamorphic rocks of all pressure and temperature grades. Because of its high mechanical and chemical durability, orthite, in spite of its quantitative insignificance in primary rocks, is a very frequent constituent of the heavy mineral fraction of sediments.

Pyroxene minerals are generally quite unimportant among the heavy minerals, since among the primary minerals, they are less common than the amphiboles and in addition are not especially mechanically or chemically resistant. Orthopyroxenes occur in basic and ultrabasic igneous intrusives and their corresponding extrusives, as well as in contact metamorphic rocks of the pyroxene-hornfels facies and in regional metamorphic rocks of the granulite facies. Augite is found in syenite, diorites, gabbros and the corresponding volcanic equivalents. Aegirine is typical of alkali-rich igneous rocks, especially nepheline syenites and phonolites. Diopside occurs in different igneous plutonics (pyroxene granites, syenites, diorites, gabbros), in basalts, in contact metamorphic rocks of the pyroxene-hornfels and hornblende-hornfels facies, as well as regional metamorphic rocks of the amphibolite and granulite facies. Spodumene is found primarily in granites and granite pegmatites, but also in crystalline schists.

Rutile occurs in igneous rocks, accessory to and most likely as a secondary alteration product of magnetite and ilmenite. It is found in pegmatites and especially in crystalline schists, granulites and eclogites, as well as in contact metamorphic rocks. It is especially resistant to weathering.

Sillimanite is found in contact metamorphic rocks of the pyroxene-hornfels facies, as well as in regional metamorphic rocks of the amphibolite facies, corresponding to the highest temperature of formation.

The spinel minerals are high temperature phases which occur in low SiO_2-igneous rocks, in contact metamorphic rocks of the highest temperature grade (sanidinite facies, calc-silicate fels) and in regional metamorphic rocks of the catazone.

Staurolite occurs in regional metamorphic rocks of the almandine-amphibolite facies, as a phase in the low temperature grades.

Sphene is distributed in granitic, dioritic, syenitic and nepheline syenitic plutonic rocks and their extrusive equivalents. It occurs, moreover, in regional metamorphic as well as contact metamorphic rocks.

Topaz occurs in pneumatolytic modified granites and in their country rocks.

Tourmaline occurs in diorites, syenites, granites, and is especially abundant in pneumatolytic altered granites and country rocks. It is rarely found in SiO_2-rich volcanic rocks. It is widely distributed in contact metamorphosed shales. It is found also in regionally metamorphosed crystalline schists. Tourmaline can form diagenetically in sediments.

Vesuvianite is found in contact metamorphic Ca-rich rocks, especially in the highest temperature pyroxene-hornfels facies and occasionally also in metamorphic rocks of the greenschist facies.

Wollastonite occurs primarily in contact metamorphic, Ca-rich rocks of the hornblende-hornfels and pyroxene-hornfels facies. It occurs also in limestone members of crystalline schists.

Xenotime occurs in granites, granite pegmatites, and nepheline syenites, and analogous gneisses. Because of its high mechanical and chemical stability, it is not an uncommon constituent of the heavy mineral fraction of sediments.

Zircon is widespread in granites, syenites, and diorites, as well as volcanic equivalents of these rocks. Frequently it is found in metamorphic rocks, probably usually as relicts from igneous and sedimentary rocks. It is especially resistant to weathering. According to HOPPE (1962, 1963) the morphology of zircon can serve to differentiate its origin: zircons from granite develop predominantly the form $\{100\}$ and $\{111\}$ and to some extent $\{310\}$. In quartz porphyries the dominant combination is $\{110\}$ and $\{111\}$. Rounded zircons also occur in igneous rocks, and are typical for metamorphites. The zircons in granulites are most highly rounded.

2. Subaerial Weathering

2.1 General

The term weathering includes all changes that rocks and minerals undergo in contact with the atmosphere and hydrosphere. By subaerial weathering we mean all processes which take place at the rock-atmosphere interface. Subaquatic weathering includes weathering processes at the rock-water interface and will be dealt with in a later section (3.172).

The level at which subaerial weathering processes take place is bounded above by the free atmosphere and extends downward to where the action of atmosphere and surface water is still effective. Especially in sedimentary rocks which contain continuous pore space, it is in certain cases difficult to sharply define this lower boundary. That ground water zone in which there is still noticeable exchange of material with surface water and atmosphere, which participates in fluid movement, and which is taxed and fed by precipitation or evaporation, must also be included in the weathering zone. The deeper zone which is isolated from moving ground water belongs to the zone of diagenesis, which extends in turn only as far at depth as continuous pore space exists.

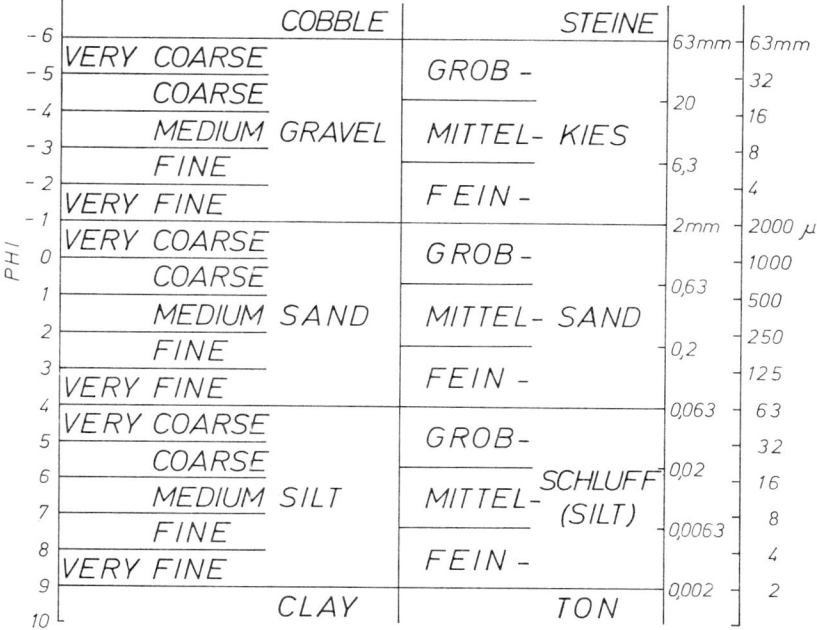

Fig. 2-1. Grain size nomenclature. At left the WENTWORTH scale with modifications by DOEGLAS (1968), at right the DIN 4022 scale used in Germany.

We distinguish between the breakdown of primary rocks into loose materials which takes place without significant change in chemical composition, and breakdown resulting in substantial change of primary minerals by chemical reaction. In each actual weathering situation both processes are active, but to variable extents. In further chapters the newly formed minerals which result from weathering and the most important soil types produced will be considered. Fig. 2-1 summarizes the grain size classification which will be utilized in subsequent discussion throughout this book.

2.2 Breakdown of primary rocks and minerals into loose materials

Where solid rock is exposed at the surface, the compact mass breaks down into fragments, ranging in size from boulders to fine gravel. The manner and intensity of disintegration depends upon climatic relations, so that characteristic relationships can be established between weathering-determined morphology and climatic zones (WIL-HELMY 1958).

In the humid climate of the tropics and in the cooler latitudes very thick zones of loose debris often form. This material is significantly altered in its chemical make-up from the composition of the parent rock by the action of continually renewed precipitation. In this case chemical processes play an important role in the observed disintegration. Quite different effects are observed in year round arid desert regions and in nival climates of the high latitudes and altitudes. Rocks exposed there become covered during the course of time with a mantle of boulders, rocks, or fine gravel, although fluid water is absent or only rarely present. Fragmentation of solid rock in this case can be attributed only rarely to the solution and decomposition action of precipitation permeating into the weathering zone.

The breakdown of compact rock into a loose mass is, therefore, a general weathering phenomenon in all climatic zones. All plutonic igneous rocks, all metamorphic rocks, and most sedimentary rocks have acquired their coherence and strength in the deeper parts of the crust where high pressures and temperatures prevail. Volcanic rocks likewise crystallize at high temperatures. At the earth's surface all solid rocks find themselves subject to conditions quite different from those under which they originally formed. Under these new conditions, they are no longer stable and break down into loose masses. The question arises as to what forces and mechanisms bring about these changes.

First of all rocks experience mechanical effects in that they are brought from conditions of higher temperature and higher hydrostatic pressures to those which prevail at the earth's surface. Stresses arise, in the case of plutonic rocks from relief of pressure, and in the case of volcanics as a result of cooling. This causes internal fracture, both on a large scale and microscopically. The frequently observed sheeting of rock masses, which follows the earth's contours, especially on mountain and valley slopes, is the result of relief forces which give rise to fracture running parallel to the surface. Also the surface parallel exfoliation, which led to the unusual crystalline rock formations of Sugarloaf Mountain and Glockenberg of Brazil, has been attributed to relief from pressure (WILHELMY 1958). Frequently, laminated rock breakdown as a result of pressure relief is observed in mines and tunnel excavations.

The macroscopic joints produced by relief of pressure and cooling have their microscopic counterparts in fine fractures and joints. It can be presumed that most, perhaps all primary rocks occurring at the earth's surface, even when they outwardly appear solid, are permeated by a more or less dense network of internal fracture surfaces before the onset of weathering.

Each fracture surface in a solid material has a corresponding specific free surface energy (erg/cm^2), which is equal to the mechanical work which must be expended to produce 1 cm^2 of surface. In a homogeneous volcanic glass the specific surface energy is uniform and independent of direction (isotropic). On the other hand, for a polycrystalline rock the specific surface energy is different for each mineral and dependent upon direction within each mineral grain. Cleavage planes are planes of especially low surface energy. Grain boundaries are also generally surfaces of lower surface energy. As a result of mechanical stress, fracture surfaces of low surface energy are preferentially formed. Therefore, fractures formed by unloading and weathering develop preferential orientation parallel to cleavage planes of minerals (feldspars, micas) or to grain boundaries. The loose mantle resulting from rock disintegration consists, therefore, largely of monomineralic grains. The excellent cleavage of feldspar results in finer grain sizes than non-cleavable quartz, even though feldspar grains in primary rocks are often larger than their quartz grains.

Numerical values for the specific free surface energies of the rock-forming minerals are not available. In any case they lie in the range of several hundred to a few thousand erg/cm². They are considerably higher than the surface energy of water (76 erg/cm²). Water molecules are absorbed, therefore, on fresh fracture surfaces of these rock-forming minerals, where, as in all cases of adsorption, the specific free surface energy is diminished. A limiting situation would result when the surface energy of the boundary layer is equal to that of water.

Adsorption of water on the internal fracture surfaces of rocks is capable of producing mechanical stresses, and is, therefore, an important factor in weathering. On a clean surface the adsorbed water layers spread out in all directions like a two-dimensional gas, with a definite pressure, called the spreading pressure. In this way adsorbed water can penetrate wedge-shaped fractures, causing them to be forced open farther. Solid bodies with high internal surface experience measurable deformation as a consequence of the mechanical effects of expanding surfaces upon adsorption. This deformation has been measured in the case of charcoal substances by BANGHAM and co-workers (1932, 1934) and in the case of porous glass rods by YATES (1954) and AMBERG & McINTOSH (1952). The expansion of these substances by adsorption of water vapor and other gases corresponds to the well-known swelling of many organic and inorganic gels (for example, clays). Although no such measurements have been made on rocks, it can be presumed with some certainty, that similar mechanical effects do occur, resulting from the adsorption of water vapor.

Values for spreading pressures can be measured, based on a simple relationship which applies to spreading thin films on fluid surfaces (see, for example, WOLF, I., p. 171):

$$\pi = \sigma_0 - \sigma_a \; \{dyn/cm\} \tag{2.1}$$

The spreading pressure π is equal to the difference between the specific free surface energy of the clean surface (σ_0) and the reduced surface energy of the contiguous adsorption layer (σ_a). It is assumed that the specific free boundary surface energy between the adsorption layer and crystalline substrate can be ignored. The so-defined spreading pressure, like surface energy, has the dimensions work/distance, relating to 1 cm along the boundary between the adsorbed layer and the clean surface. The actual pressure, which must prevail, for example, in the wedge-shaped termination of a fissure, is obtained by dividing by the thickness δ of the adsorbed layer.

$$p = \pi/\delta \tag{2.2}$$

Exact values which can be applied to rocks are not known, either for σ_0 and σ_a or for the thickness of the adsorption layer. Nevertheless an estimate of the magnitude of the expected pressure is of interest. $\sigma_0 = 1000$ erg/cm² and $\sigma_a = 200$ erg/cm² are plausible values for silicates and an adsorbed water layer. For a layer thickness of $\delta = 10$ Å we obtain the considerable pressure of 8000 kg/cm². We can probably conclude that this is sufficient to force open further existing fissures.

Still stronger effects are encountered when the rock contains swelling or expandable clay minerals, such as montmorillonite. Water is adsorbed then not only on the external mineral surfaces, but is imbibed in innercrystalline positions. As a result considerable volume effects and swelling pressures are encountered. According to VAN OLPHEN (1963) the swelling pressure for the first layer of water molecules adsorbed on completely dehydrated montmorillonite amounts to about 5400 atmospheres; for the second layer 2500 atmospheres.

We may conclude, therefore, that dry, highly fractured rocks, considerably broken up as a result of adsorption of water vapor, can be completely disintegrated under certain circumstances. For progressive destruction it is necessary for the rock to be subjected to repeated adsorption and desorption. Each single adsorption cycle at the site of previously formed fractures can produce only limited enlargement of the fracture, corresponding to the amount of energy liberated by the adsorption of water molecules. Further and continuing rock decay can take place only if the first adsorbed water molecules are repeatedly evaporated (desorption). Then each new adsorption cycle can be newly effective, enlarging the fractures further.

These kinds of conditions prevail in arid climates, where great temperature differences give rise to a change in saturation of the air with water vapor. At times of low temperature and high water vapor saturation (at night) water can be adsorbed and expansion exerted. At times of high temperature and low saturation (in the day time) the adsorbed water can be completely or partially evaporated.

Thus rock disintegration in desert regions, which is called insolation, is due not only to the temperature changes resulting from solar radiation, but also to adsorption of atmospheric moisture on mineral surfaces (hydration). Various investigators have come to this conclusion, either by direct observation (see for example, WILHELMY 1958) or from general theoretical considerations (see REICHE 1950). Similar conclusions were reached by GRIGGS (1936), whose experiments will be mentioned later.

If atmospheric humidity rises sufficiently in arid climates, in addition to adsorbed water, water in the fluid state can also bring about rock disintegration. This effect is little mentioned in published works. As a result of the phenomena of capillary condensation, fluid water collects in fine capillaries at water vapor pressures below the normal saturation vapor pressure. For a completely wetted cylindrical capillary of radius r, the following expression relates the normal saturation pressure

where:
$$\ln \frac{P_o}{P_r} = \frac{2\sigma V}{rRT} \tag{2.3}$$

σ = specific free surface energy of water
V = molecular volume of water
R = gas constant
T = absolute temperature

For a temperature of 15 °C this gives the equation (r in cm):

$$\lg \frac{P_o}{P_r} = \frac{0.48 \cdot 10^{-7}}{r} \tag{2.4}$$

2.2 Breakdown of primary rocks and minerals into loose minerals

Therefore, in unsaturated air water can occur in very fine hairline cracks. In the case of a concave meniscus (contact angle $<90°$) within water-filled capillaries, a negative pressure prevails, whose magnitude with complete wetting (contact angle $= 0°$) in a circular capillary is expressed according to the equation:

$$p_c = \frac{2\sigma}{r} \tag{2.5}$$

The filling of capillaries with water acts to hold the rock together and thus works against disintegration. On the other hand, in filled capillaries chemical reaction of water on mineral surfaces can begin. This will be dealt with in the next section. To the extent that a gradual removal of reaction products is provided for, chemical decomposition of minerals occurs below the saturation water vapor pressure in the finest hairline cracks. The cracks are extended and widened, therefore, leading to destruction of the rock without significant change in the bulk chemical composition of the rock.

The observed rock weathering in arid climates was attributed earlier exclusively to the rock stresses which develop as a result of daily temperature oscillations. For brittle substances like rock, the tensile strength is less than shear or compressive strengths. Whether temperature changes can give rise to rupture depends, therefore, on the prevailing tensile stress. A somewhat spherical rock body which is periodically exposed to heating by solar rays and to cooling by heat radiation, experiences the maximum tensile stress at its surface during cooling and at its center during heating. Since in nature disintegration is observed predominantly at the surface, it is interesting to inquire as to the maximum tensile stress occurring upon cooling at the surface of a rock body, and as to whether this exceeds the strength of the rock. Unfortunately a theoretical study of this question is fraught with great difficulties and, therefore, has not yet been carried out.

Calculations are possible only for models based on homogeneous materials. Rocks are aggregates of mineral grains, whose heat conductivity, thermal expansion, elastic constants, and tensile strengths vary considerably. They depend on direction and are not adequately represented by a homogeneous model.

BLACKWELDER (1933) and GRIGGS (1936) have carried out experiments on the behavior of rocks subjected to rapid temperature changes. BLACKWELDER found that all of the rock samples investigated by him withstand sudden quenching of 200—300°C without noticeable change. Fractures form upon sudden quenching only when temperature differences of 300° and 375°C prevail. Upon slower cooling differences up to 600°C could be imposed without damage. GRIGGS tested the effect of repeated heating and cooling and found no change in a granite after 89,400 temperature cycles between 30° and 140°C.

On the other hand, a roughening of the surface and, as shown in photomicrographs, a widening and deepening of fractures occurred after only about 900 cycles, if the rock sample was cooled each time by a spray of water. These studies, therefore, provide no basis for the view that temperature changes in desert regions alone can bring about rock disintegration: On the contrary, the rock disintegrates only when water is present, loosening the bonds between grains, either by solution or by adsorption and desorption.

In humid climates solid rocks disintegrate in the continued presence of fluid water. At temperatures above the freezing point, chemical reactions between water and different minerals provide sufficient energy to promote the breakdown of a solid rock into a loose mass.

Since the dissolving and decomposing chemical reactions preferentially follow the boundaries between different mineral grains, polymineralic rocks, such as granite or gneiss, are more readily broken down than monomineralic rocks, such as limestones, quartzites, or volcanic glasses. In this way, by quantitatively insignificant chemical reaction, large masses of polymineralic rocks can be thoroughly disintegrated.

Water enclosed in the pores of rocks can produce considerable pressure, if the temperature falls below the freezing point, since ice occupies a greater volume than fluid water. The contribution of this frost weathering to rock disintegration is of great importance at high latitudes and elevations. In temperate zones it also plays a role during the winter months, especially where solid rock crops out without a protective cover of weathering debris. The disintegration of solid rock under a cover of soil or rubble also proceeds in cooler climatic zones without participation of frost weathering, if this protective cover is deeper than the frost line in winter.

The bursting action of freezing water can be effective only under certain conditions. The pores of the rock must be completely filled with water. In addition the configuration of the pores and the nature of the freezing process must guarantee that the entrance is closed to the outside before all of the water is frozen. The maximum pressure attainable for an isolated pore completely filled with water can be deduced from the melting curve of the H_2O-phase diagram. At $-5\,°C$ this pressure amounts to somewhat over 500 kg/cm², at $-10\,°C$ about 1100 kg/cm²; at $-22\,°C$ about 2100 kg/cm². This is the highest pressure that freezing water can produce. At higher pressures and lower temperatures we pass into the stability field of ice III, whose specific volume is less than that of water.

In addition to ice crystallization, other crystallization processes within the rock pores can, under certain conditions, loosen the texture. With CORRENS (1939) we distinguish three different processes: volume effects of crystallizing solutions in pore spaces, effects on the solid surroundings of growing crystals, and the volume effects of salt hydration.

If a supersaturated solution has a smaller volume than the sum of the volumes of saturated solution and precipitating crystals, crystallization from the supersaturated solution under certain conditions can produce expansion. As was the case for frost weathering, it must be assured that the exit from the rock pore is blocked immediately by the formation of the first crystals, so that no solution can escape. An alum solution, supersaturated by a factor of 2, upon crystallization experiences a volume expansion of about 0.58%. Taking into account the compressibility of water, this corresponds to a pressure of 127 atm (CORRENS & STEINBORN 1939). The expansion effects of alum or Glauber's salt is put to use in the breaking up and working of shales. All salts that dissolve in water with a volume decrease show effects similar to alum. That is, the sum of the volumes of solid salt and pure water is greater than the volume of the saturated solution formed from them. In addition to alum, Na_2CO_3, $NaCl$, Na_2SO_4 and $CaCl_2 \cdot 6H_2O$ behave in this way.

Growing crystals can exert a pressure, since the solubility of a crystal increases with pressure. If a crystal is in equilibrium with its supersaturated solution when growing, it can withstand or exert a higher pressure, which is proportional to the supersaturation. The following equation (CORRENS & STEINBORN 1939) interrelates the pressure P, saturation concentration c_s and concentration c (both in mol/l) of the solution:

$$P = \frac{RT}{v} \ln \frac{c}{c_s} \qquad (2.6)$$

(R = gas constant; T = abs. temperature; v = mole volume).

As a prerequisite for a growing crystal to exert a pressure on another solid phase with which it is in contact, the solution must be able to penetrate between both solid phases. This depends on the specific free surface energies of the two solids and the solution.

Finally, an expansion effect can occur under certain conditions, when anhydrous salts take up crystalline water with a volume increase (hydration). As an example, the formation of gypsum from anhydrite and water is most often cited. 74 cm³ of gypsum forms from 46 cm³ anhydrite and 36 cm³ of water. Therefore, an expansion effect can occur only if the water has access to the anhydrite, somehow through porous layers, so that the newly formed gypsum cannot take the place of the water. MORTENSEN (1933) has attributed rock disintegration in the Egyptian desert to this kind of hydration. He has calculated the following pressures:

$$Na_2CO_3 \cdot H_2O \rightarrow Na_2CO_3 \cdot 7H_2O \approx 150 \text{ atm}$$
$$Na_2SO_4 \rightarrow Na_2SO_4 \cdot 10H_2O \approx 240 \text{ atm}$$
$$Na_2CO_3 \cdot H_2O \rightarrow Na_2CO_3 \cdot 10H_2O \approx 300 \text{ atm}$$

All three forms of rock disintegration by salt crystallization can take place only in those rocks whose pores are filled with highly concentrated solutions or with solid salts. Crystallization or recrystallization is brought about by strong temperature changes or by a change in relative humidity of the air. Such conditions prevail most commonly in arid or semi-arid climates.

Plant roots, penetrating joints and fine cracks in rocks, can exert mechanical effects which lead to rock disintegration. The forces originate from swelling pressure, resulting from water uptake of organic substances. This is the same mechanism which is utilized in quarrying, where a dry wooden wedge, driven into a joint, is wetted to cause it to expand and loosen blocks of rock.

In summary the following processes are instrumental in reducing solid rocks to loose masses:

A. In all climate zones:
   1. Unloading and cooling of rocks which solidified at high temperatures and pressures.
B. Chiefly in arid climates:
   2. Repeated adsorption and desorption of water.
   3. Solution effects of water condensed in hairline cracks (capillary condensation).
   4. Expansion effects due to crystallization and hydration of salts.
C. Chiefly in humid climates:
   5. Solution effects of water along grain boundaries and cleavage planes.
   6. Expansion effects of plant roots.
D. In cooler climates:
   7. Expansion effects of freezing water.

2.3 Chemical decomposition

2.31 General

The agents of chemical decomposition of minerals and rocks in the zone of subaerial weathering are the atmospheric gases and aqueous solutions occurring at the earth's surface.

The free atmosphere has an approximately constant composition containing the following major constituents (RANKAMA & SAHAMA 1950):

	Vol. %
Nitrogen	78.09
Oxygen	20.95
Carbon dioxide	0.03
Argon	0.93

Within the soil and dependent upon the activity of plants and soil organisms, greater amounts of carbon dioxide and less oxygen may occur. This is especially so for air contained in the upper horizons of the soil. Here soil atmospheres usually contain 0.2 to 0.7 volume-% carbon dioxide and about 20% oxygen. In fine-grained, relatively impermeable soils the carbon dioxide content can climb up to 5 volume-% and the oxygen content fall below 10% (SCHEFFER & SCHACHTSCHABEL 1966).

The water budget of the earth's surface is determined by the equilibrium between precipitation and evaporation. At the surface of the ocean all together more water is lost through evaporation than is returned by precipitation. On the continents in contrast an excess of precipitation occurs over evaporation. Therefore, through the atmosphere the continents are supplied by water from the ocean, which in turn flows back to the ocean on the surface. The ratio of precipitation to evaporation on the continents depends in particular on the local climate and is subject to strong variations. As far as material balance of the whole earth's surface is concerned, it is, however, of great importance that for all of the continents collectively precipitation exceeds evaporation. Only because such a water surplus exists, are the primary rocks exposed at the land surface subject to the continual processes of chemical weathering. As a result fresh solvent is supplied repeatedly so that an equilibrium state is never reached. The surplus precipitation permeates the weathering zone and reacts with minerals. In this way solutions are formed which combine with the ground water or migrate as surface runoff. The greater portion of these solutions reach by means of streams and lakes or through ground water the collecting basin of the ocean. A smaller amount flows into internally drained basins on the continents.

Table 2-1 summarizes a compilation by DIETRICH & KALLE (1957) of the amounts of water which are supplied to and lost from the ocean and land surfaces as a result of evaporation and precipitation.

Table 2-1. Evaporation and precipitation over the ocean and land surfaces (after DIETRICH & KALLE 1957).

	km^3/yr.	cm/yr.
Evaporation from the ocean surface	351 000	97
Precipitation on the ocean surface	324 000	90
Evaporation from the land surface	74 000	50
Precipitation on the land surface	101 000	68

According to these data, which of course are only estimates, the solid surface of the earth is supplied on the average annually with 27,000 km^3 more water than is lost by evaporation. This surplus, about 27% of the total precipitation, is drained off as ground water and stream run-off.

In table 2-2 data from DURUM, HEIDL & TISON (1960) are summarized. These represent recent estimates of precipitation and run-off for different regions of the

Table 2-2. Precipitation and run-off in the United States (after DURUM, HEIDL & TISON 1960).

	1 Catchment basin km^2	2 Average annual Precipit. cm	3 Average run-off 10^{3}m^3/sec	4 Average run-off per km^2 10^{9}m/sec	5 Average annual run-off cm	6 5:2 in %
North Atlantic slope	383 311	102.1	594.7	1.55	49.02	48.0
South Atlantic slope	372 951	124.5	404.9	1.09	34.04	27.3
Eastern Gulf of Mexico	362 592	132.1	515.4	1.42	44.96	34.0
Mississippi River Basin	3 221 885	73.7	1795.3	0.557	17.53	23.8
Western Gulf of Mexico	1 069 645	62.5	158.6	0.148	4.83	7.7
Colorado River Basin	637 125	32.0	155.7	0.244	0.76	2.4
Pacific slope of California	303 022	67.8	226.5	0.747	23.37	34.5
Columbia River Basin	782 162	80.3	1186.5	1.52	48.01	59.8
Total U.S.A.	7 132 695	76.2	4897.4	0.687	21.84	28.7

United States. The run-offs for the large river systems vary considerably in the percent of total precipitation which they represent in their respective catchment areas. The influence of climate is very apparent. In the dry catchment area of the Colorado River only 2.4% of the precipitation reaches the ocean while in the northern part of the Atlantic coastal region 48%, and in the catchment basin of the Columbia River 60% of the total precipitation flows into the sea. In questioning the large possible sources of error in making worldwide run-off estimates, it is noteworthy that the calculated percent for the entire United States of 29% agrees very well with the value for all continents, calculated at 27% from table 2-1.

Rainwater contains atmospheric gases and various dissolved inorganic substances. Gases dissolved in rainwater consist of up to 63% nitrogen, 34% oxygen, and up to 3% carbon dioxide. Their proportions are quite different with respect to the free atmosphere, oxygen and carbon dioxide being enriched in rainwater. The high carbon dioxide content is of considerable importance in weathering.

The electrolyte content of precipitation is subject to great variation, depending upon the origin of the dissolved salts. Repeatedly very small solid salt particles serve as condensation nuclei in forming raindrops and thus are incorporated into rainwater. The most important anions are chloride and sulfate; the commonest cation is sodium, but in addition calcium and magnesium, as well as potassium, are observed.

Chloride and sodium originate primarily from sea water and pass into the atmosphere from the sea water spray in the breaker zone. It has been established that the sodium chloride content of rainwater decreases away from the coast and is especially high at certain localities where the wind blows in from the sea. The average chloride content usually lies between 0.3 and 3.0 mg/l. However, values between 100 and 1000 mg/l occur along the coasts. Sulfate appears always to occur in rainwater. On the continents, far from the coast, the ratio of sulfate to chloride is about one or greater. Therefore, sulfate must originate from sources other than sea spray. A portion of the sulfate can be traced back to the natural oxidation of hydrogen sulfide or other sulfides. Higher sulfate contents, which occur especially in densely populated regions, stem mainly from artificial sources and through human activity. In the vicinity of cities and industrial centers the presence even of free sulfuric acid has been established. Since systematic rainwater studies over large areas of the earth's surface and analyses before the age of modern industrialization are lacking, it cannot be ascertained to what extent the composition and especially the acidity of precipitation has been altered by human activity in recent times. It is probable, because of this, that chemical weathering at the present time, at least in heavily populated regions, proceeds differently than in the geologic past.

The absolute as well as the relative amount of soluble matter in precipitation varies greatly. Since little systematic investigation has been carried on, even average values cannot be state. Distance from coasts and large cities or industrial centers, wind direction, and duration and amounts of precipitation all have a strong influence on the dissolved content of rainwater, so that even for a specific locality very great variations are encountered. For a region in mid-England GORHAM (1957) has assembled the results of analyses of rain samples collected during the course of one year.

Table 2-3. Summary for rain samples collected during the course of a year in the Lake District, England (GORHAM 1957).

Dissolved ions	0.054 – 0.439 millieq./liter
pH	4.0 – 5.8
Na	0.2 – 7.5 mg/kg
K	0.05 – 0.7 mg/kg
Ca	0.1 – 2.0 mg/kg
Mg	0.0 – 0.8 mg/kg
HCO_3	0.0 – 2.8 mg/kg
Cl	0.2 – 12.6 mg/kg
SO_4	1.1 – 9.6 mg/kg
Soot-index	0.0 – 4.4

From this study the dependence of electrolyte content on weather conditions can be clearly deduced. A summary of all his data is found in table 2-3.

Soot-free rain, that is, rainfall not influenced by industrial and urban contamination, gave on the average very low values for sulfate, higher carbonate content, lower values for Ca and K, and a somewhat higher pH, than sooty rain.

At the earth's surface not only are minerals and rocks exposed to the very dilute electrolyte solutions in precipitation. Bacteria, fungi, and lichens which populate their surfaces play an important role in the chemical decomposition and mechanical disintegration of rocks. From numerous studies summarized by KRUMBEIN (1972), it has been established that the vital activities of microflora bring about the dissolution of different minerals. Carbonates are especially affected, but micas and feldspars are dissolved, as is even quartz. Lichens are especially active agents in producing a large number of organic acids.

At depth in the weathering zone soil solutions are derived from chemical decomposition of minerals and rocks, from the decomposition of organic matter, and the activities of soil organisms. Higher contents of CO_2 prevail in these solutions. In addition, sulfuric acid, nitric acid, and different organic acids occur. Thus soil solutions are usually acid (to pH 3). Less often, as a result of NH_3-content, soil solutions are alkaline (to pH 11). The oxidation potentials of soil solutions vary over a wide range of Eh, from $+750$ mV to -350 mV (BAAS, BECKING, KAPLAN & MOORE 1960). Therefore, in weathered soils oxidizing as well as reducing conditions prevail. Fig. 2-2 gives the oxidation potentials and hydrogen ion concentrations for different natural waters at the earth's surface. The dashed lines represent the boundaries for the stability fields for divalent and trivalent iron ions as well as for hematite and magnetite. These are taken from fig. 2-6.

Which acids play the major role in weathering, is best determined from the composition of river waters, which include the soil solutions resulting from chemical weathering processes. Table 2-4 gives the calculated world wide average composition of surface waters, taken from LIVINGSTONE (1963).

From these data the dominant role of carbonic acid is apparent. It originates in part from the atmosphere, but predominantly from the activity of organisms. Sulfuric acid also plays a considerable role. It originates from the inorganic and bacterial

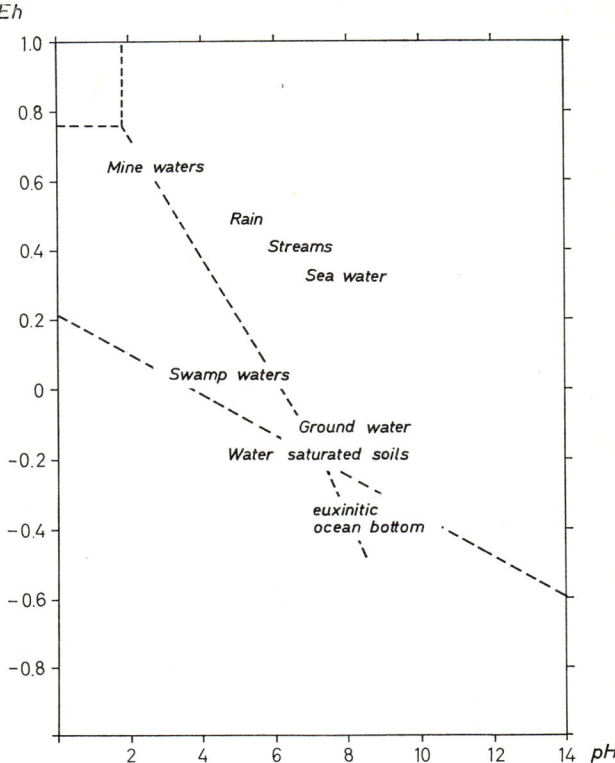

Fig. 2-2. Oxidation potential (Eh) and hydrogen ion concentration (pH) of natural waters (BAAS, BECKING, KAPLAN & MOORE 1960).

oxidation of primary sulfides and biogenic protein, as well as from artificial burning processes. Chloride originates predominantly from the chloride content of precipitation.

Table 2-4. World-wide average composition of surface waters (after LIVINGSTONE 1963).

	mg/kg
HCO_3	58.4
SO_4	11.2
Cl	7.8
NO_3	1.0
Ca	15.0
Mg	4.1
Na	6.3
K	2.3
Fe	0.67
SiO^2	13.1
Total	120.0

The chemical reactions which take place between the solid phases and the solutions and gases (free air and soil air) penetrating the weathering zone are of two types:

1. Mineral dissolution, when active solutions are undersaturated with respect to the various respective phases.

2. New minerals form from solution either by reconstitution or by neoformation. Reconstitution occurs when partially degraded minerals take into their structures components from solution. Neoformation occurs when minerals crystallize anew from dissolved components.

In the following sections, solution processes, the important equilibrium conditions for reconstitution and neoformation, and how these processes operate in the weathering zone, will be considered. In a special section, soils in which the processes of subaerial weathering take place, will be considered.

2.32 Solution processes

In the very dilute solutions which permeate the upper horizons of the weathering zone and are replenished by rainfall, every mineral is soluble, even though often sparsely so. The weathering events take place here essentially by very slow dissolution in an open system. In such cases thermodynamic equilibrium generally cannot be obtained. What happens to a particular mineral depends upon kinetic factors, that is, the nature and velocity of the solution process. Since all minerals are soluble, a mineral series representing increasing stability (resistance to weathering) is nothing more than a series of decreasing solution velocities. In order to understand weathering, knowledge of the general theory of crystal dissolution as well as that of the dissolution of rock-forming minerals in weathering solutions are especially essential preconditions.

According to the older theory on dissolution of crystals by NERNST & BRUNNER (1904) a static boundary layer of thickness δ forms on the crystal surface. Within this layer saturation concentration C_s falls off to that concentration which prevails in the solution agent (C). If we assume that establishment of equilibrium is infinitely rapid, the dissolution velocity L (referred to 1 cm² of surface) is determined by diffusion alone in the boundary layer. The following relation applies:

$$L = \frac{D}{\delta}(C_s - C) \tag{2.7}$$

D is the diffusion coefficient of the substance going into solution.

The simple theory requires some elaboration. If the solution is not static, but flowing over the crystal, according to VIELSTICH (1953) the thickness of the boundary layer is related as follows:

$$\delta \simeq \gamma^n \cdot D^{1/3} \tag{2.8}$$

γ is the kinematic viscosity of the solution. The exponent n has the value of about $1/6$ in the case of laminar and of $1/3$ to $1/2$ with turbulent flow.

2.3 Chemical decomposition

In addition, from experiments on crystal dissolution, the assumption of instantaneous establishment of equilibrium at the crystal surface is not correct. The chemical constituents are dissolved from the structure rather with definite finite velocities, which depend, among other factors, on the orientation of the crystal face undergoing dissolution (BERTHOUD 1912).

Under such circumstances the dissolution of a crystal face by a pure solvent ($C = 0$), which flows with constant velocity is defined as follows

$$L = A \cdot v^{-n} \cdot D^{2/3} \cdot C_s P \tag{2.9}$$

A is a constant dependent upon experimental conditions. The dissolution velocity thus depends on viscosity, diffusion coefficient, saturation concentration, and a factor P which is less than 1 and takes into account the fact that the crystal constituents do not pass freely into solution, but do so only after overcoming a lattice energy barrier. The value of P depends for a particular crystal on its crystallographic orientation and in addition is strongly dependent on temperature.

Experiments by W. MÜLLER (1965) show that the crystal dissolution velocities in moving solutions, under like conditions, are not simply proportional to solubility, as could be expected from the simplified equation (2.7). MÜLLER's data are tabulated in table 2-5. Potassium sulfate, sodium nitrate, and sodium chloride behave similarly

Table 2-5. Relative dissolution velocity (L) of crystals of differing solubility (l) (MÜLLER 1965).

Crystal	l	L	$L/l \cdot 10^3$
K_2SO_4	0.0637	0.327	5.1
$NaNO_3$	1.03	4.12	4.0
NaCl	0.620	3.26	5.2
K-Alum	0.0234	0.0592	2.5
Sucrose	0.597	0.175	0.29

L: dissolved in 45 sec from a planar crystal surface (122.5 mm²), by a 100 ml stream of distilled water (amounts in moles).
l: solubility in water in moles/100 gr.
Temperature: 20 °C.

inasmuch as they all give L/l ratios around 4-5. On the other hand alum and more particularly sucrose, which have very different solubilities, dissolve relatively slowly. This is especially apparent for sucrose, which has somewhat the same molar solubility as sodium chloride, but dissolves about 18 times slower. Here the factor P of equation (2.9) plays a role related, to detaching the crystal constituents from the structure. This detachment apparently requires a greater activation energy the more complicated the structure. Thus the complex alum and sugar crystal structures dissolve comparatively more slowly than the simpler salts.

The dissolution rates of minerals in the weathering zone are dependent, therefore, primarily on their water solubility, but also on the as yet undetermined rates with which individual mineral species react with natural solutions. The course of this

2.32 Solution processes

dissolution either can be observed in nature or investigated experimentally in the laboratory.

In the case of many minerals, one can observe directly complete dissolution by rainwater. Dissolution is immediately evident for the comparatively easily water-soluble minerals such as halite, anhydrite, gypsum, calcite, and dolomite. Their dissolution progresses relatively rapidly at the earth's surface. The CO_2-content of the effective solution is especially important in the dissolution of carbonates. In chapter 4 carbonate and sulfate reactions and equilibria will be discussed.

Microscopic observations of soil profiles teach us that in principle in moving ground water sparingly soluble minerals dissolve no differently than readily soluble ones. One can conclude something about the solution process from comparisons of the mineralogy of the higher leached soil horizon (A-horizon) and less affected subsoil (C-horizon). The mineral grains themselves, by their shapes and surface characteristics, provide clues which are in turn characteristic of dissolving crystals. Examples are: typical solution forms, which develop as a result of more rapid and advanced dissolution along certain directions or planes (cleavage planes); the formation of etchpits, which are sometimes polygonal-crystallographically shaped or aligned along specific crystal directions; dissolution may proceed preferentially in certain parts of the crystal, from imperfections, zones of varying composition, from inclusions or other irregularities, resulting in irregular embayments and tubular indentations and finally, perforated or skeletal residues.

In many soils one finds feldspar grains in the state of dissolution, which clearly is usually progressing preferentially along the main cleavage planes. In addition the grains are attacked irregularly from the borders. Typically they occur as tattered granules, indented or ragged along cleavage cracks. Since larger grains are reduced by this dissolution to smaller grains, one finds frequently in the upper soil horizons less and finer-grained feldspar than is found in the parent material (v. ENGELHARDT 1937a, 1940b; GADOW 1965).

The so-called heavy minerals are dissolved also in permeable soils. Characteristic solution- and etch-phenomena can be observed with many minerals.

Garnet grains frequently exhibit on their surfaces etchpits which usually are irregularly circular, but sometimes polygonal in shape. Often the surface contains rounded etching ridges. In advanced stages of dissolution the grains are broken up by deep corrosion pits, so that finally irregular perforated skeletons remain.

In the case of hornblende grains preferential solution along cleavage planes produces serrated boundaries with fine points parallel to the long crystal axis. Continued dissolution reduces the grains to narrow fibers or to perforated skeletons. Similar solution forms have been observed for augite.

With epidote shallow etchpits or extensive corrosion at the grain boundaries is observed.

On staurolite grains shallow etchpits or crystallographically oriented etch grooves and ridges are found. More intense solution proceeds from the crystal edges, enlarges the otherwise obscure cleavage cracks and produces serrated boundaries and finally perforated skeletal forms.

In the case of apatite which is relatively soluble in an acid milieu, diagnostic solution features are not so apparent, since the dissolution usually proceeds very uniformly. From even the most elongate crystals, rounded grains form. Occasionally etchpits are observed also on apatite.

Descriptions and illustrations of solution phenomena observed on the heavy minerals can be found in the works of EDELMAN & DOEGLAS (1932, 1934), ZÖBELEIN (1940) and GRIMM (1957).

Individual heavy minerals show variable resistance to solution in the weathering zone. A group of stable minerals, rutile, zircon, tourmaline, and sphene, is especially resistant under all conditions. In soils which have been exposed to continuous and extensive weathering, this group of resistant minerals is concentrated, because the other heavy minerals are more or less completely dissolved. Examples are the deeply weathered lateritic tropical and subtropical soils and the Tertiary residual kaolinite-rich soils of central Europe.

The resistance of the less stable heavy minerals depends in particular on the chemical properties of the solutions permeating the soil. pH may be of particular importance. In temperate climates the dissolution of heavy minerals in the acid solutions of podsols is especially intensive.

The heavy minerals of calcareous soils are less intensively dissolved. The weathering resistance series for heavy minerals, proposed by different authors, do not agree always with one another, because special local factors were not taken into account (GOLDICH 1938; WEYL 1950, 1952; VAN ANDEL 1952, 1959; DRYDEN & DRYDEN 1946). There is general agreement that olivines and pyroxenes show little resistance to weathering. Amphibole, garnet and epidote are more resistant, although differences in their behavior are found, depending on local conditions. For example, in the weathered calcareous sands of the Obere Suesswassermolasse (upper Miocene), garnet is especially intensively dissolved, whereas epidote and hornblende is concentrated (LEMCKE et al. 1953). On the other hand WEYL (1950) found that under podsolic weathering conditions in the Tertiary sands of Schleswig-Holstein, hornblende is intensively dissolved, and garnet and epidote concentrated, with garnet more stable than epidote. Staurolite, kyanite, sillimanite and andalusite in general appear to be more resistant than hornblende, garnet and epidote. Apatite is especially sensitive with respect to the pH of the solution: Where lime is present, as, for example, in the south German Molasse sands, it is barely attacked, while in non-calcareous soils it is one of the most highly soluble minerals.

While the minerals mentioned so far are dissolved gradually in the weathering zone without leaving an insoluble residue, there are also minerals which behave differently. Mica is a good example. In soils and sediments micas of the muscovite and biotite groups occur. From these certain constituents may be dissolved and replaced by new constituents. The result is pseudomorphs with modified composition, which in turn gradually fall prey to dissolution. The latter process happens, however, much more slowly. RIMSAITE (1957) gave a recent description of this kind of altered mica, as have earlier investigators.

Biotite loses potassium, magnesium, and iron preferentially. Either the iron is

2.32 Solution processes

transported away completely, or, after oxidation to the trivalent state, deposited as hydrated oxide on or between flakes. The color of the biotite flakes changes as a result from dark brown or green to golden yellow, bright yellow to bright green tones; also colorless or opaque white flakes occur. Indexes of refraction and birefringence decrease continuously and X-ray interferences become increasingly weaker and broader.

Similarly there develops from muscovite a potassium-deficient weathering product with reduced indexes of refraction and birefringence.

From these mica weathering products, alkali-free and alkali-deficient clay minerals form as a result of reconstitution. These will be discussed in detail in section 2.332.2.

Repeated attempts have been made to clarify the nature of rock-forming silicate decomposition in water and dilute solutions, utilizing laboratory experiments. It is easy to show, for example, that feldspars are attacked by water. If fine, powdered feldspar is placed in water, the solution rapidly becomes alkaline, as alkalis go into solution.

As a result of studies by different investigators, it has been shown that many freshly pulverized silicates, reacting with a limited volume of solvent, dissolve incongruently. That is, the constituents in solution do not occur in the same proportions as in the mineral undergoing dissolution. The solutions always contain relatively more alkalis, alkaline earths, or iron than alumina or silica, giving an alkaline pH. As an example, the results of a study by KELLER, BALGORD & REESMAN (1963) are given in table 2-6. In this study, during grinding with definite amounts of distilled water, different rock-forming minerals undergo dissolution:

From dissolution experiments of this kind it can be wrongly concluded that silicates are decomposed in water by hydrolysis, forming on the one hand a solution containing predominantly Na, K, Mg, Ca, and Fe, and on the other hand a hydrated alumina-silica residue. Direct observations for all minerals, however, speak against this, since in soil solutions they dissolve completely and without perceptible residue. This contradiction is best explained by noting that natural weathering does not take place in a closed system.

The progress of mineral dissolution as it takes place in the weathering zone can, therefore, be clarified only by experimenting with open systems, in which, as in nature, repeatedly renewed, fresh solvents act on the minerals. In this way the dissolution always proceeds forward and is not interrupted by discontinuous equilibrium. Investigations of this kind are very tedious and time-consuming because of the slowness of the process and the large amounts of necessary reaction solution. Therefore, not much data of this kind are available. Nevertheless a few important rock-forming minerals have been studied in this way, so that some general conclusions can be drawn (CORRENS 1940, 1961, 1963).

The experiments simulating open systems were carried out in the following manner: Fresh, dry-ground powders of pure minerals of defined particle size were exposed to the action of a stream of fresh solvent. The solution, after permeating the powder, was separated by passage through a collodian or cellophane membrane,

Table 2-6. Decomposition of rock-forming silicates during grinding in distilled water in a closed system (after KELLER, BALGORD & REESMAN 1963).

	Mineral composition (cations)	Dissolved cations mg/l	Solution pH	Cationic composition of dissolved constituents
Quartz	Si	3.5	6.5	Si
Microcline	0.725 K · 0.24 Na · 0.05 Ca · 1.05 Al · 2.95 Si	66.06	8.0	11 K · 8.6 Na · 0.4 Ca · 0.36 Al · 2.95 Si
Labradorite	0.53 Ca · 0.47 Na · 1.53 Al · 2.47 Si	28.96	8.0	0.68 Ca · 0.17 Na · 0.23 Al · 2.47 Si
Nepheline	0.82 Na · 0.18 K · 1.0 Al · 1.0 Si	114.86	8.8	7.0 Na · 0.57 K · 0.15 Al · 1.00 Si
Muscovite	1.7 K · 0.23 Na · 0.10 Ca · 0.25 Fe · 5.4 Al · 6.1 Si	32.89	8.0	13 K · 4.4 Na · 0.22 Ca · 0.27 Fe · 0.57 Al · 6.1 Si
Biotite	1.9 K · 0.13 Na · 2.2 Fe² · 1.6 Mg · 0.16 Ca · 1.0 Fe³ · 2.98 Al · 5.3 Si	62.64	8.5	18 K · 3 Na · 5.3 Fe · 0.61 Mg · 0.53 Ca · 0.36 Al · 5.3 Si
Enstatite	0.78 Mg · 0.073 Fe · 0.086 Ca · 0.005 Na · 0.046 Al · 1.0 Si	37.46	8.6	0.83 Mg · 0.002 Fe · 0.063 Ca · 0.41 Na · 0.001 Al · 1.0 Si
Diopside	0.45 Ca · 0.46 Mg · 0.047 Fe · 0.19 Al · 1.84 Si	30.64	9.0	0.21 Ca · 0.013 Mg · 0.005 Fe · 0.004 Al · 1.84 Si
Augite	0.87 Ca · 0.07 Na · 0.34 Mg · 0.60 Fe² · 0.01 Fe³ · 0.08 Al · 1.99 Si	34.62	8.8	0.35 Ca · 1.1 Na · 0.30 Mg · 0.057 Fe · 0.002 Al · 1.99 Si
Hornblende	0.45 K · 0.48 Na · 1.4 Ca · 1.7 Mg · 3.7 Fe · 1.13 Al · 6.9 Si	24.31	8.9	5.7 K · 16 Na · 1.8 Ca · 2.9 Mg · 0.1 Fe · 0.37 Al · 6.9 Si
Olivine	1.86 Mg · 0.14 Fe · 1.0 Si	20.55	8.8	2.2 Mg · 0.02 Fe · 1.0 Si

2.32 Solution processes

Table 2-7. Dissolution of 1.5 g powdered K-feldspar ($< 2\,\mu$) in a weakly acid solution (pH = 3.0) in an open system (after Correns & v. Engelhardt 1938).

ml Total solvent	ml Filtrate analyzed	Dissolved constituents mg/l		Dissolved constituents Millimole			Composition of dissolved constituents			Composition of residue		Amt. of residue
		mg		SiO_2	Al_2O_3	K_2O	SiO_2 :	Al_2O_3 :	K_2O	SiO_2 :	Al_2O_3	
1842	1842	84.0	45.5	0.263	0.233	0.473	0.56 :	0.49 :	1.00	10.7 :	1.00	179 mg
4742	2900	79.7	27.5	0.762	0.190	0.166	4.59 :	1.14 :	1.00	13.0 :	1.00	179
9078	4336	47.2	10.9	0.516	0.0715	0.0940	5.49 :	0.76 :	1.00	12.0 :	1.00	190
Totals:		210.9	23.2	1.541	0.495	0.733	2.10 :	0.68 :	1.00			196

and individual portions analyzed. In this way it could be ascertained how the amounts and composition of dissolved constituents changed during the solution process.

Two important results of these studies were to be anticipated. First of all they showed that all of the silicate components pass through the membrane, and thus go into solution in ionic or molecular-dispersed form. It showed secondly, that the nature and rate of dissolution changes with time, so that a long time passes before a steady state of dissolution is attained. Experiments in closed systems simulate only the initial incongruent starting reaction, while in nature weathering is the result of steady state processes.

The manner of decomposition and dissolution is characteristic for each mineral group, so they will be discussed individually.

Quartz dissolves congruently. Details of the dissolution of quartz and amorphous SiO_2, as it relates to temperature, pH and other factors are considered in section 2.331.

Of the minerals of the feldspar group, K-feldspar (CORRENS & v. ENGELHARDT 1938) and albite (BIEGER 1952) dissolution have been studied in open systems. As a typical example, the decomposition of K-feldspar by means of a weak acid solution of pH = 3 (0.001 n H_2SO_4) is described in some detail. The solution flowed through fine powdered feldspar ($< 2\,\mu$), was filtered through an ultrafilter (collodian membrane), and successive portions of filtrate were dried by evaporation. Fig. 2-3 shows the course of the solution process. We can see how the dissolution

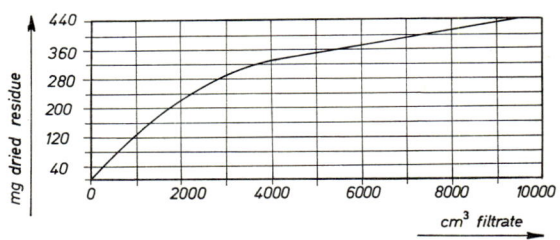

Fig. 2-3. Dissolution of potassium feldspar in aqueous solution at pH 3 (20 °C) (CORRENS & v. ENGELHARDT 1938).

rate is at first rapid and then finally decreases to a somewhat constant value. The dissolved substances in the filtrates were analyzed in 3 successive portions. The results are reproduced in table 2-7 and fig. 2-4. The composition of dissolved substances changes with time. The initial rapidly reacting solution contains an excess of potassium: the dissolution clearly takes place incongruently. Later it changes, and the solution collected at the end contains components in proportions approaching those of the initial mineral (6 SiO_2 . Al_2O_3 . K_2O). Therefore, a gradual transition from incongruent to congruent solution occurs, with simultaneous decrease in the solution rate to a constant value.

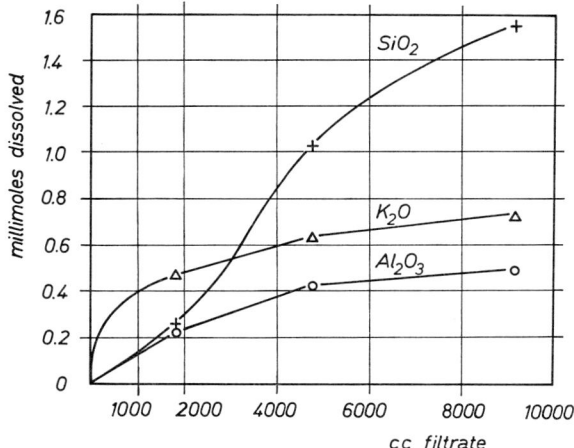

Fig. 2-4. Amount of silica, alumina, and potassium in aqueous solution at pH 3 (20 °C) during dissolution of K-feldspar (CORRENS & v. ENGELHARDT 1938).

Apparently during the initial solvent attack, the chemical constituents go into solution at different rates. K-ions are preferentially dissolved, followed by Al, while SiO_2 goes into solution the slowest. Naturally this is possible only by exchange with H-ions which take the place of dissolved K- and Al-ions dissolved from the structure. As a result a K-free and Al-deficient residual layer is formed on the surface of the crystals. During further dissolution K and Al must diffuse through this layer in order to get into solution. The $SiO_2:Al_2O_3$-ratio of the residual layer depends on the relative rates at which Al and Si go into solution. In solution of pH $=$ 3, Al is dissolved more rapidly than Si, so that a residual layer forms, whose $SiO_2:Al_2O_3$-ratio is higher than in feldspar. From the analysis of the dissolved constituents the composition and amount of the residual layer can be calculated as well. This shows that the amount of residual layer tends toward a constant value, while at the same time the dissolution rate becomes constant and the dissolving components go into solution in proportions approaching those in the mineral. Apparently the particles are then surrounded by a residual layer whose thickness is just such that the dissolution of K and Al is inhibited sufficiently, so that all three components (K, Al, Si) continue to dissolve in the ratios corresponding to the initial mineral composition. Then further solution occurs congruently and the thickness of the residual layer remains constant.

The same solution mechanism was found for K-feldspar, as well as albite by the action of other solutions in the pH-range 0—11. The thickness and composition of the residual layer depends upon the different experimental conditions, especially the solution pH. The lowest ratio, $6\,SiO_2:Al_2O_3$, was observed for near neutral solutions. Acid and alkaline solutions produce residual layers richer in SiO_2. At pH $=$ 0 the residual layer from K-feldspar is essentially pure SiO_2. Residual layer thicknesses give estimated values between 0.03 and 0.2 µ. It has not yet been possible to make these layers optically visible or to otherwise observe them directly.

According to these experiments the dissolution rates of feldspars clearly depend on pH. This is illustrated by the data summarized in table 2-8.

No similar experiments have been carried out as yet for Ca-plagioclases. It may be presumed, however, that these feldspars behave basically like the alkali feldspars.

Table 2-8. Dissolved constituents produced by 2000 ml of solvent from 1.5 g powdered K-feldspar ($< 2\,\mu$) at 25 °C (open system) (CORRENS & v. ENGELHARDT 1938).

pH of Initial solution	Total amt. in % of 1.5 g	Dissolved constituents Oxides in % of amount in feldspar		
		Al_2O_3	SiO_2	K_2O
3	5.9	9.4	2.0	18.2
6.6	3.8	0.8	1.4	9.6
11	5.5	4.3	3.9	14.2

The results of these feldspar weathering experiments are in good agreement with observations in the leached horizons of soils. The obviously incongruent, hydrolytic decomposition characteristic of closed systems is an initial reaction, which, after the formation of a SiO_2—Al_2O_3 protective layer, changes over to much less rapid congruent dissolution. In this main phase of weathering solutions are produced which contain feldspar components in the same ratio as the crystal structure. The amount of silica-alumina residue remaining in the last formed residual layer depends on the original grain size of the weathering feldspar.

Of course dissolution proceeds differently, if, as a result of mechanical action, repeatedly fresh crystal surfaces are produced and the inhibiting residual layer is destroyed. One can accomplish this experimentally by carrying out the dissolution in a ball mill. Then one observes no interference and obtains dissolution at a constant rate.

Among the **feldspathoids** the decomposition of **leucite** has been investigated in an open system (KRÜGER 1939). In this case also incongruent dissolution is observed initially in all solutions between pH = 0 and pH = 11: with respect to dissolved SiO_2, significantly more K and Al are dissolved than corresponds to the mineral composition. From attack by the solution a residual layer surrounds the mineral grain, whose $Al_2O_3:SiO_2$-ratio depends primarily on pH. At pH = 0 a residual layer of pure SiO_2 forms, as it did with feldspar. At higher pH the Al-content increases; near neutrality a composition of about $3\,SiO_2 \cdot Al_2O_3$ was found. Because of the higher K- and Al-content of leucite ($K_2O \cdot Al_2O_3 \cdot 4\,SiO_2$) the K-free residual layer is much more permeable to K and Al than with feldspar. Thus in strongly acid solution of pH = 0, all K and Al can be dissolved, while only minor SiO_2 goes into solution. The mineral grains dissolved at pH = 0 were altered to pure amorphous SiO_2 pseudomorphs. Also in the weakly acid and neutral range, the inhibiting action of the residual layer is less than with feldspar, so that constant rate

2.32 Solution processes

was not attained experimentally. Therefore, the incongruent dissolution of leucite perseveres longer than for feldspar and small grains may be converted to SiO_2-rich pseudomorphs in weakly acid solutions. Experiments at pH = 11 indicate approach to constant residual layer thickness and congruent dissolution.

Also for leucite the solution rate is less in neutral than in acid or alkaline solutions (somewhat similar to K-feldspar).

Among the micas, biotite, in strongly acid solutions (MEHMEL 1937) and muscovite, in the range pH = 3—11 (BÖHMEKE 1952) have been studied in open systems. Over the entire pH-range micas behave differently from the feldspars, in that incongruent dissolution is not hindered by a residual layer, and, therefore, persists in forming a K-deficient or K-free mineral residue, such has been shown to occur in soils and sediments. As a result of reaction of muscovite with water, much more potassium and somewhat more SiO_2 than aluminum appears in solutions, in comparison to the mineral composition. The remaining K-free residue, formed between pH of 3—11, has approximately constant composition with a SiO_2:Al_2O_3-ratio between 1.8 and 1.5. The amount of mineral residue increases constantly with continuing dissolution, and the solution rate experiences essentially no reduction, since the exchange of K- by H-ions along (001)-planes of the layer structure can take place unhindered.

The rate of dissolution is strongly dependent upon pH. Thus at pH = 3 it is approximately double and at pH = 11 about quadruple the decomposition rate at pH = 6.

The decomposition of biotite until now has been investigated only at pH = 0 (1 normal H_2SO_4) (MEHMEL 1937). In this strongly acid solution the mineral loses all other constituents and is transformed finally into a pure SiO_2-pseudomorph. Natural observations show that biotite is decomposed more readily than muscovite. This is probably because of the greater solubility and ease of solubility of Fe^{+2} and Mg in comparison to Al over the pH-range 3—9 (see fig. 2-5).

As an example of a chain silicate investigated in an open system, the dissolution of tremolite ($2CaO.5MgO.8SiO_2.H_2O$) was studied by TUNN (1940). This mineral also is subject to severe incongruent dissolution between pH 0 and 11. In strongly acid solution (pH = 0) its residual layer consists of pure SiO_2; at pH 3 it has a composition of about $2SiO_2.MgO$; at pH 5.8 of about $SiO_2.MgO$. At pH = 11 the residual layer is silica-free with a CaO:MgO-ratio of approximately 1:8. The residual layer formed from tremolite has a relatively high permeability, so that during the course of dissolution the dissolution rate diminishes only very slowly. The decomposition rate in the case of tremolite also clearly depends upon solution pH. At pH 3 or 11 the dissolution rate is almost double that at pH 6.

From the inosilicates an iron-rich olivine was studied in open system experiments by HOPPE (1947). Its residual layer consists at pH = 3, 6, and 9 of a mixture of iron and magnesium hydroxides, with the highest MgO:FeO-ratios forming in alkaline solutions. The residual layer is relatively permeable, so that high layer thicknesses were observed and steady state conditions (constant layer thickness and congruent solution) barely could be attained in laboratory experiments.

As an example of weathering of natural glasses, of interest is a study by HOPPE (1947) on the decomposition of sideromelane in an open system at pH-values between 3 and 11. Basaltic glasses (SiO_2-content $= 52\%$) were shown by these studies to decompose incongruently. All residual layers are alkali-free. At pH $= 3$ SiO_2 and Al_2O_3 are concentrated; at pH $= 5.8$ the residual layer contains more SiO_2, Al_2O_3, and Fe_2O_3 than the parent material. At pH $= 11$ it contains an excess of Fe_2O_3 and MgO. At all values of pH the residual layers formed are so permeable that little or no reduction in dissolution rate of the glasses was observable. The residual layer thickness increased continuously at almost constant rate. The dissolution rate is lowest near neutrality and was found to be about equal at pH $= 3$ or pH $= 11$.

Therefore, it can be concluded that basaltic type glasses in contact with dilute solutions, on the one hand go completely into solution, but on the other at particular pH lose preferentially the more rapidly soluble constituents. Thus larger fragments of natural glasses can, as a result of weathering, experience significant changes in chemical composition: through the action of acid solutions alkali-deficient to alkali-free glasses form, with higher SiO_2- and H_2O- and lower alkaline earth contents. Neutral solutions produce alkali-free or -deficient products with increased Al-, Fe^{+3}- and SiO_2- and lowered alkaline earth contents. Alkaline solutions result in low alkali or alkali-free glasses with increased Fe^{+3}- and Mg- and decreased Al- and Ca-contents. Natural palagonite is produced by the action of weakly alkaline solutions (sea water) on a basaltic glass (sideromelane). It contains relatively less alkalis, alkaline earths, Al_2O_3, and SiO_2, but more Fe_2O_3, than the parent sideromelane. In addition it has a high water content.

Whether SiO_2-rich glasses react similarly in dilute solution has not yet been studied. Because of the difficulty of solution of SiO_2, resistance to solution should increase with increasing SiO_2-content. It is likely that SiO_2-rich glasses, at least in acid milieu, would form less permeable residual layers, which would exert a protective action analogous to the feldspars.

It is noteworthy, that according to these experiments in open systems, the initial incongruent reaction rates for feldspar, leucite, and tremolite are approximately equal. Since natural dissolution proceeds at very different rates (feldspar is certainly much more resistant than leucite and tremolite), it follows, that during natural weathering the main congruent phase of decomposition, greatly affected by the inhibition of the residual layer, is decisive or rate controlling. In experiments we only begin to see its effects. Thus from such experiments we are unable still to deduce quantitative measures for the weathering resistance of individual minerals.

Qualitatively it can be said that silicate dissolution rates must be lower, and thus weathering resistance higher, the higher the SiO_2-content. The residual layer is the more permeable and thus less inhibiting, the lower the SiO_2-content of the mineral. Thus the alkali-feldspars are more stable than the plagioclases, and all feldspars more stable than the alkali-rich feldspathoids. The feldspars should be more resistant than the chain silicates and these more resistant than the soro- and inosilicates. The phyllosilicates are a special group. They are easily decomposed, since, because of their structures, no surrounding protective residual layer forms.

2.33 Formation of new minerals

In the weathering environment new minerals can be formed in place of the dissolved and decomposed minerals of the parent rock. On the one hand we have the iron and aluminum oxides and hydrated oxides, as well as varieties of SiO_2; on the other hand the special layer silicates, which we call clay minerals. The oxides and hydrated oxides form as neoformations from the weathering solutions. Clay minerals can be formed either by reconstitution of primary layer silicates or crystallize authigenically from the weathering solutions. In the following sections we shall discuss the different means of new mineral formation during weathering.

2.331 Oxides and hydrated oxides

The simplest reaction products which can precipitate from weathering solutions, are the oxides and hydrated oxides of Fe and Al, as well as various forms of SiO_2.

Oxides and hydroxyoxides of aluminum (GARRELS & CHRIST 1965): The following minerals occur as neoformations in the weathering zone:

γ-Al(OH)$_3$ . . . . gibbsite
α-AlO(OH) . . . . boehmite
γ-AlO(OH) . . . . diaspore

Aluminum hydroxide forms with water, either Al^{+3}-cations or AlO_2^--anions:

$$Al(OH)_3 + 3H^+ = Al^{+3} + 3H_2O \quad \quad \text{I} \quad \quad (2.10)$$
$$Al(OH)_3 = H^+ + AlO_2^- + H_2O \quad \quad \text{II} \quad \quad (2.11)$$

For these reactions the following equilibrium constants apply (at 25 °C):

$$K_I = \frac{[Al^{+3}]}{[H^+]^3} = 10^{5.7} \quad \quad (2.12)$$

$$K_{II} = [H^+][AlO_2^-] = 10^{-14.6} \quad \quad (2.13)$$

From these one obtains the following equations for the activities (equal to concentrations in very dilute solution) of the aluminum ionic species:

$$\lg Al^{+3} = 5.7 - 3\,pH \quad \quad (2.14)$$
$$\log AlO_2^- = pH - 14.6 \quad \quad (2.15)$$

In fig. 2-5 the solubility of gibbsite in pure solution, based on these equations, is plotted as a function of pH. Below pH = 5.1, Al^{+3}-ions form predominantly; above pH = 5.1, predominantly AlO_2^--ions. At pH = 5.1 gibbsite has its lowest solubility. Since Al^{+3} has the tendency to form complex ions with different anions, this diagram applies only for solubility in pure water.

Based on their free energy values, diaspore is more stable than boehmite. The difference, however, is very small (0.1 kcal/mol), so that the occurrence of both forms in nature is understandable. In the presence of free water, gibbsite is more stable than diaspore. The difference in their free energies is 6 kcal/mol. The reaction

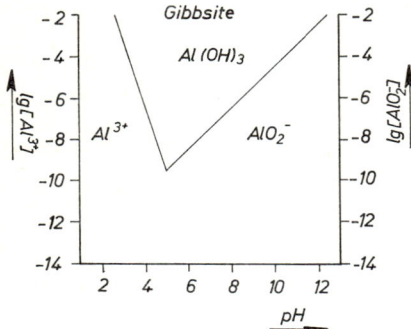

Fig. 2-5. Solubility of Al(OH)$_3$ in aqueous solution (GARRELS & CHRIST 1965).

rate, however, is very low at low temperatures. Gibbsite should, therefore, precipitate first from solution. Upon drying the soil it can transform to diaspore or boehmite, which in turn, because of the sluggishness of the reaction, are able to persist metastably during following wet periods.

Oxides and hydroxyoxides of iron (GARRELS & CHRIST 1965):
The following minerals occur as neoformations in the weathering zone:

α-FeO(OH) . . . . goethite
γ-FeO(OH) . . . . lepidocrosite
Fe$_2$O$_3$ . . . . hematite
Fe$_3$O$_4$ . . . . magnetite

Freshly precipitated "ferrihydroxide" is also unstable in the presence of water and always transforms to FeO(OH) or hematite. Similarly in the absence of oxygen, ferrohydroxide precipitates unstably and transforms to magnetite:

$$3\,Fe(OH)_2 \rightarrow Fe_3O_4 + H_2 + 2\,H_2O \tag{2.16}$$

Therefore, at normal pressure ferrohydroxide persistantly converts to magnetite.

The stability fields for the different Fe-minerals have been determined from solution oxidation potentials and pH's. Since thermodynamic data are lacking for goethite and lepidocrosite, only hematite and magnetite are considered as precipitates in the following discussion, which relates to the published diagram of GARRELS & CHRIST (1965). In the hematite stability field FeO(OH)-minerals can occur also.

For the oxidation of magnetite to hematite we have the following equation:

$$2\,Fe_3O_4 + H_2O = 3\,Fe_2O_3 + 2\,H^+ + 2\,e^- \tag{2.17}$$

For equilibrium between these phases, from free energies of the separate components, the following relation between oxidation potential Eh (volts, with reference to standard hydrogen electrode) and pH can be calculated:

$$Eh = 0.221 - 0.059\,pH \tag{2.18}$$

From this equation the boundary between the stability fields of magnetite and hematite is drawn in fig. 2-6.

2.33 Formation of new minerals

In solutions in contact with magnetite and/or hematite the following ions occur:

Fe^{+3}, $Fe(OH)^{+2}$, $Fe(OH)_2^+$
Fe^{+2}, $Fe(OH)^+$, $HFeO_2^-$

For the summary desired here it is sufficient only to consider the Fe^{+3}- and Fe^{+2}-ions, since the other ionic species occur only in exceptionally low concentrations (exact values in GARRELS & CHRIST 1965). The following equilibria are to be considered.

$$Fe_2O_3 + 6H^+ = 2Fe^{+3} + 3H_2O \quad \ldots \quad I \quad (2.19)$$
$$Fe_3O_4 + 8H^+ = 3Fe^{+3} + 4H_2O + e^- \quad \ldots \quad II \quad (2.20)$$
$$2Fe^{+2} + 3H_2O = Fe_2O_3 + 6H^+ + 2e^- \quad \ldots \quad III \quad (2.21)$$
$$3Fe^{+2} + 4H_2O = Fe_3O_4 + 8H^+ + 2e^- \quad \ldots \quad IV \quad (2.22)$$

For these four reactions the following constants and oxidation potentials can be calculated from thermodynamic data.

$$\lg K_I = \lg \frac{[Fe^{+3}]^2}{[H^+]^6} = -1.45 \quad (2.23)$$

$$Eh_{II} = 0.337 + 0.177 \lg [Fe^{+3}] + 0.472 \, pH \quad (2.24)$$
$$Eh_{III} = 0.728 + 0.059 \lg [Fe^{+2}] - 0.177 \, pH \quad (2.25)$$
$$Eh_{IV} = 0.980 + 0.0885 \lg [Fe^{+2}] - 0.236 \, pH \quad (2.26)$$

For these considerations we obtain the following dependence of Fe^{+2}- and Fe^{+3}-ion concentrations in equilibrium with hematite and magnetite:

For hematite:

$$\lg [Fe^{+3}] = -0.72 - 3 \, pH \quad (2.27)$$

$$\lg [Fe^{+2}] = \frac{1}{0.059} (Eh - 0.728 + 0.177 \, pH) \quad (2.28)$$

For magnetite:

$$\lg [Fe^{+3}] = \frac{1}{0.177} (Eh - 0.337 - 0.472 \, pH) \quad (2.29)$$

$$\lg [Fe^{+2}] = \frac{1}{0.0885} (Eh - 0.980 + 0.236 \, pH) \quad (2.30)$$

In fig. 2-6, using these equations, lines are drawn for some different concentrations (10^{-8}, 10^{-6}, 10^{-4}, 10^{-2} mol/l) of Fe^{+3} and Fe^{+2}. In the strongly stippled area of fig. 2-6 the concentration of Fe^{+3}-ions is higher than Fe^{+2}-ions. It can be seen that trivalent Fe can be transported in significant amounts only in strongly acid solutions with high oxidation potentials. Under all other conditions Fe occurs essentially only as the divalent ion.

At high oxidation potentials precipitation of hematite [or FeO(OH)] occurs under relatively acid conditions (about pH = 3). At lower oxidation potentials, in

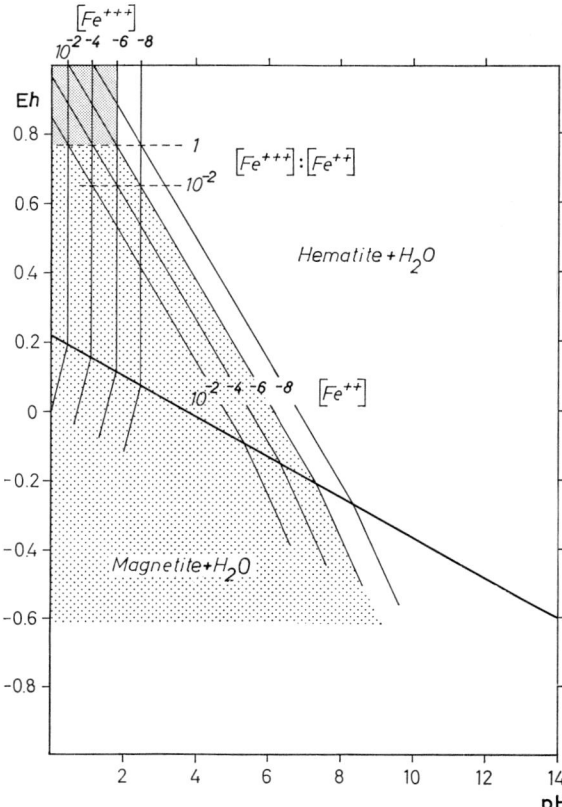

Fig. 2-6. Stability fields of magnetite and hematite in aqueous solution as a function of oxidation potential (Eh) and hydrogen ion concentration (pH). Concentration of Fe^{+2}- and Fe^{+3}-ions in saturated solutions (GARRELS & CHRIST 1965).
Heavy stippling: $[Fe^{+3}] > [Fe^{+2}]$. Light stippling: $[Fe^{+3}] < [Fe^{+2}]$; $[Fe^{+2}] > 10^{-6}$.

weakly acid solutions, divalent Fe is in equilibrium with hematite. Only when the oxidation potential falls below the hematite-magnetite boundary, does magnetite occur as a stable precipitate.

By comparing fig. 2-6 with the pH- and Eh-values of naturally occurring solutions (fig. 2-2), one can conclude under what conditions which Fe-ions and minerals are to be expected. Of course we must bear in mind that new phases occur and other equilibria prevail when such additional components as sulfur and carbon dioxide are added. Then phases such as FeS_2 and $FeCO_3$ are possible, but stable only at low oxidation potentials (Eh < 0.1) and near neutral pH (see GARRELS & CHRIST 1965). Further complication can arise as a result of organic matter, which forms Fe-complexing compounds, inhibiting precipitation as oxides or hydroxyoxides.

2.33 Formation of new minerals

SiO_2 (KRAUSKOPF 1959; SIEVER 1962)

SiO_2 occurs in soils and sediments in different amorphous forms (SIEVER 1957), as finely crystalline chalcedony, and as crystalline quartz. Neoformation and dissolution of the different forms of SiO_2 are characterized by very slow reactions between solutions and solid SiO_2.

Experimental investigations by different authors (summarized by KRAUSKOPF 1959 and SIEVER 1962) at temperatures between 0° and 200 °C have demonstrated that amorphous SiO_2 has a well-defined water solubility, which is relatively independent of the kind of amorphous SiO_2. Fig. 2-7 shows the temperature dependence of the experimentally determined solubility.

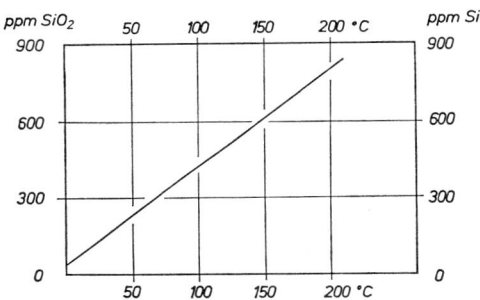

Fig. 2-7. Solubility of amorphous SiO_2 in water as a function of temperature (KRAUSKOPF 1959).

The experimentally determined solubility of amorphous SiO_2 in pure solutions of differing hydrogen ion concentration is given in fig. 2-8. At pH-values below 8, the solubility is essentially independent of pH. Above $pH = 8$ it increases with increasing pH. This development of solubility results from the fact that the solution contains monomeric H_4SiO_4, which is dissociated to a very small extent. According to ROLLER & ERVIN (1940) the first dissociation constant at 30 °C has the value $10^{-9.8}$. If the concentration of undissociated silicic acid is essentially constant over the entire range, then, as a result of dissociation, the limited increase in solubility with increasing pH only begins to be apparent at about $pH = 8$. The experimentally determined solubility curve corresponds to a constant concentration of undissociated acid of about 120 ppm (ALEXANDER, HESTON & ILER 1954).

According to measurements by GREENBERG & PRICE (1957) the solubility of amorphous SiO_2 is influenced little by the presence of salts. In a normal solution of NaCl the solubility decrease amounts to somewhat less than 10%. On the contrary, ions like Al^{+3}, which combine with silica to form insoluble silicates, clearly decrease the solubility.

The solubility of quartz is difficult to determine in the laboratory, since its dissolution rate is extraordinarily low. From measurements of different investigators, SIEVER (1962) determined its temperature dependence, as shown in fig. 2-9. According to SIEVER (1962) the solubility in water can be calculated from the following equation:

$$\log c = 4.829 - 1.132 \cdot \frac{1}{T} \cdot 10^3 \tag{2.31}$$

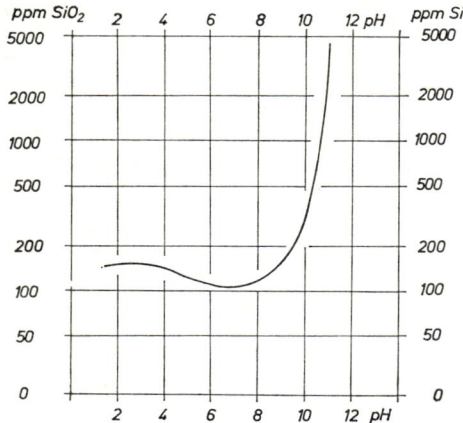
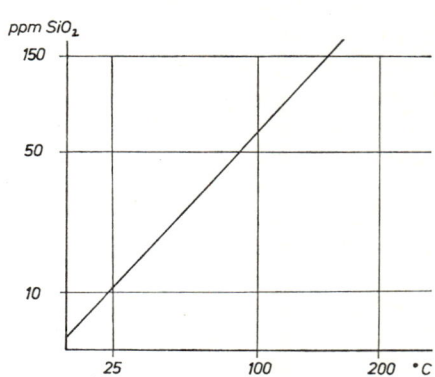

Fig. 2-8. Solubility of amorphous SiO₂ in aqueous solution as a function of pH (KRAUSKOPF 1959).

Fig. 2—9. Solubility of quartz in water as a function of temperature (after different investigators, SIEVER 1962).

where c is the concentration in ppm SiO_2 and T is the absolute temperature. On the average the solubility of quartz is about one-tenth that of amorphous SiO_2. With increasing temperature, the difference in solubility decreases. It can be assumed that for quartz the dependence of solubility on pH takes the same course as for amorphous SiO_2.

According to measurements by PELTO (1956), chalcedony has a higher solubility than quartz (51 ppm after 17 hours at 95 °C).

For all natural processes, not only the solubilities, but especially the reaction rates are important. These are very low for all the different forms of SiO_2, and different for each. The reaction rate for quartz is so low, that solutions formed by dissolution of amphorous SiO_2 can maintain supersaturation which respect to quartz for a very long time. The precipitation of quartz from supersaturated solutions is facilitated by certain substrate surfaces onto which quartz can precipitate with reduced nucleation energy. HARDER & FLEHMIG (1967) found that Al-, Fe-, Mn-, and Mg-hydroxide precipitates promote SiO_2-precipitation from dilute SiO_2-solutions. This SiO_2 changes in a short time at 20 °C to finely crystalline quartz.

The reaction rate of amorphous SiO_2 is very much higher than that of quartz. It is dependent, however, on numerous factors, and thus diverse kinds of amorphous SiO_2 occur. High specific surface raises its solution rate. Organic matter can greatly reduce the rate.

In summary, it follows that by increasing temperature and/or pH above 8, the solubility of all forms of SiO_2 is augmented. The very low reaction rate of quartz permits the metastable persistance of supersaturated SiO_2-solutions, which have formed from amorphous SiO_2. Ultimately, in a system containing co-existing quartz, amorphous SiO_2 and solution, all of the amorphous SiO_2 disappears at the expense of quartz.

2.332 Clay minerals

2.332.1 Classification of clay minerals

In tables 2-9—12, the clay mineral classification and nomenclature used in this book are summarized. For all details the reader is referred to the specialized publications of BROWN and co-workers (1961) and JASMUND (1955). In the first volume of this series (MÜLLER 1964) data for the identification of the different clay minerals can be found.

Basic to the classification is the manner in which oxygen-tetrahedral layers and oxygen-hydroxy-octahedral layers are articulated.

The two-layer minerals are condensed from a single tetrahedral and a single octahedral layer.

Table 2-9. Two-layer silicates (1:1 type).

T-sheet \ O-sheet	Non-hydrated Al dioctahedral	Non-hydrated Mg trioctahedral	Hydrated Al dioctahedral
Si	Kaolinite group $Al_4(OH)_8Si_4O_{10}$ 7 Å	Serpentine group $Mg_6(OH)_8Si_4O_{10}$ 7 Å	Halloysite group H: $Al_4(OH)_8Si_4O_{10} \cdot 4H_2O$ 10 Å Meta-H: $Al_4(OH)_8Si_4O_{10}$ 7 Å

Table 2-10. Non-hydrated three-layer silicates (2:1 type).

T-sheet \ O-sheet	predominantly Al dioctahedral	predominantly Mg, Fe^{+2} trioctahedral
Si $x = 0$	pyrophyllite $Al_2(OH)_2Si_4O_{10}$ 9.3 Å	talc $Mg_3(OH)_2Si_4O_{10}$ 9.4 Å
Si. Al $0 < x < 1$	deficient dioctahedral mica dioct. illite, glauconite 10 Å	deficient trioctahedral mica trioct. illite 10 Å
Si_3Al $x = 1$	dioctahedral mica muscovite $K Al_2(OH)_2Si_3Al O_{10}$ 10 Å	trioctahedral mica biotite $K Mg_3(OH)_2Si_3Al O_{10}$ 10 Å

x = layer charge

The three-layer minerals are condensed from a single octahedral and two tetrahedral layers.

The four-layer minerals have an additional discrete octahedral layer between two three-layer packets.

Dioctahedral minerals contain Al exclusively or predominantly in two-thirds of the available octahedral sites. Trioctahedral minerals have all of the available octahedral sites occupied completely or predominantly by Mg and Fe^{+3}.

A layer charge (x) can occur if Si in tetrahedral coordination is replaced partially by Al.

Hydrated layer silicate minerals contain interlayer water between the silicate packets.

Table 2-11. Hydrated three-layer silicates (2:1 type).

T-sheet \ O-sheet	predominantly Al dioctahedral	predominantly Mg, Fe^{+2} trioctahedral
Si, Al $0 < x < 0.5$	montmorillonite $(Al,Mg)_2(OH)_2(Si,Al)_4O_{10}$ $(Ca,Na\ldots)_x \cdot m\,H_2O$	saponite $(Mg,Fe^{+2},Al)_3(OH)_2(Si,Al)_4O_{10}$ $(Ca,Na\ldots)_x \cdot m\,H_2O$
Si, Al $0.5 < x < 0.8$	dioctahedral vermiculite $(Al,Mg)_2(OH)_2(Si,Al)_4O_{10}$ $(Ca,Na\ldots)_x \cdot m\,H_2O$ 14 Å	trioctahedral vermiculite $(Mg,Fe^{+2},Al)_3(OH)_2(Si,Al)_4O_{10}$ $(Mg,Ca\ldots)_x \cdot m\,H_2O$ 14.4 Å

x = layer charge

Table 2-12. Four-layer silicates (2:2 type).

O-sheet	Al dioctahedral	predominantly Mg, Fe^{+2} trioctahedral
	sudoite $Al_2(OH)_2Si_4O_{10} \cdot Al_2(OH)_6$ 14 Å	chlorite $(Mg,Fe,Al)_3(OH)_2(Si,Al)_4O_{10} \cdot (Mg,Fe)_3(OH)_6$ 14 Å

Remarks regarding tables 2-9 to 2-12:

The kaolinite group consists of the following polytypes:

        Kaolinite = 1 T kaolinite
        Dickite   = 2 M kaolinite
        Nacrite = 4 M kaolinite

The various kaolinite species are distinguished by the number of elementary layers per unit cell (1, 2, 4) and the resulting symmetry (T: triclinic; M: monoclinic). 1 T kaolinite, which occurs in sediments, forming at low temperatures, is probably

2.33 Formation of new minerals

the most stable form. There is also a disordered variety of kaolinite (fireclay-mineral), the disordering resulting from random orientations in the basal plane of contiguous layers.

The minerals of the **serpentine group** are antigorite (platy variety) and chrysotile (fibrous variety). Like the mineral **halloysite**, chrysotile fibers consist of individual hollow tubes.

Not tabulated are the minerals of the **berthierine-cronstedite group**. These are trioctahedral minerals with a layer thickness of 7 Å, which differ from serpentine by greater substitution of tetrahedral Si by Al and Fe^{+3}. All trioctahedral two-layer minerals, because they are chemically similar to the chlorites, are called collectively septachlorites.

Dioctahedral deficient mica differs from muscovite in that it contains less Al in tetrahedral and more Fe^{+2} and Mg substituting for Al in octahedral positions. Its layer charge is less than that of muscovite, and accordingly it contains less K. **Illite** is a deficient mica, with reduced tetrahedral Al, predominantly Al in octahedral positions, and also reduced K-content (HOWER & MOWATT 1966). **Glauconite** is a deficient mica with lower tetrahedral Al-content, with much Fe^{+3} in addition to Fe^{+2}, Mg, and Al in octahedral sites, and also with reduced K-content.

Trioctahedral illites correspond to the dioctahedral varieties, but are much rarer.

The **hydrated three-layer silicates** swell in water, as a result, increasing interlayer separation from water imbibition. **Montmorillonite** and **saponite** belong to the more extensive montmorin- or smectite-group of clay minerals. These minerals have lower but relatively constant negative layer charge, resulting in part from tetrahedral substitution of Al, and in part from octahedral substitutions. This layer charge is balanced by loosely bonded, exchangeable cations between the layers.

Vermiculite differs from the smectites by its higher layer charge. There is probably no sharp boundary between the vermiculites and smectites. The layer thicknesses indicated in table 2-11 for vermiculite correspond to air-dried material.

For **sudoite** an ideal formula is given in table 2-12, which corresponds to the composition of kaolinite. The few analyses available for sudoite (MÜLLER 1963) show that tetrahedral Si is partially replaced by Al. In addition divalent ions (Mg) occur in octahedral sites, giving total octahedral occupancy in excess of dioctahedral.

As the composition of sudoite corresponds to that of kaolinite, so does **chlorite** correspond chemically to the berthierine group.

2.332.2 Alteration of primary layer silicates to clay minerals

Clay minerals can form by alteration of primary layer silicates, when individual components are dissolved out of the primary structure and replaced by other constituents, while at the same time a remnant of the old structure remains to serve as a solid scaffold.

The especially common alteration of **micas**, resulting in loss of potassium and reduction of layer charge, is called degradation (GRIFFITHS 1952; MILLOT 1964). The opposite process, aggradation, that is, K-uptake and layer charge increase in a deficient mica plays a role in sedimentation and diagenesis. Degradation begins at the edge of each individual mineral platelet, progressing inward along the basal planes. As a result heterogeneous reaction products occur, whose characterization and classification cause considerable difficulty. The following types of heterogeneous alteration minerals can be distinguished.

In the so-called interstratified or mixed-layer minerals (MACEWAN, RUIZ AMIL & BROWN 1961) layers of differing composition alternate with each other (for example, mica and montmorillonite layers). Mixed-layering of alteration products resulting from weathering is usually likely to be random in nature (LUCAS 1962).

In minerals with so-called intergradation structures (JACKSON 1963a) isolated domains or "islands" of different composition occur between the same layers. An example would be chlorite-islands between montmorillonite layers.

Both mixed-layering and intergradational structure may occur together in combination.

The trioctahedral micas (biotites) lose potassium as a result of weathering. Charge balance can be maintained by exchange of K by hydronium-ions, by the oxidation of divalent Fe and by exchange of tetrahedral Si by Al. These trioctahedral illites occur with diminished K-content. They are much less stable than the dioctahedral illites and thus are much rarer than the latter. Illites are, for the most part, heterogeneous minerals displaying mixed-layer and intergrade structures. Mica layers which have lost their layer charge to a substantial degree, become expandable like the smectite clay minerals. In many weathered soils in humid climates, this sort of illite-smectite mixed-layering is observed, resulting from the weathering of biotite (WALKER 1950; GJEMS 1960; DROSTE, BHATTACHARYA & SUNDERMAN 1962).

In numerous soils and weathered rocks, one can recognize members of an alteration series, ranging from biotite, through mixed-layer minerals to vermiculite (examples: French podsols, MILLOT & CAMEZ 1963: Germany, MATTHES 1950; Scotland, WALKER 1950, MITCHELL 1955, 1963; Siberian platform, KOSSOVSKAYA, DRITS, & ALEXANDROVA 1965; North America, BASSETT 1963). Alteration of biotite to vermiculite takes place with only slight reduction in layer charge, indicating, therefore, less severe change than alteration to smectite. Since all gradations between trioctahedral vermiculite and saponite appear to occur, their characterization in any particular case is difficult (WALKER 1957). As indicated by the chemical composition of vermiculite, its formation form biotite proceeds in the following manner (FOSTER 1963): First of all divalent Fe in the octahedral layer is oxidized to the trivalent state and a corresponding amount of K is expelled. This process requires a pH less than 7 and produces free hydrogen on the one hand, and an alkaline reaction on the other. Secondly K is replaced by strongly hydrated Mg-ions, which are exchangeable in contrast to Mg in the octahedral layers. The degree of hydration of Mg-ions depends on water vapor pressure and restricts the intercrystalline swelling of vermiculite. The weakly acid, Mg-bearing solutions necessary for the transformation of biotite to vermiculite occur in humid climates in soils developed on biotite rich rocks.

Also chlorite minerals can alter to vermiculite in humid climates, forming chlorite-vermiculite mixed-layer intermediates. In this case the cations of the interlayers, which are octahedrally coordinated by hydroxyls in chlorite, are progressively replaced by hydrated cations, especially Mg (HARRISON & MURRAY 1959; DROSTE, BHATTACHARYA & SUNDERMAN 1962). Ultimately smectite interlayers can be formed. In this way, strongly swelling minerals occur, which are recognized as randomly interstratified chlorite and saponite or vermiculite and saponite. Various occurrences of these have been described as swelling chlorite (STEPHEN & MACEWAN 1951; MARTIN-VIVALDI & MACEWAN 1957; ROY & ROMO 1957; VENIALE & VAN DER MAREL 1963).

Dioctahedral micas undergo alteration by weathering with removal of K and reduction in negative layer charge. Dioctahedral illites result. A recent summary of chemical analyses of carefully purified illites from sediments of Cambrian through Eocene age (HOWER & MOWATT 1966) indicates the following characteristic differences from muscovite. While in ideal muscovite tetrahedral sites are occupied by Si and Al in the ratio 3:1, in illites this ratio varies between 3.4:0.6 and 3.9:0.1. The negative interlayer charge is less and the K-content correspondingly lowered. In addition a high water content results, water molecules presumably filling K-sites between silicate sheets. Many K-deficient members of the group have been proven by X-ray studies to be illite-montmorillonite mixed-layer minerals (MACEWAN et al. 1961). All gradations exist between non-swelling illite, illite-montmorillonite mixed-layers, and montmorillonite. Since, as already mentioned, dioctahedral illites are much more stable than trioctahedral ones, they are found widely distributed in soils of humid climates and in sediments derived from them.

Analogous to the formation of trioctahedral Mg-vermiculite from biotite is the frequent deposition of $Al(OH)_3$ between silicate layers in K-deficient or K-free derivative dioctahedral micas in soils of humid climates (JACKSON 1963 a, b). In this way minerals form, which can be described as dioctahedral vermiculites and which can likewise occur as mixed-layer montmorillonite-sudoite or as intergradational structures with islands of sudoite within montmorillonite layers. These kinds of minerals are found in different soils (BROWN 1953; DIXON & JACKSON 1962; RICH & COOK 1963; GLEN & NASH 1964). An end member of this alteration series is sudoite, which ideally consists of neutral dioctahedral three-layer packets with intercalated $Al(OH)_3$-layers. Sudoite has been found in several soils (MÜLLER 1963; BRYDON, CLARKE & OSBORNE 1961; PAWLUCK 1963; KELLER 1964). The precipitation of Al in the interlayers of illite or montmorillonite has been called the "anti-gibbsite effect" by JACKSON (1963 a, b), since the Al freed by the dissolution of feldspars and other minerals is fixed by this process. Thus gibbsite cannot form as long as montmorillonite layers are available.

It has been possible also to produce sudoite-like substances in the laboratory, by fixation of $Al(OH)_3$ in montmorillonite (LONGUET-ESCARD 1950; SLAUGHTER & MILNE 1960; BRYDON & KODAMA 1966).

WILSON (1966) described the deposition of island-like $Al(OH)_3$-layers in a weathered trioctahedral mica in a Scottish soil developed on gabbro. Therefore, we must consider also the possibility of gradations between pure dioctahedral and tri-octahedral clay minerals in weathered materials.

2.332.3 Neoformation of clay minerals

Clay minerals originate by neoformation when they form in the weathering zone from solutions generated by the decomposition of primary minerals.

From the younger geologic formations (Cretaceous-Pleistocene) bentonite occurrences are known at many places throughout the world. These consist mainly of montmorillonites, arising from the alteration of volcanic ash (glass). To these belong the bentonites of Wyoming (Cretaceous), California (Tertiary), New Zealand (Tertiary) and North Africa (Tertiary). Many of these montmorillonites certainly do not result from weathering, but rather as a result of hydrothermal action. Some have formed by the action of lake or sea water on the volcanic ash (GRIM 1953; MILLOT 1964). Nevertheless, montmorillonite does originate under certain conditions by neoformation in the weathering domain. Developed on Hawaiian basalts, at places where rainfall is especially great, is a thick lateritic weathering layer containing gibbsite and halloysite. In drier localities in Hawaii the same rocks form soils containing montmorillonite, which can be accepted definitely as having been formed as neoformations from the soil solutions (BATES 1962). MILLOT (1964) has reported on other instances where neoformations of montmorillonite developed by the weathering of volcanic ash. In all of these instances montmorillonite occurs when the soil is not strongly drained by rainfall, so that highly concentrated solutions of neutral to alkaline pH can develop.

That minerals of the smectite group can form at low temperatures is confirmed by laboratory synthesis of crystalline saponites. This is relatively easy to accomplish at low temperatures by reaction of very dilute solutions of H_2SiO_4 and $MgCl_2$. The pH of the reaction solutions must lie in the range pH 8 to 12 (HENIN & CAILLERE 1961; WEY & SIFFERT 1961).

It still has not been clarified whether minerals of the illite group can form anew in the weathering milieu. At first sight it would seem that the great bulk of illite and mixed-layer illites which occur in sediments could not be derived from primary rocks. Thus neoformation must be assumed for the great abundance of dioctahedral mica, either forming in certain soils, during sedimentation, or during diagenesis. However, it must be recalled that sedimentary micas stem not only from igneous and metamorphic rocks, but also from reworked, older sediments. Like quartz and the especially resistant heavy minerals (zircon, rutile, tourmaline) dioctahedral micas also are able to survive several weathering and sedimentation cycles, thereby experiencing repeated substantial alteration, that is, repeated degradation and aggradation.

Minerals of the kaolinite and halloysite group form where large amounts of fresh but weakly acid precipitation permeate the weathering zone. Kaolinite or halloysite forms in the soils of all climatic zones, when the rainfall is sufficiently high. They are especially prevalent in the tropics and subtropics. In the lateritic soils of these two climatic zones, developed from different parent rocks under the prevailing high temperatures and high rainfall, mixtures of kaolinite and iron and aluminum oxides and hydroxyoxides occur. The most extensive kaolin deposits of the world originate from tropical soils of earlier geologic epochs. Examples are the Bohemian Massif of Germany and Czechoslovakia, the Mesozoic kaolin deposits of Russia (PETROV 1958) and the Cretaceous kaolins of the southeastern United States (BATES 1964; JONAS 1964).

In cooler climates kaolinite neoformation is less extensive. Yet there are various known examples of soils in which the dissolution of primary silicates and kaolinite neoformation are observed to take place hand in hand (Northern Germany: v. ENGELHARDT 1940b; Alabama: GRANT 1963; BRYANT & DIXON 1964; New South Wales: BAYLISS & LOUGHNAN 1964).

In many cases halloysite neoformation has been observed in soils. Yet up to the present time it has not been possible to define the special conditions which sometimes leads to the formation of halloysite, rather than kaolinite. SUDO (1954) found halloysite along with amorphous allophane as the most important neoformed minerals resulting from the weathering of acid and intermediate volcanic glasses of the Japanese Islands. Large amounts of halloysite form under lateritic soil-forming conditions on Hawaii (BATES 1962; PATTERSON 1964). At places of high rainfall thick lateritic soils develop in which predominantly gibbsite and goethite are concentrated near the surface, and mainly halloysite at depth. In addition amorphous gels of aluminum hydroxide and allophane occur. At places where the ground water is obstructed and cannot run off freely, thick deposits of almost pure halloysite form.

Based on an estimate of the free energy of formation of kaolinite, GARRELS & CHRIST (1965) have calculated the stability field of kaolinite in relation to Al-concentration (Al^{+3} or AlO_2^-), H_4SiO_4-concentration, and pH and compared it to the stability field for gibbsite (fig. 2-5). This shows that alumina is dissolved from kaolinite when the SiO_2-concentration of the solution is greater than about 10^{-5} mole/liter (0.6 mg/l). At SiO_2-contents above this amount, the kaolinite field has about the same dimensions as the gibbsite field shown in fig. 2-5 (the pH-range is somewhat broader along both the acid and alkaline sides). Therefore, kaolinite will form from solutions of around $pH = 5$ as long as the SiO_2-content is sufficient. This is compatible with natural observations.

Although kaolinite and halloysite occur most frequently of all the clay minerals as neoformed minerals, it has proven extremely difficult to synthesize these minerals at low temperatures in the laboratory. BRINDLEY, DE KIMPE and GASTUCHE (BRINDLEY & DE KIMPE 1961; DE KIMPE, GASTUCHE & BRINDLEY 1961, 1966) have succeeded by the action of pure dilute SiO_2-solutions on aluminum hydroxide between $pH = 4.5$ and 5 (at 40 °C) in producing very small amounts of a crystalline substance which they identified by electron-microscopy as kaolinite. Earlier WEY & SIFFERT (1961) had obtained a product by reacting dilute solutions of Na-Al-oxalate and SiO_2, which gave kaolinite-like electron- and X-ray diffraction patterns. From the experience of these syntheses, it is concluded that Al must maintain 6-fold coordination and not unite in 4-fold coordination with Si. In the presence of SiO_2 this is assured only under weakly acid conditions.

2.4 Soils

2.41 General

The unconsolidated layer of rock material, which, as a result of superficial influences covers unaltered rock, and serves as the habitat and milieu for plant and animal life, is called soil. If we broaden this pedological definition to include all superficial materials, even those of the desert and polar regions which contain little organic constituents, then the term soil includes that upper layer of the earth's crust in which all of the previously described weathering processes take place. In the following discussion soils will be considered in this context.

As a result of weathering the soil consists both of the loose materials from which sediments form, and solutions which make their way into streams and ground water, and ultimately into the sea. Thus transport processes come into play which act from the surface downward, and sometimes in reverse. Through their action, in most undisturbed soils typical profiles develop; that is, a sequence of more or less sharply defined parallel horizons separated from one another. The uppermost humus-bearing layer is called the A-horizon (upper soil); unaltered bedrock constitutes the C-horizon. Usually it can be established that in the A-horizon decomposition and dissolution of primary minerals has taken place. In many soils another B-horizon (subsoil) lies between A- and C-horizons. Characteristic of the B-horizon is the precipitation and enrichment of some of the constituents of the primary rock, mobilized in the upper soil.

To a lesser extent than all other geologic formations a soil can be viewed as the end-product of a completed process. On the one hand the uppermost soil layers are more or less rapidly eroded and transported by streams, forming new sediments. On the other hand weathering proceeds continuously, leading to more and more severe alteration of primary materials, and deeper and deeper weathering of the parent rock. Each soil in its immediate state is in a temporary stage of a long lasting process. Climatic changes or changes in vegetation can steer the once initiated development of weathering and soil formation in new directions. Therefore, it is often difficult to ascribe the observed tendencies of soil development and weathering to well-defined external conditions.

As soil science has shown, it is possible to reduce the soils of the world to a few main types. Understanding these soil types and weathering processes related to them is of considerable importance in understanding sediments. We can deduce from the mineral content of a sediment the climatic conditions which prevailed at the time of formation, only when we know how soil formation and weathering depend in turn on climatic factors.

For a detailed description of soils and their development, we must refer to pedology texts (for example, SCHEFFER & SCHACHTSCHABEL 1966, and other literature). In the following discussion the most important soil types are considered, mainly from the viewpoint of process of mineral decomposition and synthesis, so far as is today possible.

2.42 Factors in soil formation

Various factors determine the nature and tempo of soil development and weathering.

By far the most important factors are climatic, especially the mean annual temperature and the ratio of precipitation to evaporation. This ratio can be measured and expressed as the N/S-quotient or as saturation.

The N/S-quotient (MEYER, see SHEFFER & SCHACHTSCHABEL 1966, p. 333) is the ratio of the average annual precipitation to average saturation deficit of the air. The saturation (LAATSCH, see SCHEFFER & SCHACHTSCHABEL 1966, p. 333) is the excess annual precipitation, after subtraction of surface water runoff, divided by annual evaporation (in mm). The overwhelming importance of climate arises from the fact that surface soils are distributed according to major climatic zones. These are expressed especially well on the large continents (North America, Eurasia). For example, in the European part of the U.S.S.R., the following soil types follow in sequence from north to south: Arctic wasteland soils, tundra soils, podsols, parabrown earths, black loams or czernozem, and chestnut colored soils. The same zonation is repeated in mountanous areas, higher elevation soils corresponding to those of the northern latitudes.

High N/S-quotients or high saturation values characterize soils of humid climatic zones. There large amounts of fresh rain water constantly flow downward and through the soil. In dry climates such inundation of the soil is decreased in significance. In arid climates downward flow in the soil occurs only during occasional rain downpours. During dry periods solutions at depth rise upward and evaporate in the upper horizons. The influence of variable water penetration of soils is modified by the influence of average annual temperature, since the rate of all chemical weathering reactions increases markedly with temperature.

The influence of parent rock makes itself noticeable during early stages of soil development or when very contrasting rock compositions are involved. If the weathering is very intensive, as under tropical rain forests, uniform soil types are developed from different rocks. Under less extreme conditions, weathering of Ca- and Mg-rich basic rocks, which inhibits the formation of strongly acid soil solutions, proceeds more strongly than for SiO_2-rich rocks. Special soil types develop on carbonate rocks. As long as carbonate persists and is not completely dissolved, little silicate mineral decomposition takes place.

The real soil lies above the ground water table, that is, above the zone in which pore spaces are continuously filled with water. If, at certain times the upper edge of the ground water extends up into the soil profile, this zone can be characterized by certain precipitates and chemical reactions. Such soils are included in the hydromorphic soil group.

The influence of plants and animals living in and on the soil is very important and cannot be dealt with in detail here. Meriting special emphasis is the role of vegetation in producing, from different plant forms, under different climatic conditions, different types of humus. This collects in the upper soil, forming acid solutions which accelerate decomposition of primary minerals. Chemical complexes can be formed between inorganic constituents (Al, Fe, SiO_2) and humic substances, which facilitate their transport in solution. Low oxidation potentials may be produced ultimately due to the easily oxidized humic substances.

Surface relief is an important factor in weathering and soil formation. In terraines with diversified relief different subclimates prevail side by side, leading to different soil development within a limited area. Also the extent of erosion, that is, the removal of the uppermost soil horizons, depends greatly on the slope of the land. A region with high relief will supply much loose mantle material which is little altered from the initial parent rock. On the contrary on a flat continental segment, even when weathering is slow and of low intensity, thick soils and weathering horizons develop, which consist of highly altered material.

An important factor in soil formation is time. Under constant climatic conditions a soil experiences successively different stages of development which typify a respective climate. The time span necessary to reach a certain stage depends, under like climatic conditions, very much on parent rock and vegetation.

2.43 Soil types

Podsols are characteristic of extreme types of soil development in humid-cool to temperate climates. Soils of this type are distributed in the northern parts of Asiatic and European Russia, northern and north-central Western Europe, on the Canadian shield, and in the northeast United States. Beneath a humus layer a leached eluvial horizon (A) follows, and then a dark colored illuvial horizon (B, hard pan). This passes transitionally into unweathered parent rock (C). The A-horizon extends to a depth of 20—60 cm; the thickness of the B-horizon amounts to about 10—20 cm. Podsols are developed especially easily on permeable rocks, containing little Ca and Mg (sands, sandstones, granites, gneisses). Certain plant varieties which supply an acid humus, such as conifers and heather, promote development of these soil types. The abundant rainfall produces a downward flow of solutions which have a low pH (approx. 3—4) in the upper soil. In these solutions primary minerals are dissolved, especially the feldspars. Thus characteristic feldspar solution forms and a reduction of feldspar content is observed in the A-horizon. Micas experience varied transformations, which lead to illites, illites with intercalated swelling layers, vermiculites and to montmorillonites. Small amounts of kaolinite may be observed as a neoformed mineral.

Humic substances flushed out of raw humus form complexed compounds with Al and Fe. In these Fe is present predominantly in the divalent state. These complexes can be transported over considerable distances. SiO_2 may be carried away also under the influence of organic components, since observed dissolution of feldspar is accompanied by only slight kaolinite neoformation, and since mica transformation requires little SiO_2.

Some of the components removed from the upper soil are enriched in the B-horizon. The continuous increase in pH, as soil solutions percolate downward, is primarily responsible for precipitation of new phases. Clay minerals, especially those that formed in the upper soil, are mechanically filtered out and concentrated in the B-horizon. The dark color originates from organic matter which is deposited here. Part of the Al is deposited as gibbsite and Fe as goethite. Al is introduced also into K-deficient illite interlayers, so that sudoite or sudoite-bearing mixed-layer phases occur.

Under milder conditions the brown earths (brown forest soils, sol brun) develop in humid-temperate climates. They are widely distributed in central Europe. An approximately 20 cm thick A-horizon is grey brown in color and passes gradationally into a brown B-horizon, whose thickness can vary between 20 and 150 cm. Transition into the C-horizon is gradational also. Soil solutions are distinctly acid, although less so than in podsols. The brown pigment, released from dissolved silicates, results from the oxidation of Fe to goethite.

Parabrown earths (grey brown podsols, sol brun lessive) are very widely distributed soils in the humid-temperate climatic regions of Eurasia and America. They are distinguished from brown earth soils by the distinct differentiation of the A- and B-horizons. The A-horizon is impoverished in clay and lighter in its lower portion and can be up to 60 cm thick. An enrichment in clay is characteristic of the lower 40—400 cm thick B-horizon.

To the extent to which they have been studied, dissolution of primary minerals, especially feldspars, is observed in brown earth and parabrown earth soils. If the parent rock contains carbonates, these are completely dissolved in the upper soil. Like the podsols these soils contain illite, illite with swelling layers, vermiculite, and sudoite, or sudoite-bearing mixed-layers, all forming from micas. Formation of kaolinite and gibbsite in these soils is at most minor.

Soils containing predominantly or only montmorillonite clay minerals are rare. They form in humid-temperate and in warm climates from certain special parent rocks and under special conditions. Volcanic rocks, especially glass-rich ejecta, produce montmorillonite-rich soils. In addition, to obtain montmorillonite it is necessary that the soil be poorly drained. This can be achieved by low soil permeability and low relief (poorly drained depression). In any case, montmorillonite is permanently stable only in neutral or weakly alkaline milieu. A high Ca- and Mg-concentration is desirable.

Rendzina soils develop on carbonate rocks in humid-temperate climates. These consist of a dark humus-rich upper horizon (A) lying without a B-horizon directly on unweathered rock (C). The upper soil is impoverished in carbonate relative to the parent rock, but still contains enough carbonate that the soil solution is alkaline or at least weakly acid. Apart from carbonate dissolution chemical alteration is minor. Primary silicates are only slightly dissolved and neoformed and transformed minerals play a minor role.

Widely distributed on the grassy plains or steppes of Eurasia and America are the black earth soils (czernozems). A black, humus-rich A-horizon up to over 100 cm thick lies directly on a C-horizon, whose upper layer is enriched in concretionary $CaCO_3$. The black earths form preferentially from loose sediments (loess) with finely divided $CaCO_3$ under steppe vegetation and, therefore, in relatively dry climates. The soil solution of the upper soil is weakly alkaline to weakly acid. In the upper soil the $CaCO_3$ is completely or partially dissolved.

The hydromorphic soils have a zone of fluctuating ground water level beneath a humus-bearing A-horizon. In the lower zone the dissolved Fe which occurs in the divalent state in ground water is oxidized and precipitated as goethite. This results in the formation of either rust colored mottling or solid limonitic crusts or concretions. Also $CaCO_3$ is deposited if Ca-bearing ground water loses CO_2 in this zone.

In the humid climatic zones of the tropics (central Africa and South America, India, Indonesia, Pacific Islands) soil formation is governed by widely distributed kaolinitic-lateritic soils (see MILLOT 1964). Under the influence of high precipitation and high average annual temperature, intensive dissolution of all minerals of primary rocks takes place. Only a few minerals, which precipitate together by neoformation from the weathering solutions, form the constituents of mature lateritic soils. These are gibbsite, boehmite, diaspore, goethite, hematite, and kaolinite. The widespread reconstitution of three-layer silicates typical of cooler climates is insignificant or absent in laterites.

For lateritic decomposition of primary rocks it is desirable to have low relief and good drainage, so that ground water can penetrate deep into the bedrock and easily flow off. Under the most favorable conditions a decomposition zone of much greater thickness forms than can occur in temperate and cool climates. For example, the lateritic zones of decomposition under the rain forests of equatorial Africa are 10 to 20 m thick.

The details of formation of lateritic soils and the mineral distribution in profile are very numerous. The amount of rainfall, the nature of water movement in the profile, the amount and kind of organic substances, the parent rock, and other factors exert modifying influences, which have not been studied sufficiently.

According to MILLOT (1964) drainage above all plays an important role. If the soil is quickly and completely permeated by a deep-lying ground water level, primarily gibbsite, boehmite, diaspore, goethite, and hematite occur. If the drainage is less adequate and the ground water level stands higher, kaolinite forms. Since, during the course of development of a soil profile or as a result of annual changes, these conditions can change, diverse proportions and distributions of individual minerals are found which often cannot be explained by a single, simple sequence of weathering events. A good example of the dependence of drainage conditions is soil formation on Hawaii, which was mentioned earlier.

The relations are complicated further if ground water flow and changes in ground water table occur within the lateritic decomposition profile. Below the ground water table precipitation of solid crusts and concretions of Al and Fe oxides and hydroxyoxides occurs, if the ground water is oxidized. Hard, red colored horizons originate which, after subsequent erosion of the overlying layers, appear as lateritic "armour". Bauxites are low-Fe concentrations of Al minerals enriched in this way.

The conditions of formation of gibbsite and kaolinite, as observed naturally are qualitatively in accord with the theoretical stability relations of these minerals, which were discussed earlier. It must be assumed that gibbsite formations occurs when copious amounts of fresh ground water are supplied so rapidly that soil solutions formed achieve insignificant SiO_2-concentrations. Under poorer drainage conditions or in deeper soil horizons, the critical boundary value for SiO_2-concentration is exceeded, permitting formation of kaolinite instead of gibbsite. Of course it is possible that as conditions change, earlier deposited gibbsite may transform to kaolinite, or the reverse may also happen. Therefore, in lateritic soils different sequences of mineral genesis may be observed; for example, feldspar → gibbsite → kaolinite in the bauxites of Arkansas (GORDON et al. 1958) and in lateritic soils of Nyassaland, Africa (STEPHEN 1963) or feldspar → halloysite → gibbsite on Hawaii (BATES 1962).

The question as to what precise conditions lead to formation of gibbsite or diaspore or boehmite is not yet answered. The supposition of some authors (KELLER 1964), that gibbsite originates above the ground water table, and diaspore below it, is not in agreement with thermodynamic consideration. Since in some bauxites all three minerals occur together, it must be presumed that these compounds can exist metastably for long times as a result of very low reaction rates.

Transport and deposition of Fe in lateritic soils are determined by pH and oxidation potential. High organic content and corresponding low oxidation potential facilitate the removal of Fe in the divalent state. Under conditions of lateritic weathering a substantial part of the Fe is oxidized by atmospheric oxygen and deposited as hematite, giving the intensive red color to these soils.

Even when weathering is not extensive enough to lead to extreme laterization, as in humid tropic and subtropical regions, soils are more or less red-colored by hematite or goethite. When developed on silicate rocks, so-called red loams are formed; on carbonate rocks, the terra rossas.

Montmorillonitic soils also form in tropical and subtropical climates if the water permeability of the soil is inhibited in any way and if the parent rock (volcanic rocks) composition favors the formation of neutral or alkaline, highly concentrated soil solutions. Hawaii provides an excellent example, where, in dryer regions montmorillonite forms instead of the usual wide spread lateritic deposits (BATES 1962).

Soils of arid regions are very low in organic matter. Two important factors of soil genesis are absent here: organic acids resulting from humus development and the low oxidation potential important for Fe transport. In addition downward flushing of the soil by ground water occurs only seldomly. Prominent profile development is usually lacking. In warm regions oxidation of Fe produces red colors. As a result of evaporation of upward moving ground water, deposits or crusts form at the surface or deeper, which are enriched in soluble minerals such as gypsum or calcite. At the surface, brown or black Fe- and Mn-oxide coatings form from evaporating solutions (desert varnish). Also on desert surfaces deposition of quartz, chalcedony or opal can occur, from evaporating solutions circulating through quartz sands (MILLOT, LUCAS & WEYL 1963).

Soils of the vegetation-barren cold regions (arctic, antarctic, high mountains) consist of a A-horizon lying directly on the C-horizon. There is little chemical alteration and mechanical processes predominant in reducing parent rock to loose material. Loess is the fine fraction of such soils, transported by the wind.

In summary, it follows that subaerial weathering yields different end products in different climatic zones.

In humid-cool to humid-temperate climates complete weathering leads to a soil consisting of quartz as a insoluble weathering residue (also minor zircon, rutile, tourmaline), and predominantly dioctahedral three-layer silicates, such as illite, vermiculite, and montmorillonite. In addition sudoite and the corresponding mixed-layer minerals occur along with some kaolinite.

In humid-tropic and humid-subtropic climates primary minerals are destroyed more rapidly and intensively. Weathering is characterized by a stronger removal of SiO_2 and leads to an end product containing kaolinite (or halloysite), gibbsite, boehmite, diaspore, goethite, and hematite in quite variable proportions.

In arid and extremely cold regions, soils develop that in general are mechanically reduced in size and which differ little chemically from their parent material.

3. Transport and Deposition of Clastic Constituents

3.1 Transport and deposition in water

3.11 General

The comminuted, altered and newly formed material, resulting from continental subaerial weathering, is extensively eroded by surface waters and carried into streams. The base of the layer continues to develop downward into the primary rocks, while at the same time material at the surface is being removed. The extent of removal depends on local conditions, such as annual rainfall and evaporation, surface slope, bedrock conditions, and other factors. Streams carry both fine and course sediment particles into the collecting basins of continental depressions and the sea. As a consequence of this transport particles are sorted according to size and shape, they are altered physically in size and shape, and chemically altered by dissolution and reconstitution. Within continental and marine basins additional transport processes generally join in before deposition and sedimentation come to an end. Especially in the ocean, transported mineral and rock particles are exposed to further chemical alteration, until at the site of deposition they are buried by new sediment layers and are isolated from direct effects of the depositional environment.

The following chapter considers these processes in particular. In the first section mechanisms of transport in streams and in the ocean are described. Additional sections deal with changes which sedimentary materials experience during transport and until covered with sediment. We shall distinguish between the mechanical effects of transport and subaquatic weathering, that is, the chemical effects of the transport and depositional medium on mineral and rock particles. A last section considers the mineral composition and grain size distributions of clastic sediments formed in rivers, deltas, lakes, and in the oceans, as they reflect these sedimentary processes.

3.12 Laminar and turbulent flow[1]

At low velocities water flows in a laminar fashion. Thin layers of water slide over one another without mixing. This process is illustrated by means of a simple model (fig. 3-1): The space between two parallel plates, separated by distance a, is

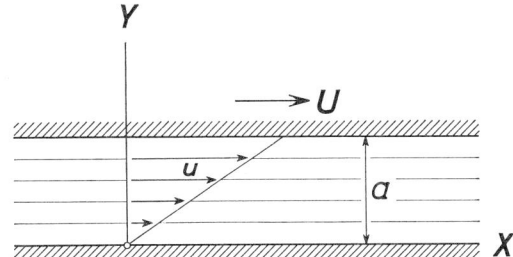

Fig. 3-1. Laminar flow (explanation in text).

filled with water. If the upper plate is moved parallel with a velocity U with respect to the lower plate, the fluid layer immediately in contact with the plate has the same velocity as the plate (it adheres to the plate). The intermediate water layers slide over one another with a velocity u, which is proportional to the distance y from the stationary plate.

$$u = U \frac{y}{a} \qquad (3.1)$$

Therefore, a linear velocity gradient prevails between the plates. All real fluids possess a certain viscosity, which makes it necessary to exert a certain force in order to maintain the motion of the upper plate and fluid. If τ is the necessary shear stress (force : area), then

$$\tau = \mu \frac{du}{dy} \qquad (3.2)$$

The constant μ is viscosity. It is expressed in poise = $gcm^{-1}sec^{-1}$ or centipoises = 1/100 poises. Table 3-1 gives some values for the viscosity of water at different temperatures.

---

[1] For details see hydrology texts and special publications, for example, PRANDTL (1965).

Table 3-1. Viscosity of water at different temperatures.

Temperature	Viscosity (in centipoises)
0°	1.789
5°	1.516
10°	1.306
15°	1.141
20°	1.005
25°	0.894
30°	0.802

As a model for a stream, we can consider a semicircular open channel of radius r_m, inclined in the direction of flow at an angle β with respect to the horizontal. The water is flowing in this channel as a result of its own weight. Thus water bodies of semicircular cross section glide past one another. In contact with the channel surface one such layer of radius r produces the shear stress τ_r. (Since in natural streams, β is very small, we can substitute tan β for sin β).

$$\tau_r = 1/2 \, r \, g \, \varrho_1 \, \text{tg} \, \beta \tag{3.3}$$

where g = acceleration due to gravity, and
ϱ_1 = specific weight of water.

Analogous to equation (3.2):

$$\frac{du}{dr} = \frac{\tau}{\mu}$$

Taking into account the boundary conditions and integrating, the flow velocity μ becomes zero, when $r = r_m$; for the velocity of a cylindrical layer at distance r from the midpoint of the water surface:

$$u_r = \frac{1}{4\mu} (r_m^2 - r^2) g \, \varrho_1 \, \text{tg} \, \beta \tag{3.4}$$

Water particles along the centerline of the stream (r = 0) have the highest velocity. The velocity decreases with the square of the distance from the center line, becoming zero at the bottom of the channel. On the other hand the shear stress increases from the water surface downward, reaching a maximum value at the bottom.

$$\tau_0 = \frac{1}{2} r_m \, g \, \varrho_1 \, \text{tg} \, \beta \tag{3.5}$$

As we shall see later, this maximum shear stress acting on the stream bed, is an especially important factor for transport processes in a stream. If the stream channel does not have a semicircular cross section, but rather some sort of more generalized shape, a more generalized expression of equation (3.5) applies for the bottom shear stress:

$$\tau_0 = m \, g \, \varrho_1 \, \text{tg} \, \beta \tag{3.6}$$

where m is the hydraulic radius and is equal to the cross sectional area/wetted perimeter. The free water surface is not included in the wetted perimeter. Therefore, for a semicircular open channel,

$$m = 1/2\ \pi\ r_m^2 : \pi\ r_m = r_m/2. \qquad \text{(see equation 3.6)}$$

The mean velocity U is the greater, the greater the stream gradient and hydraulic radius. Theoretically, an equation of the following type applies:

$$U = C\ \sqrt{g\ m\ \text{tg}\ \beta} \qquad (3.7)$$

where C depends principally on the roughness factor of the stream bed, but also on hydraulic radius.

With increasing velocity laminar flow passes into turbulent flow. This happens when the Reynolds number, R,

$$R = \frac{U\ \varrho_1\ m}{\mu} \qquad (3.8)$$

exceeds a critical value. This critical value depends strongly on the character of the wetted perimeter. It is greater for smooth than for rough channel surfaces. It is not possible to assign an exact and universally valid value for natural streams and currents. It probably lies in the range

$$500 < R_{\text{Crit.}} < 1500$$

Using these values for a semicircular channel of 10 meters radius, the critical velocity obtained would fall between 0.1 and 0.3 mm/sec; for a 5 meter radius, between 0.2 and 0.6 mm/sec. We can see from these values that the flow of natural streams probably always will be turbulent.

Turbulent flow is characterized by irregular, swirling, mixed motions, superimposed on the general stream flow. From these vortex motions all of the energy content or components of the flowing fluid are transferred as impulses, as heat content, to dissolved and suspended components. If the content of these constituents or the magnitude of the fluid properties varies spatially, the turbulent mixing results in transport from sites of higher to lower content.

At each site of turbulent flow, flow velocities prevail which vary in direction and magnitude. Indeed at every point, at any given time mean values of the velocity components can be resolved into three directional components $\bar{u}$, $\bar{v}$, $\bar{w}$. At any instant variations u', v', w', apply so that the actual velocities are:

$$\begin{aligned} u &= \bar{u} + u' \\ v &= \bar{v} + v' \\ w &= \bar{w} + w' \end{aligned} \qquad (3.9)$$

Turbulent flow may develop parallel to a channel wall in the direction of the x-axis (velocity component u). The y-axis (component v) runs perpendicular to the wall. Let s be some property of a constituent of the fluid (per cm³), which, as a result of turbulent mixing motions, is transported transverse to prevailing direction of flow. The following exchange equation, analogous to a diffusion process, is valid:

$$\frac{ds}{dt} = -A\frac{ds}{dy} \tag{3.10}$$

ds/dt measures the transport of s per unit area in the y-direction (perpendicular to the wall); ds/dy corresponds to the momentary distribution of s in the y-direction. A is the exchange coefficient which increases with increasing velocity.

If s is the impulse $\varrho_1 u$, transported in the y-direction, ds/dt has the significance of a shear stress τ which must be overcome by flowing in the x-direction. Theoretically this shear stress is proportional to the mean of the product of the velocity variations.

$$\tau = \varrho_1 \, u'v' = -A\varrho_1 \frac{d\bar{u}}{dy} \tag{3.11}$$

The quantity $A\varrho_1$ is the "turbulent viscosity". In the case of fully developed turbulence this quantity takes the place of the fluid viscosity μ of laminar flow. The turbulent viscosity is very much greater than normal viscosity. In contrast to the latter, it is not a material constant, but is dependent on the mean velocity, on geometric dimensions, and on other conditions which influence the turbulent mixing motions and therewith the exchange coefficients.

According to the turbulence theory of PRANDTL (1965) the following expression defines the exchange coefficient:

$$A = l^2 \left| \frac{d\bar{u}}{dy} \right| \tag{3.12}$$

l is the so-called mixing length, that is, that distance which the uniformly agitated fluid particles travel in the y-direction, before they loose their individuality. From equation (3.11) we obtain the relation

$$\tau = \varrho_1 \, l^2 \left(\frac{d\bar{u}}{dy}\right)^2 \tag{3.13}$$

$$\sqrt{\frac{\tau}{\varrho_1}} = 1 \, \frac{d\bar{u}}{dy} \tag{3.14}$$

The radical term has the dimension of velocity. If one substitutes the maximum shear stress τ_0 (at the wall) for the shear stress τ (varies with distance from the wall), we obtain:

$$u_\tau = \sqrt{\frac{\tau_0}{\varrho_1}} = 1 \, \frac{d\bar{u}'}{dy} \sqrt{\frac{\tau_0}{\tau}} \tag{3.15}$$

The quantity u_τ is called the shear velocity.

3.12 Laminar and turbulent flow

Based on dimensional considerations, KÁRMÁN (1930) derived the following expression for the mixing length l:

$$l = -K \frac{(d\bar{u}/dy)}{d^2\bar{u}/dy^2} \sqrt{\frac{\tau_0}{\tau}} \qquad (3.16)$$

From empirical measurements in cylindrical tubes near the tube walls the constant K has the value 0.4, so that from equation (3.15), we obtain:

$$u_\tau = -0{,}4 \frac{(d\bar{u}/dy)^2}{d^2\bar{u}/dy^2} \cdot \frac{\tau_0}{\tau} \qquad (3.17)$$

In the vicinity of the wall $\tau_0 \cong \tau$, so that at distances not to far from the wall we can write:

$$u_\tau = -0{,}4 \frac{(d\bar{u}/dy)^2}{d^2\bar{u}/dy^2} \qquad (3.18)$$

By integration of this equation KÁRMÁN obtained the following expression for the velocity of a turbulent stream in the vicinity of a wall (as a function of y):

$$\bar{u}_y = \frac{u_\tau}{0{,}4} \ln \frac{y}{y_0} \qquad (3.19)$$

or

$$\bar{u}_y = 5{,}75 \, u_\tau \lg \frac{y}{y_0} \qquad (3.20)$$

From this equation it turns out that the velocity distribution in turbulent flow near the tube wall is interesting. The flow velocity falls off to zero, not only at the wall, as in the case of laminar flow, but also falls off to zero, or at least very low values at some distance y_0 from the wall. This distance y_0 depends on the wall roughness. By experiments with sand-coated walls PRANDTL (1965, p. 190) determined that $y_0 \cong D/30$ where D is the diameter of the largest sand grains.

The value (K = 0.4) for the KÁRMÁN constant of equation (3.20) was derived from measurements in smooth-walled tubes. In natural streams the testing of equation (3.20) leads to divergent values. For example NORDIN & DEMPSTER (1963) found values of K varying between 0.15 and 1.20 for the middle Rio Grande (New Mexico, U.S.A.). It appears that K becomes less than 0.4 with increasing concentration of suspended matter. The presence of sandbanks in the river bed appears to increase K. Also the value of y_0 is subject to marked variations in natural streams and cannot be calculated simply from the grain size of the bed load.

For these reasons the numerical application of equation (3.20) to natural streams leads only to approximate values for the velocity distribution above the stream bed. However, the general nature of velocity distribution, that is, increase in velocity with the logarithm of height above the stream bed, has been well established in natural streams. This is shown in fig. 3-2 for some measurements from the Rio Grande.

5 Sedimentary Petrology III

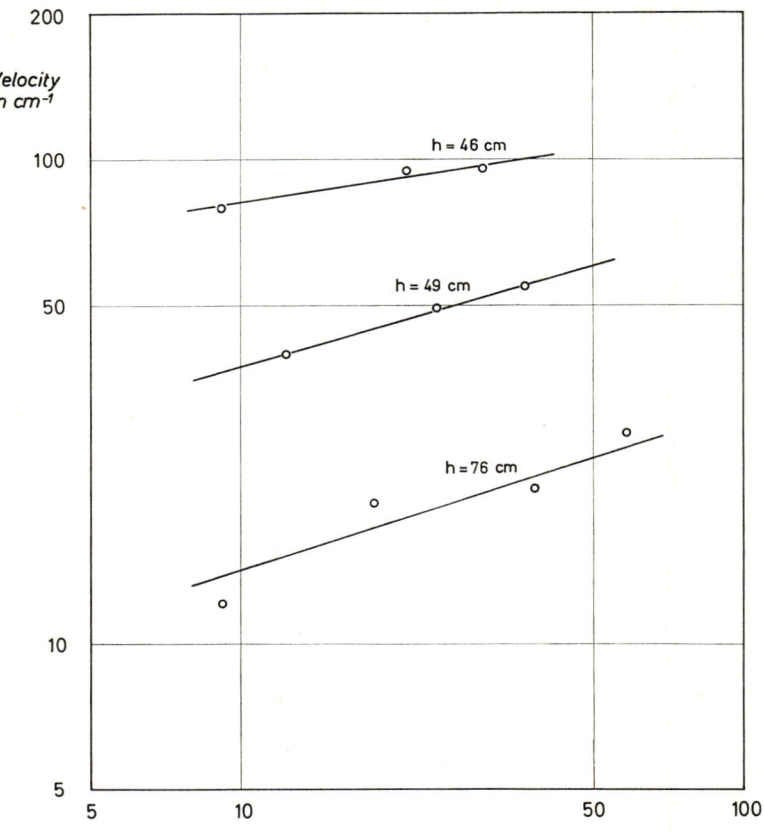

Fig. 3-2. Flow velocity distribution in the Rio Grande. Measurements taken at Bernalillo (water depth h = 29 cm) and at Socorro (water depths h = 46 and 76 cm) (after NORDIN & DEMPSTER 1963).

3.13 Transport in suspension (suspended load)

The mineral particles transported in suspension have a higher specific gravity (ϱ_2) than water (ϱ_1). Therefore, as a result of their weight they experience a downwardly directed acceleration. On the other hand the motion of particles through the water acts in opposition to a frictional force, which increases with velocity. The opposing action of both forces finally gives rise to a constant particle settling velocity, which depends on particle size and mass.

For the frictional force or resistance to particle motion through the water one can generally note

$$W = \frac{1}{2} c \varrho_1 F v^2 \qquad (3.21)$$

3.13 Transport in suspension (suspended load)

F is particle cross section perpendicular to the direction of motion (for a sphere of radius r, $F = \pi r^2$). ϱ_1 is the specific gravity of water and v the relative velocity between particle and water. c is the dimensionless resistance coefficient. If V is the volume and ϱ_2 the specific gravity of the particle, the gravitation force acting on the particle is

$$G = (\varrho_2 - \varrho_1) g V \qquad (3.22)$$

The particle will assume a settling velocity such that the gravitational and frictional forces are just in balance:

$$\frac{1}{2} c \varrho_1 F v^2 = (\varrho_2 - \varrho_1) g V \qquad (3.23)$$

If the resistance coefficient is known, the settling velocity for particles of defined form and dimensions can be ascertained from this relation:

$$v^2 = \frac{2 (\varrho_2 - \varrho_1) g V}{c \varrho_1 F} \qquad (3.24)$$

c can be expressed as a function of Reynolds number:

$$c = f(R) \qquad (3.25)$$

with

$$R = \frac{\varrho_1 v D}{\mu} \qquad (3.26)$$

where D is the particle diameter.

In fig. 3-3, after SCHILLER (1932), the empirically determined dependence of resistance coefficient on Reynolds number is shown for spherical particles. For Reynolds number up to $R = 1$,

$$c = \frac{24}{R} \qquad (3.27)$$

From (3.21), (3.26) and (3.27) we obtain the resistance law for laminar flow around a sphere, derived theoretically by STOKES:

$$W = 3 \pi \mu D v \qquad (3.28)$$

Equating (3.28) with the gravitational force acting on a spherical particle (3.22), leads to the constant settling velocity:

$$v_S = \frac{(\varrho_2 - \varrho_1) g D^2}{18 \mu} \quad \text{(STOKES)} \qquad (3.29\text{ a})$$

For the diameter of a sphere with settling velocity v we obtain:

$$D_s = \sqrt{\frac{18 \mu v}{(\varrho_2 - \varrho_1) g}} \quad \text{(STOKES)} \qquad (3.29\text{ b})$$

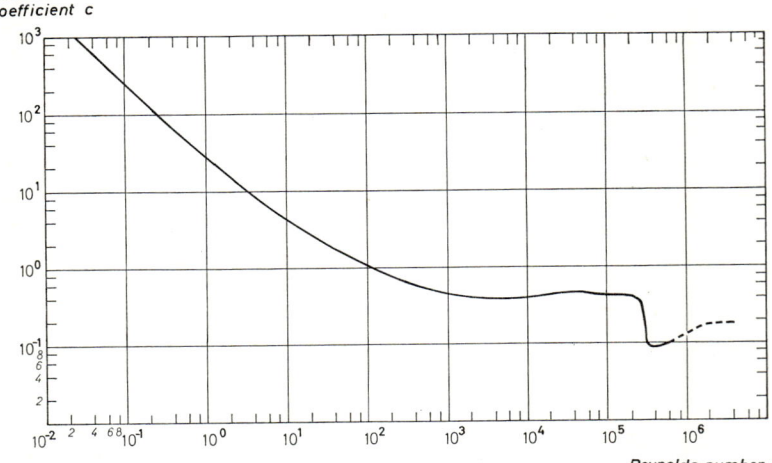

Fig. 3-3. Relation between resistance coefficient for the movement of spheres in a viscous medium and Reynolds number (SCHILLER 1932).

For the range $1 < R < 5$, OSEEN (1927) calculated a refinement of the STOKES equation:

$$c = \frac{24}{R}\left(1 + \frac{3}{16}R\right) \tag{3.30}$$

$$W = 3\pi\mu Dv\left(1 + \frac{3 D \varrho_1 v}{16 \mu}\right) \tag{3.31}$$

Equating (3.31) to the gravitational force acting on a spherical particle (3.22) gives for the settling velocity:

$$v_0 = \frac{-8\pi}{3 \varrho_1 D} + \sqrt{\frac{8(\varrho_2 - \varrho_1) gD}{27 \varrho_1} + \frac{64 \mu^2}{9 \varrho_1^2 D^2}} \quad \text{(OSEEN)} \tag{3.32 a}$$

For the diameter of a spherical particle of settling velocity v:

$$D_0 = \frac{27}{16}\frac{\varrho_1 v^2}{(\varrho_2 - \varrho_1) g} + \sqrt{\frac{18 \mu v}{(\varrho_2 - \varrho_1) g} + \frac{729 \varrho_1^2 v^4}{256 (\varrho_2 - \varrho_1)^2 g^2}} \quad \text{(OSEEN)} \tag{3.32 b}$$

For high Reynolds numbers ($R > 10^3$) the resistance coefficient decreases to a relatively constant value ($c \simeq 0.4$; the approximately horizontal portion of the curve, fig. 3-3). Here turbulence prevails and the settling velocity of the spherical particle is no longer dependent on the viscosity. For this range the resistance equation given first by NEWTON applies:

$$W = 0.05 \pi \varrho_1 D^2 v^2 \tag{3.33}$$

3.13 Transport in suspension (suspended load)

Equating (3.33) to the gravitational force (3.22), gives for the settling velocity:

$$v_N = \sqrt{\frac{10\,(\varrho_2 - \varrho_1)\,gD}{3\,\varrho_1}} \quad \text{(NEWTON)} \qquad (3.34\text{ a})$$

For the diameter of a sphere of settling velocity v we obtain:

$$D_N = \frac{3\,\varrho_1\,v^2}{10\,(\varrho_2 - \varrho_1)\,g} \quad \text{(NEWTON)} \qquad (3.34\text{ b})$$

For any pair of values of resistance coefficient and Reynolds number from fig. 3-2, one can calculate the corresponding sphere diameters and settling velocities from the following two formulae. These are derived from equations (3.24) and (3.26):

$$D = \sqrt[3]{\frac{3\,c\,R^2\,\mu^2}{4\,(\varrho_2 - \varrho_1)\,\varrho_1\,g}} \qquad (3.35)$$

$$v = \sqrt[3]{\frac{4\,R\mu\,(\varrho_2 - \varrho_1)\,g}{3\,c\,\varrho_1^2}} \qquad (3.36)$$

From equation (3.35) are calculated for quartz spheres ($\varrho_2 = 2.65$) in water the values shown in table 3-2 for the range of applicability of the STOKES, OSEEN, or NEWTON formulae for settling velocities.

Table 3-2. Range of applicability of different formulae for settling velocities of spherical quartz particles in water.

Reynolds number	Diameter in cm	Applicable formula
		} STOKES
1	$1 \cdot 10^{-2}$	
		} OSEEN
5	$2.4 \cdot 10^{-2}$	
		} Transition zone
1000	$2.6 \cdot 10^{-1}$	
		} NEWTON

For the transition between the ranges covered by the OSEEN- and NEWTON-formulae, the corresponding sphere diameters and settling velocities can be ascertained by use of the empirical curve (fig. 3-3) with the aid of equations (3.35) and (3.36).

Fig. 3-4 shows the settling velocities of spherical quartz grains in water. The previously derived formulae are strictly applicable only for individual spheres in an infinitely extended fluid. In order to use the equations for real suspensions, one must

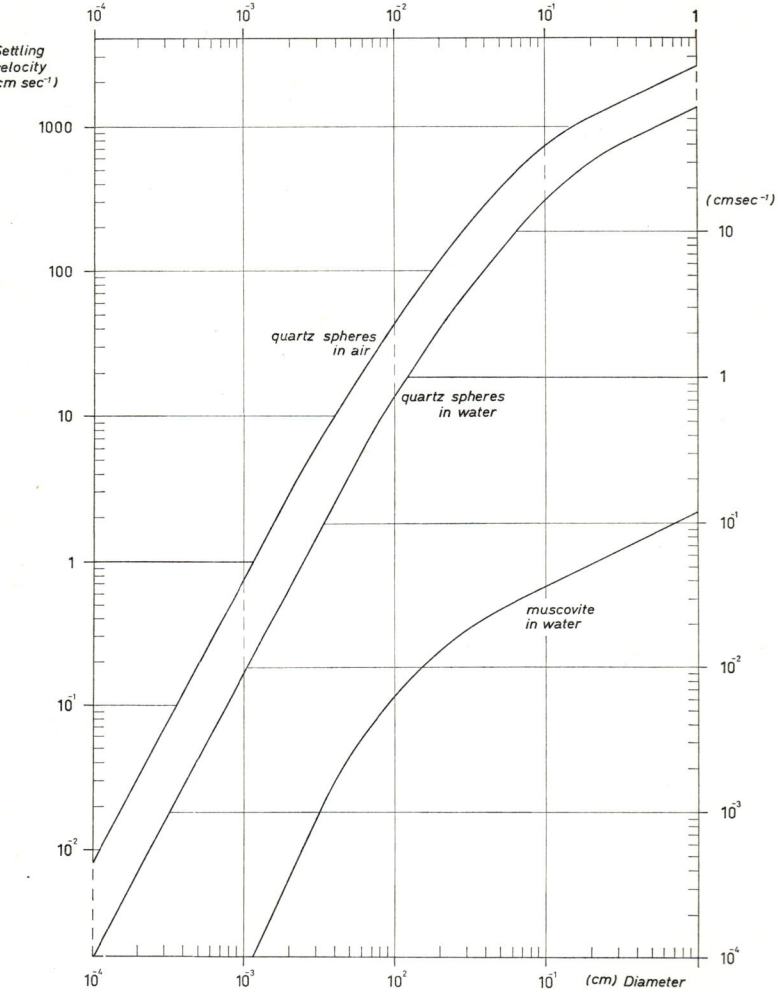

Fig. 3-4. Settling velocities of quartz spheres in water and air and of muscovite in water (20 °C).

take into account deviations from spherical shape as well as the mutual interactions of numerous particles. In highly concentrated suspensions the effect of particle interaction leads to an increase in the resistance coefficient c and with it a decrease in settling velocity. In fig. 3-5 (after BAGNOLD 1957) resistance coefficients for spheres in suspensions of different volume concentrations (C) are plotted as a function of Reynolds number.

Deviations from spherical shape can have a significant influence on settling velocities. Of all possible particle morphologies, for particles of the same weight,

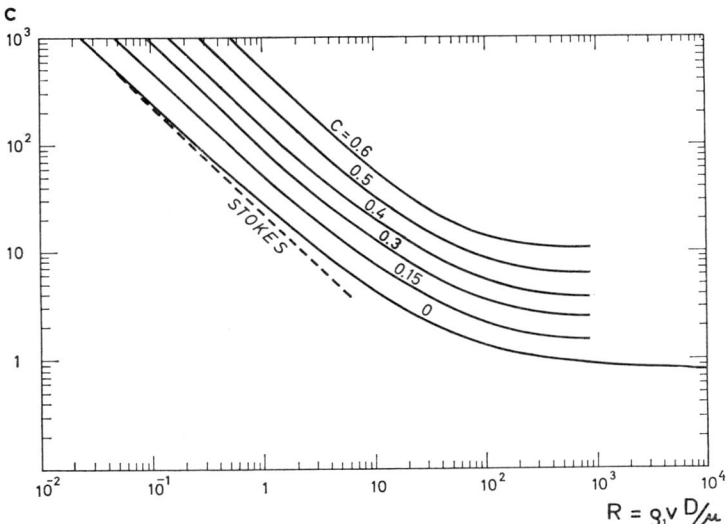

Fig. 3-5. Resistance coefficient for the movement of spheres in suspensions of different concentration (c = volume concentration) (BAGNOLD 1957).

spherical ones settle most rapidly. Fall velocity is more strongly hindered, the lower the sphericity. Similarly rough surfaces or sharp corners and edges decrease the settling velocity.[2]

For the formation of sedimentary rocks the relations of platy minerals, such as mica and clay mineral particles, are of special importance.

The resistance of a infinitely thin circular disk of radius r to a flow directed perpendicular to the disk's surface can, at small Reynolds numbers, be defined by the following expression (OBERBECK, after SCHILLER 1932):

$$W_I = 16\,\mu\,r\,v \tag{3.37}$$

If d is the thickness of the disk (d ≪ r), one obtains the following settling velocity for motion perpendicular to the disk's surface:

$$v_I = \frac{\pi\,(\varrho_2 - \varrho_1)\,gdr^2}{16\,\mu} \quad \text{(OBERBECK)} \tag{3.38}$$

For motion parallel to the disk surface, according to OBERBECK, the following expressions apply:

$$W_{II} = 10.67\,\mu\,v\,r \tag{3.39}$$

$$v_{II} = \frac{\pi\,(\varrho_2 - \varrho_1)\,gdr^2}{10.67\,\mu} \quad \text{(OBERBECK)} \tag{3.40}$$

---

[2] BRIGGS, McCULLOCH & MOSER (1962) have published results of an experimental study of the influence of particle shape on the settling velocity of natural grains of different minerals.

These equations apply for Reynolds numbers up to about 0.05. For somewhat higher values to about $R = 1$, OSEEN (1927) modified the expression as follows:

$$W_I = 16\,\mu\pi\,rv\left(1 + \frac{\varrho_1\,rv}{\pi\mu}\right) \qquad (3.41)$$

Derived from this, the settling velocity is:

$$v_I = \frac{-\mu\pi}{\varrho_1\,r} + \sqrt{\frac{\pi^2(\varrho_2 - \varrho_1)\,gc}{16\,\varrho_1} + \frac{\mu^2\,\pi^2}{4\,\varrho_1^2\,r^2}} \quad \text{(OSEEN)} \qquad (3.42)$$

By means of settling experiments, these equations have been verified to values of $R = 1$ (SCHMIEDL after SCHILLER 1932). At these small Reynolds numbers experimental disks maintain their original orientation. At higher Reynolds numbers (up to $R = 80$) they had a tendency to orient themselves with the flat surface perpendicular to the settling direction. At R-values between 80—300 oscillations about an axis in the plane of the disc were observed. Thereby the center of gravity did not move linearly downward, but oscillated back and forth in the plane of oscillation.

As an example of the influence of grain shape, the settling velocity of mica platelets in water is shown also in fig. 3-4. The specific gravity of mica was assumed to be $\varrho_2 = 2.80$. The mica platelets would be circular discs of radius r_M and thickness $d = 0.01 r_M$. It was assumed that they were introduced perpendicular to the settling direction. It is concluded from fig. 3-4 that the diameters of mica platelets are about 10 times and their volumes about 8—11 times as large as the respective diameters and volumes of quartz spheres with the same settling velocity.

The distance which suspended particles can be transported in laminar flow depends on the ratio of the mean flow velocity to settling velocity and also on the water depth. If L is the length of the transport path, h the water depth, u the flow velocity, and v the settling velocity of the particles, then:

$$L = \frac{hu}{v} \qquad (3.43)$$

Table 3-3 gives some values for the transport distance of slowly settling particles with a mean flow velocity of 1 mm/sec and in different water depths. These data show that with laminar flow suspension transport over geologically significant

Table 3-3. Particle transport distance for different settling velocities (v) in laminar flow of velocity $u = 1$ mm/sec^{-1} and at different water depths (h). The diameter (D) refers to spherical quartz particles ($1\,\mu = 10^{-4}$ cm).

$v =$	10^{-2} cm/sec ($D = 10\,\mu$)	10^{-3} cm/sec ($D = 4\,\mu$)	10^{-4} cm/sec ($D = 1\,\mu$)
$h = 1$ m	10 m	10^2 m	10^3 m
$h = 10$ m	10^2 m	10^3 m	10^4 m
$h = 100$ m	10^3 m	10^4 m	10^5 m

3.13 Transport in suspension (suspended load)

distances is possible, only when the settling velocity is very small and when water depth is considerable. Such conditions can be realized, for example, for the finest clay detritus suspended in ocean currents. However, as soon as the flow becomes turbulent, coarser material can also be transported long distances in suspension. SCHMIDT (1925) first showed that suspension transport during turbulent flow can be described and understood utilizing the concept of an exchange coefficient.

Let n equal the number of particles with a settling velocity v, which are present in 1 cm³ of flowing water. n is a function of the distance (y) above the stream bed. Past a reference plane perpendicular to the y-axis, nv particles settle per unit surface area and per unit time. In contrast turbulent exchange carries $A \frac{dn}{dy}$ particles upward.

At equilibrium both transport mechanisms must balance:

$$A \frac{dn}{dy} + nv = 0 \qquad (3.44)$$

Thus for the number of particles n_y at the height y, when n_a is the number to height $a < y$:

$$\ln \frac{n_y}{n_a} = -v \int_a^y \frac{dy}{A} \qquad (3.45)$$

If the exchange over the entire cross section of flow is constant, which we can probably assume for ocean currents, we conclude:

$$n_y = n_a e^{-\frac{v}{A}(y-a)} \qquad (3.45\,a)$$

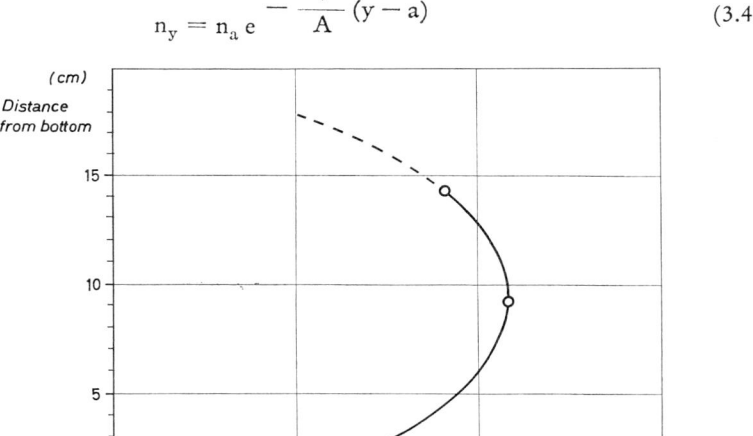

Fig. 3-6. Variation of exchange coefficients with distance above bottom for water flow in a rectangular, inclined channel. Width: 28.6 cm; depth: 18.9 cm; $\tan \beta = 0.0041$ (KALINSKE 1943).

In this simple case particle concentration decreases exponentially with height above the bottom. The slope of the curve is smaller the smaller v/A, the ratio of settling velocity to exchange. In streams the magnitude of the exchange is a function of height above the stream bed. In order in this case to be able to calculate the particle distribution, we must know how the exchange varies with distance above the stream bed. In this respect some experimental investigations are pertinent.

Using an artificial channel of 28.6 cm width, h = 18.9 cm depth, and slope of tan β = 0.0041, KALINSKE (1943) by means of diffusion experiments measured the variations in exchange coefficients. His data are reproduced in fig. 3-6. The observed distribution of sands of different grain size over the cross-sectional area of the channel is plotted in fig. 3-7. The solid curves were calculated from equation (3.45)

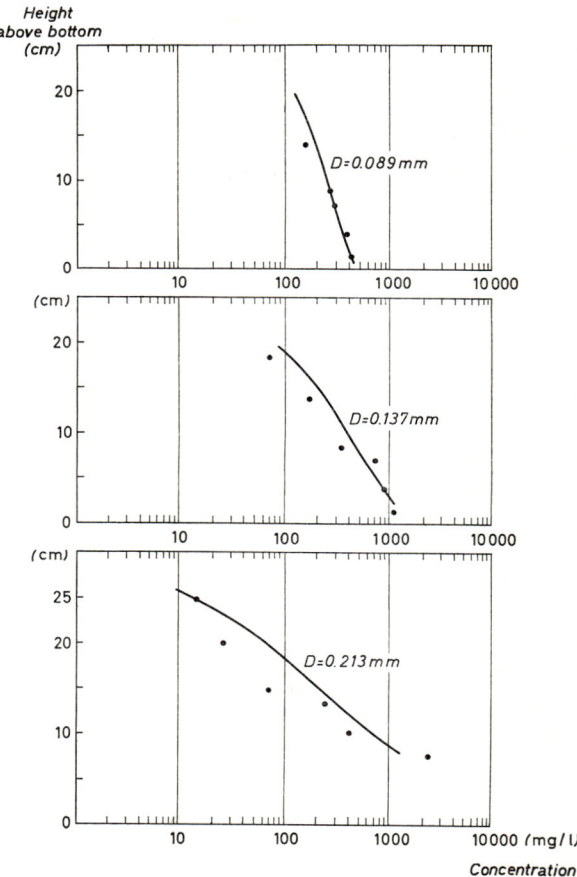

Fig. 3-7. Experimental and calculated (equation 3.45) distribution of sands of different grain size over the cross section of flow in a rectangular channel (see fig. 3—6). Curves give theoretical values (KALINSKE 1943).

using the measured exchange coefficients. As a reference point the measured concentration at h/10 above the channel bottom was used. The agreement between theoretical and experimental data is very good and the dependence on grain size distribution can be seen. For the sand of 0.089 mm diameter the concentrations at the surface and at the bottom are in the ratio of about 1 : 4; for the 0.213 mm sand the ratio is 1 : 1000. Noteworthy is the exchange value maximum at y = h/2 (fig. 3-6). HJUL-STROM (1934) likewise had found a similar maximum with his channel experiments.

Little data are available on the exchange coefficients in natural streams, and especially on the relationship to water depth. Failing this, a theoretical treatment of ROUSE (1937, after SCHEIDEGGER 1961, p. 142) must suffice. The starting point is equation (3.11):

$$\tau = -A\varrho_1 \frac{d\bar{u}}{dy} \tag{3.11}$$

The shear stress τ has its maximum value (τ_0) at the bottom $(y = 0)$ and becomes negligible at the water surface $(y = h)$. It is assumed that τ decreases linearly upward (h = water depth):

$$\tau = \tau_0 \left(1 - \frac{y}{h}\right) \tag{3.46}$$

After equations (3.12—3.15) the exchange coefficient is given by:

$$A = \frac{\tau_0}{\varrho_1} \frac{(1 - y/h)}{d\bar{u}/dy} = u_\tau^2 \frac{(1 - y/h)}{d\bar{u}/dy} \tag{3.47}$$

The dependence of velocity on distance from the channel wall is based on the KÁRMÁN equation (3.19):

$$\bar{u}_y = \frac{u_\tau}{0.4} \ln \frac{y}{y_0} \tag{3.48}$$

Then we obtain:

$$\frac{d\bar{u}}{dy} = \frac{1}{y} \cdot \frac{u_\tau}{0.4}$$

$$A = 0.4 \sqrt{\frac{\tau_0}{\varrho_1}} \left(1 - \frac{y}{h}\right) y \tag{3.49}$$

By substituting in equation (3.45) and integrating, we obtain the following expression for the distribution of particles of settling velocity v in turbulent flow as it relates to distance from the channel bottom:

$$\left. \begin{array}{c} \dfrac{n_y}{n_a} = \left(\dfrac{a\,(h-y)}{y\,(h-a)}\right)^p \\[2ex] \text{with} \quad p = \dfrac{v}{0.4 \sqrt{\tau_0/\varrho_1}} \end{array} \right\} \tag{3.50 a}$$

n_y is the number of particles per cm³ at a height y above the bottom; n_a the number at height a.

Expressing the heights y and a as fractions of the water depth h:

$$y = Yh \text{ and } a = Xh,$$

then

$$\frac{n_y}{n_a} = \left(\frac{X(1-Y)}{Y(1-X)}\right)^P \quad (3.50\text{ b})$$

If we relate the particle concentration at an arbitrary height above the bottom (y) to the concentration at half water depth, $a = \frac{1}{2}h$ ($x = \frac{1}{2}$), the following simplified formula applies:

$$\frac{n_y}{n_{1/2}} = \left(\frac{1}{Y} - 1\right)^P = \left(\frac{h}{y} - 1\right)^P \quad (3.50\text{ c})$$

Equation (3.49) infers that the exchange coefficient equals zero both at the channel bottom (y = 0) and at the water surface (y = h) and that at $y = \frac{1}{2}$ it is at a maximum. This was observed also in the experiments of KALINSKE as mentioned earlier. This maximum value is obtained:

$$A_{max} = 0.1\,h\,\sqrt{\tau_0/\varrho_1} \quad (3.51)$$

For the channel experiments of KALINSKE (1943) one obtains a hydraulic radius of m = 8.15 cm, and, based on equation (3.7), a maximum shear stress of $\tau_0 = 32.38$ gcm⁻¹sec⁻². Thus we can calculate from (3.51) a maximum exchange coefficient of 10.7 cm²sec⁻², in excellent agreement with the experimental value of $A_{max} = 10.9$ (fig. 3-6).

Fig. 3-8 reproduces the distribution of suspended particles from equation (3.50) for different values of the exponent p. If p > 3 suspension transport is of little

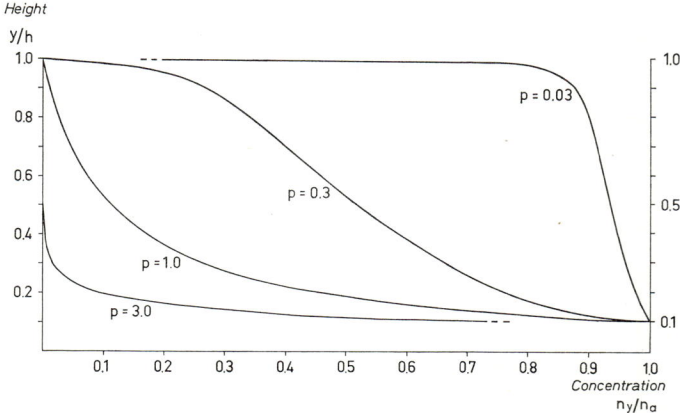

Fig. 3-8. Suspended particle distribution during turbulent flow, according to equation (3.50) (explanation in text).

3.13 Transport in suspension (suspended load)

significance; suspended particles occur only in a thin layer near the bottom. If $p < 0.03$ particles are distributed almost uniformly over the entire cross section of the stream. The suspension transport determining factor p is given by the ratio of particle settling velocity (v) to the maximum shear stress at the stream bottom (τ_0). The latter, based on equation (3.7) is determined by gradient (tan β) and hydraulic radius (m) of the stream.

From (3.50) we obtain:
$$p = \frac{0.08 \, v}{\sqrt{m \, tg \, \beta}} \quad (3.52)$$

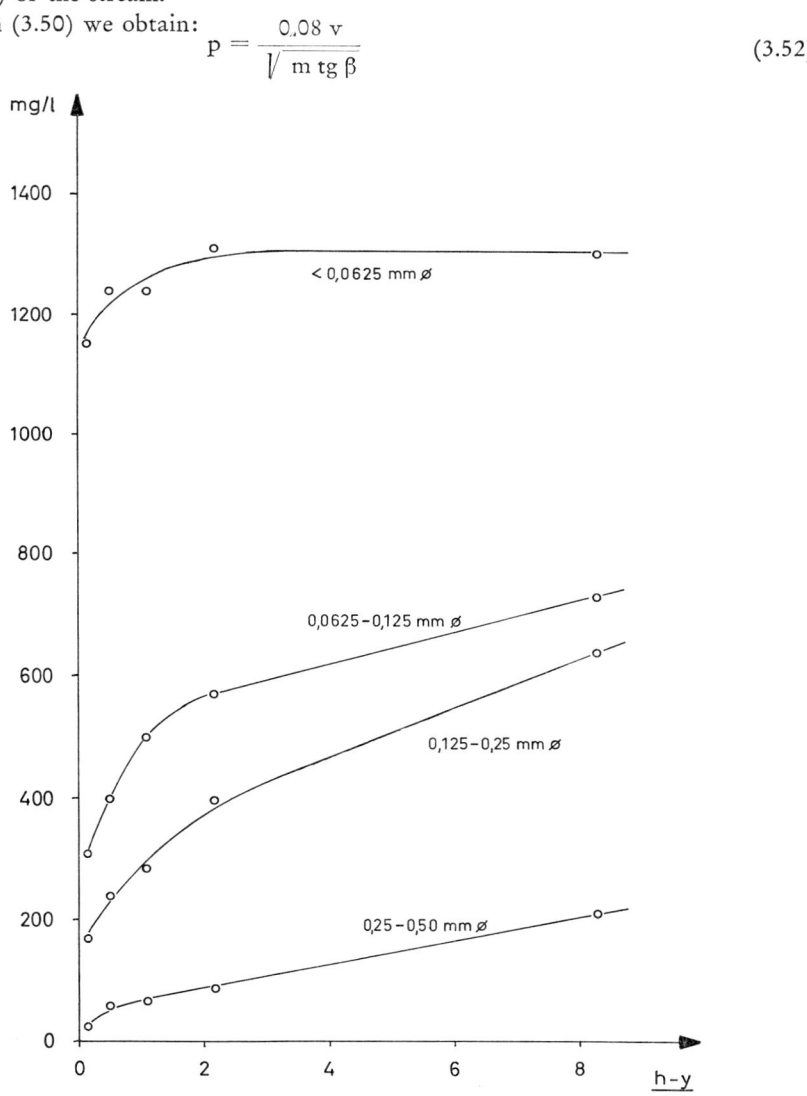

Fig. 3-9. Suspended particle distribution in a cross section of the Rio Grande (after data of Nordin & Dempster 1963).

For a stream of fixed hydraulic radius and gradient, we can, according to this theory, estimate the upper limit of the settling velocity and therewith the size of particles transported as suspension load. We should expect:

Complete transport as suspension load for $p \lesssim 0.03$, that is, for particles with

$$v \lesssim 0.375 \sqrt{m \, tg \, \beta} \qquad (3.53)$$

No transport as suspension load for $p \gtrsim 3$, that is, for particles with

$$v \gtrsim 38 \sqrt{m \, tg \, \beta} \qquad (3.54)$$

In fig. 3-13 the upper grain size limit for complete transport of quartz particles in suspension, based on equation (3.53), is noted.

Equations (3.50—3.54) can provide only approximate values for natural streams. This is because the quantitative application of equation (3.50) to natural streams is impeded by the same circumstances that stand in the way of evaluation of equation (3.20) for flow velocity distribution. Here also one must consider that the KÁRMÁN constant deviates from the value $K = 0.4$. However, measurements in streams have confirmed that the form of the distribution function of suspended particles corresponds quite well with equation (3.50). An example of such confirmation is given in fig. 3-9, which reproduces the particle distribution in the Rio Grande, as was measured by NORDIN & DEMPSTER (1963). Based on equation (3.50), a straight line must be obtained, when the log of the measured particle concentration at height y, log n_y, is plotted against log $(h - y) : y$. The slope of the curve is equal to the coefficient p:

$$p = \frac{d \, (\lg n_y)}{d \left(\lg \dfrac{y}{h - y} \right)} \qquad (3.50 \, d)$$

3.14 Bottom transport (bed load)

Transport occurring at the bottom of a natural stream, at the river bed, is difficult to observe. There the amount and kind of material in motion, as well as the flow relations prevailing, cannot be measured without disturbing the processes themselves. In this case also one must rely on experiments involving artificial channels. From such studies (see, for example, GILBERT 1914; BAGNOLD 1954, 1957; LIU 1957; DILLO 1960) the following picture emerges:

Allow water to flow over a packed layer of sand of uniform grain size. The velocity of the water, of constant depth, is gradually increased. When a certain critical velocity is exceeded, individual sand grains are set into a rolling motion. Further increase in velocity increases the movement of the sand and, if it is not too coarse grained, a regular system of current ripples form whose ridges lie perpendicular to the flow direction. These have a gentle upstream slope and a steeper downstream

3.14 Bottom transport (bed load)

slope. Sand grains are carried by the current up and over the gentle slope and deposited on the steep slope, rolling down it. In this way the gently inclined side of the ripple is eroded and the steep side built up by deposition, causing the system of ripples to migrate downstream. This migration velocity increases with, but is much less than the current velocity. From channel experiments the following empirical relation has been found (LIU 1957; DILLO 1960):

$$v_R = C u_m^B \tag{3.55}$$

v_R is the migration velocity of the ripples, and u_m the mean channel current velocity. In a 60 cm wide, 1 m deep channel DILLO (1969) found with different sands that $B = 6.4$ and that C increases with grain size. $C = 2.86 \cdot 10^{-14}$ for a mean grain diameter of $D = 0.075$ mm; $C = 1.46 \cdot 10^{-13}$ when $D = 0.185$ mm; and $C = 2.78 \cdot 10^{-13}$ when $D = 0.28$ mm. However, these values may be dependent also on the dimensions of the channel.

The slow migration of the current ripples produces a slow downstream directed transport of sand. Ripples whose separation and height vary in natural streams with water depth and current velocity are, as long as they are in motion, to be considered as transport bodies. The sand transport velocity, resulting from the migration of ripples, like the ripple migration velocity, varies exponentially with current velocity, and depends on the form, especially the height, of the ripple (KENNEDY 1964; FÜHRBÖTER 1967).

A very much stronger bottom transport sets in, when the current velocity is raised above a second critical velocity. When this limit is exceeded the ripples disappear, and the sand surface again becomes flat, with a cloud of bouncing sand grains moving over it (bed suspension). At a third critical velocity symmetrically formed regressive sand waves appear, which migrate upstream. Their amplitudes and separation distances are greater than those of current ripples. In the deeper parts of large streams sediment waves of even greater dimensions are found, which migrate downstream.

The first critical velocity at which the first grains of a sand covered substrate are set into rolling motion corresponds to a first critical shear stress τ_{01} at the base of flow. It can be postulated that this shear stress must subject the uppermost layer of solid packed sand to a definite amount of shearing action, during which individual, especially loosely bound grains are freed from the rest and caused to roll over the surface. If P is the weight of the uppermost sand layer per unit surface area, this first critical shear stress is equivalent to a force per unit area, corresponding to a certain fraction γP of this weight:

$$\tau_{01} = \gamma P = \gamma (\varrho_2 - \varrho_1) g D C_0 = \Theta_1 (\varrho_2 - \varrho_1) g D \tag{3.56}$$

C_0 is the volume concentration of sand grains in the uppermost layer. According to the afore-mentioned experiments Θ_1 depends on grain size. In fig. 3-10 values for Θ_1, determined by SHIELDS (after BAGNOLD 1957, 1966) for D-values between 0.2 and 20 mm are plotted. From these data one can calculate from equation (3.56) the necess-

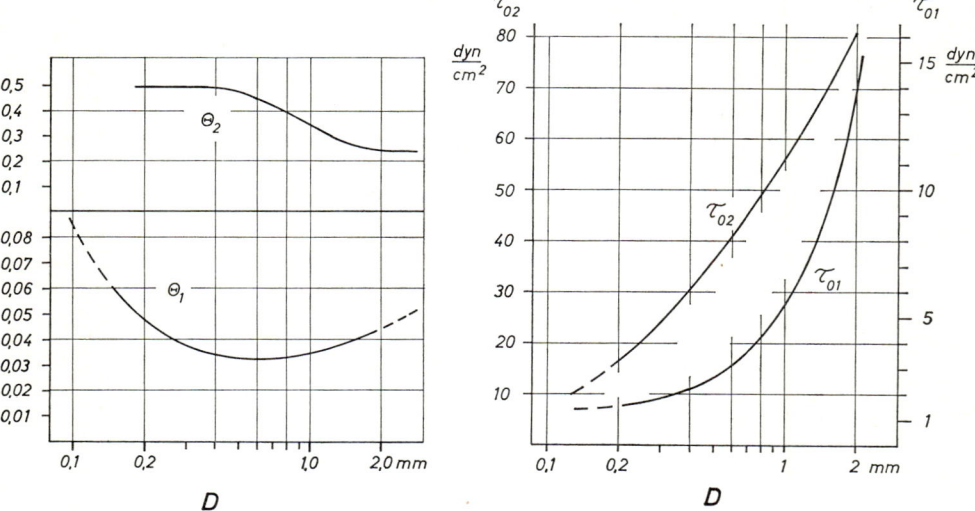

Fig. 3-10. Values of Θ_1 and Θ_2 coefficients for rolling transport in water (BAGNOLD 1957, 1966).

Fig. 3-11. Critical shear stress at the bottom of a current for rolling (τ_{01}) and saltating (τ_{02}) transport of quartz spheres in water.

ary shear stress for the first rolling motion of grains of specified specific gravity in this size range. Fig. 3-11 gives the values for spherical quartz grains. Since the underlying values of Θ_1 stem from very few measurements, we are dealing with numbers of no great accuracy. Nevertheless, one can substitute for τ_{01}, according to equation (3.6) the hydraulic radius and gradient of a stream and derive the following expression from which we can estimate the approximate grain size that a stream of given transport power (that is, of given gradient and depth) can just set into rolling motion:

$$D_1 = \frac{\varrho_1}{(\varrho_2 - \varrho_1)} \frac{1}{\Theta_1} m \, tg \, \beta \qquad (3.57)$$

Some numerical values are given in table 3-4. According to equation (3.56), when Θ_1 is approximately constant, the first critical shear stress should become steadily smaller with decreasing grain diameter. Actually this shear stress rises with decreasing grain diameter, first slowly and then more rapidly, as soon as the grains become smaller than about 0.2 mm. Unfortunately this extremely important fact of transport and deposition has not been proven adequately by means of recent and exact measurements. Thus one is referred always to a compilation of older observations which were passed on by HJULSTROM (1934). Since this time the curve, reproduced in fig. 3-12, has appeared in numerous text books. The critical shear stress, which would be of real interest, is not plotted, but rather the mean flow velocities which were observed to be necessary to initiate sand movement in different channels.

3.14 Bottom transport (bed load)

Table 3-4. Diameters of quartz sand (gravel) grains which just begin to move (D_1) or to be completely eroded (D_2) in the bed of a stream with hydraulic radius m = 5 m and at different gradients.

Gradient tan β	D_1 mm	D_2 mm
$3 \cdot 10^{-5}$	2.0	0.23
10^{-4}	6.5	0.75
$3 \cdot 10^{-4}$	20	3.6
10^{-3}	65	12

A broad minimum lies at the grain size range from 0.1—0.4 mm, with its mid-point lying near 0.2 mm. For finer particles, the velocity or shear stress necessary for mobilization rises steeply. This is because a bouncing motion is no longer possible; the hydrodynamic lift can no longer be effective because of the very small grain size, and because the adhesion between particles increases strongly with increase in their specific surface area.

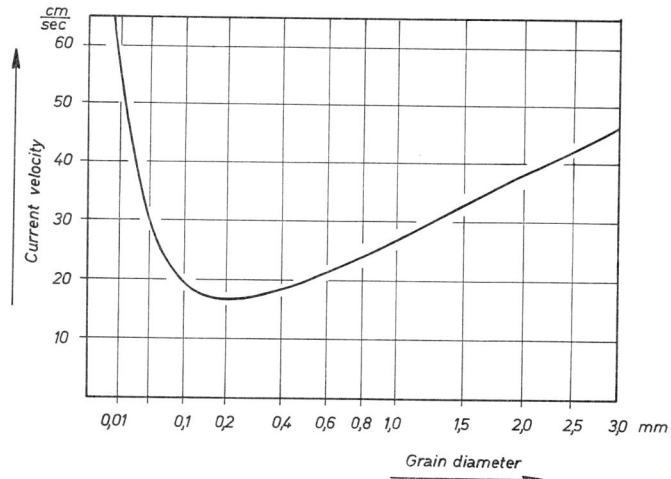

Fig. 3-12. Current velocity necessary for erosion of bottom sediments (HJULSTRÖM 1934).

As the second critical velocity, or the corresponding bottom shear stress, is approached, more and more grains are lifted from the bed into the fluid. As this velocity is exceeded, saltation becomes the predominant kind of sand grain motion at the bottom. The jumping grain is accelerated at first upwards obliquely to the current and then has its path bent in the direction of the current. It is advanced some distance and finally falls again to the bottom under the influence of gravity. The original upwardly directed acceleration probably develops primarily from a hydrodynamic lifting force, produced as a result of the current velocity gradient near the bottom. In addition a grain can be accelerated upward, if it is pressed at an inclined contact plane against a neighboring grain. The observed trajectory height of one to two grain diameters can be attained against the water friction, only if the mass of an individual grain exceeds a certain lower limit. According to the observations of BAGNOLD (1957) this limit for quartz in water lies at a diameter of about 0.2 mm. Therefore, only grains with diameters above about 0.2 mm move as saltating bed load.

At higher current velocities the number of saltating grains becomes so large that they frequently collide. As a result, a dense suspension of saltating grains forms, which moves along the bottom of the current. The suspended bed load concentration (distinct from the suspended load) and its upward expansion increase with stream velocity.

The flow of such suspensions was investigated experimentally and theoretically in a fundamental way by BAGNOLD (1954, 1957). The following picture conforms to the interpretation of BAGNOLD.

A suspension of saltating, colliding grains can be considered as a flowing assemblage of contiguous spheres. An assemblage of spheres can flow only if their rest positions with respect to each other are spread out. Spheres in contact cannot glide over one another. In a plane (shear plane) parallel to the flow direction the shear stress τ acts, and perpendicular to it a normal force P, which is equal to the pressure which the particles above this plane exert as a result of their weight. The quotient $\tau/P = \alpha$ is a frictional coefficient. For a dense packing of spheres it is equal to the tangent of the angle of repose, the steepest slope on which the spheres will rest without rolling downward. For most sands this angle is about 33°, whereas for dense packing this gives a mean value of $\alpha = 0.63$.

Also dilute suspensions of saltating grains have a frictional coefficient which determines the resistance of the suspension to flow. Instead of the close grain contacts of dense packing, collisions of saltating sand grains occur. BAGNOLD (1954, 1957, 1966) measured the coefficients of suspensions of different grain size and concentration and found that α is dependent upon grain size and shear stress. An upper limiting value of 0.75 applies for small grains, low shear stresses, and low concentrations. A lower limiting value of $\alpha = 0.375$ applies to coarse grains, high shear stresses, and high concentrations.

The formation of a suspension of saltating sand grains will begin, when the shear stress at the bottom is sufficient to suspend the first compact layer of sand grains and to support them in a dense cloud of saltating grains. This shear stress is given as the product of the force of gravity on the uppermost dense layer per unit surface and the applicable frictional coefficient for suspension movement:

$$\tau_{02} = \alpha P = \alpha (\varrho_2 - \varrho_1) g D C_0 \tag{3.58}$$

For the volume concentration C_0 of the sand grains in the bed, one can take 0.65, as a mean value corresponding to a porosity of 35%. By refining the constant values, we obtain for the second critical shear stress:

$$\tau_{02} = \Theta_2 (\varrho_2 - \varrho_1) g D \tag{3.59}$$

3.14 Bottom transport (bed load)

with $\Theta_2 = 0.65\,\alpha$. Values for α for dense suspensions ($C_0 = 0.60$), measured by BAGNOLD, are given in table 3—5 for different grain sizes. Given also are the values of Θ_2 taken from fig. 3-10.

Table 3-5. Frictional coefficients α of concentrated suspensions ($C_0 = 0.60$) and the second critical shear stress Θ for the bottom transport of spheres of different diameters (after measurements of BAGNOLD).

D [mm]	α	Θ_2
> 2.0	0.375	0.25
1.5	0.42	0.27
1.2	0.47	0.30
1.0	0.53	0.34
0.8	0.60	0.39
0.7	0.64	0.42
0.6	0.68	0.44
0.5	0.72	0.47
0.4	0.75	0.49
< 0.2	0.75	0.49

From equation (3.59) one can calculate the shear stress τ_{02} necessary to initiate saltation of grains of known specific gravity. Fig. 3-11 gives these values for quartz grains between 0.2 and 2 mm diameter. These are only approximate values.

The grain size of sand which can just be set into saltating motion by a stream of known transport power, can be calculated as follows:

$$D_2 = \frac{\varrho_1}{(\varrho_2 - \varrho_1)} \cdot \frac{1}{\Theta_2}\, m\,\mathrm{tg}\,\beta \qquad (3.60)$$

A calculation can be carried out using the values for α in table 3-5. Since α depends on grain diameter, the solution is attained, under given circumstances, only after several steps. Table 3-4 contains values calculated in this way for the limiting grain size D_2 of quartz spheres at the bottom of a stream with hydraulic radius of 5 m and at different gradients.

As table 3-5 shows, the frictional coefficient of the suspended bed load varies between $\alpha = 0.375$ (large grains) and 0.75 (small grains). The value for dense packing ($\alpha = 0.63$) falls in the middle and is attained for suspensions of grains of 0.7 to 0.8 mm diameter. The uppermost sand layer, when smaller grains are involved, can be sheared without the current being able to transport these grains in dense suspension, since a frictional coefficient $\alpha = 0.63$ is sufficient for the first process, and a higher one applies for the second. In this case, however, the planar sand layer becomes unstable. It forms into current ripples, which, by virtue of the long upstream slope angle offers greater resistance to the stream than the planar surface. According to this idea, current ripples should form in water only in sands with grain sizes below about 0.8 mm diameter. According to BAGNOLD (l. c.) this is in agreement with observations made in artificial channels.

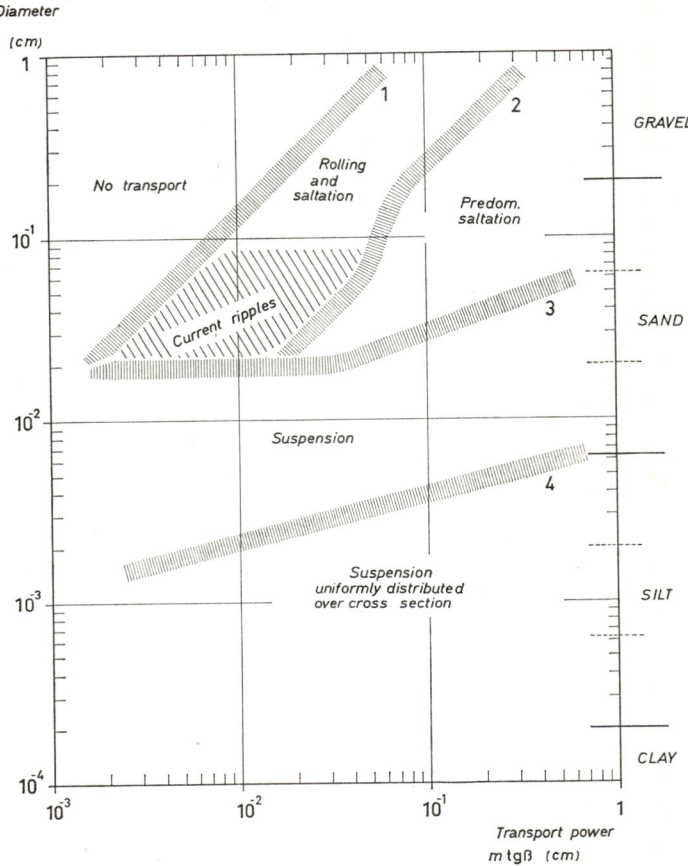

Fig. 3-13. Modes of transport of quartz spheres in water currents as related to transport power of the current (explanation in text).

Fig. 3-13 summarizes graphically the modes of transport of quartz spheres in water. The transport power of a stream can be defined by m tan β, the product of hydraulic radius (m) and stream gradient (tan β). The gradients of natural streams lie mostly between $1 \cdot 10^{-4}$ and $2 \cdot 10^{-3}$. For wide streams (depth $\ll \frac{1}{2}$ width) the hydraulic radius approximates the depth. For bed transport, two boundaries (fig. 3-13, 1 and 2) are shown, corresponding to the critical shear stresses τ_{01} and τ_{02}. Above the first boundary (1) material is not set in motion by the current. Between boundaries (1) and (2) it is transported slowly, predominantly by rolling, but in part by saltating motion. Under the same conditions grains below about 0.8 mm in size form current ripples as their mode of transport. To the right of boundary (2) the grains move predominantly by saltation as a dense bottom suspension. Boundary (3) is calculated from equation (3.52) for $p = 1$. Beneath this boundary (3) grains are distributed to a considerable extent as suspended load over the entire cross section of the stream (see fig. 3-8). For grain sizes below about 0.2 mm, independent of the transport power, bottom transport ceases to play an important role. Beneath boundary (4) (from equation 3.53 for p = 0.03) particles are distributed almost uniformly over the entire cross section (see fig. 3-8).

3.15 Transport capacity of streams

In order to understand sedimentary processes, and to obtain insight into the mechanisms of sedimentary particle transport, it is important to know the amounts of suspended and bed loads a given stream is capable of transporting. During the last century many observations from artificial flumes and natural streams have been made on the relation between transport capacity and different flow parameters. Numerous empirical formulae and semi-empirical theories developed from these observations which still have not been joined into a unified theory. Their presentation would far exceed, therefore, the space available in this book.

Instead an outline of a comprehensive theory of the transport power of streams, investigated by BAGNOLD (1966), will be reiterated. This theory has the advantage that it is based on general and simple physical assumptions and abstains a priori from empirical "laws".

The total energy released per unit of time when a quantity of water flowing in a stream travels a unit distance, is obtained as follows:

$$\Omega = \varrho_1 \, g \, Q \, \text{tg} \, \beta \tag{3.61}$$

Q ist the discharge, the volume of water passing a point in unit time, expressed in $cm^3 sec^{-1}$. The energy which is available each second for a column of water above a unit area of the stream bed is given:

$$\omega = \frac{\Omega}{b} = \varrho_1 \, g \, h \, U \, \text{tg} \, \beta = \tau_0 \, U \tag{3.62}$$

b is the width of the stream of rectangular cross section, and h its depth. $U = Q/bh$ is the mean current velocity, and τ_0 the maximum shear stress at the bottom. For wide streams one can write also:

$$\omega = \varrho_1 \, g \, m \, \text{tg} \, \beta \, U \tag{3.63}$$

If q_s is the volume of particles of suspended load (specific gravity ϱ_2) over 1 cm² of the stream bed, the force of the suspended load (in dynes), transported per second above a unit area of the stream bed is

$$i_s = (\varrho_2 - \varrho_1) g \, q_s \, U \quad (\text{dyn cm sec}^{-1}) \tag{3.64}$$

when it is assumed that the suspended particles move at the same velocity as the flowing water.

Similarly, if q_b is the volume of bed load material per unit area of the stream bed,

$$i_b = (\varrho_2 - \varrho_1) g\, q_b\, U_b \quad [\text{dyn cm sec}^{-1}] \tag{3.65}$$

where U_b is the transport velocity of the bed load. If α is the above defined frictional coefficient for bottom transport, the work required for bottom transport per unit surface area and time is

$$a_b = \alpha\, (\varrho_2 - \varrho_1) g\, q_b\, U_b \tag{3.66}$$

If v is the settling velocity of the suspended particles, the turbulent flow performs work on the suspended material, causing it in unit time to be uplifted a distance v. Accordingly the amount of work performed for suspension transport per unit surface and time is:

$$a_s = \frac{i_s v}{U} = (\varrho_2 - \varrho_1) g\, q_s\, v \tag{3.67}$$

For bottom transport a certain portion of the total applied energy ω is used, which is designated by the term e_b ($e_b < 1$):

$$a_b = \frac{e_b}{\alpha}\, \omega \tag{3.68}$$

For suspension transport another portion e_s of the applied energy $(1 - e_b)$ is used:

$$a_s = e_s\, (1 - e_b)\, \frac{U}{v}\, \omega \tag{3.69}$$

Finally we obtain for the total work involved in transport per second per unit area of the stream bed:

$$a = a_b + a_s = \omega \left(\frac{e_b}{\alpha} + e_s\, (1 - e_b)\, \frac{U}{v} \right) \tag{3.70}$$

According to a theoretical study by BAGNOLD (1966) which should be mentioned, e_b varies between the limits:

$$e_b = 0.11 \text{ to } 0.14 \quad (\text{mean, } 0.13).$$

These values are valid for quartz grains between 1.0 and 0.3 mm diameter (small values at high velocity and large diameter). α depends on grain size and can be taken from table 3-5. For quartz grains of less than 0.5 mm diameter, $e_b/\alpha = 0.17$. For suspension transport, we can estimate, according to BAGNOLD that:

$$e_s\, (1 - e_b) = 0.01$$

Therefore, the total work for transport is given:

$$a = \omega \left(\frac{0.13}{\alpha} + 0.01\, \frac{U}{v} \right) \tag{3.71}$$

3.15 Transport capacity of streams

The work applied per second and per 1 cm of flow, over the entire cross section of the stream is obtained by multiplying both sides of equation (3.71) by the stream width b:

$$A = (\varrho_2 - \varrho_1) g q_b b U_b + (\varrho_2 - \varrho_1) g q_s b U = \Omega \left(\frac{0.13}{\alpha} + 0.01 \frac{U}{v} \right) \quad (3.72)$$

The weights of bed load (M_b) and suspended load (M_s) materials (the transport capacities) per second across the cross section of the stream are given as follows:

$$M_b = \varrho_2 q_b b U_b = \frac{\varrho_1 \varrho_2}{(\varrho_2 - \varrho_1)} Q \, tg \, \beta \, \frac{0.13}{\alpha} \quad (3.73)$$

$$M_s = \varrho_2 q_s b U = \frac{\varrho_1 \varrho_2}{(\varrho_2 - \varrho_1)} Q \, tg \, \beta \, \frac{0.01 \, U}{v} \quad (3.74)$$

$M_s/Q = C_s$ is the mean concentration of suspended load. With a specific gracity $\varrho_2 = 2.65$ (quartz), one obtains for a stream:

$$C_s = \frac{M_s}{Q} = 0.016 \, tg \, \beta \, \frac{U}{v} \quad (3.75)$$

Therefore, the concentration of suspended load of a stream should depend on its gradient and the ratio of mean flow velocity to settling velocity of the suspended particles. If, as is usually the case, the suspended matter consists of particles of different settling velocities, it is necessary to calculate a mean settling velocity (BAGNOLD 1966):

$$\bar{v} = \frac{1}{100} \Sigma \, (pv_p) \quad (3.76)$$

where p is the weight percent of particles with settling velocity v_p.

Equations (3.73—3.75) give the maximum amounts of bed and suspension load that a stream may transport. The actual amounts transported can be less, if insufficient sediment is supplied.

From equations (3.73) and (3.74) the ratio of suspended to bed load is given as follows:

$$\frac{M_s}{M_b} = 0.077 \, \alpha \, \frac{U}{v} \quad (3.77)$$

For quartz particles less than 0.5 mm diameter, $\alpha = 0.75$ (see table 3-5) and, therefore:

$$\frac{M_s}{M_b} = 0.06 \, \frac{U}{v} \quad (3.78)$$

It is evident that this ratio should increase, therefore, with decreasing grain size. It should also increase with increasing current velocity. Under certain circumstances, these effects may be offset if coarser particles come into suspension, causing the mean settling velocity $\bar{v}$ to increase.

For quartz grains of 0.04 mm diameter (v = 0.13 cm sec^{-1}) the suspension load in a stream with mean current velocity U = 1 m sec^{-1}, according to equation (3.78) is about 46 times greater than the bed load. Only for particles with a settling velocity of 6 cm sec^{-1} (quartz grains with 0.4 mm diameter) would the bed and suspended loads be equal in the same stream.

For natural streams the assertions of this theory can be tested only relative to the suspended load, since measurements of bed load are not available. For a large number of measurements in North American rivers, BAGNOLD (1966) found, considering the theoretical generalizations and the uncertainties in measuring suspension load and current velocities, a relatively good agreement between theory and observation. From 146 determinations of suspension load, 49 deviate no more or less than 50% from the theoretical values; 65 measurements deviated plus or minus 100%. As examples, measurements and calculations for the Rio Grande at Bernalillo, New Mexico, are reproduced in table 3-6.

Table 3-6. Suspended load in the Rio Grande at Bernalillo (after BAGNOLD 1966).

No.	Gradient tan β	Mean velocity U (cm sec^{-1})	Mean settling velocity $\bar{v}$ (cm sec^{-1})	Suspended load concentration Observed	Calc. after eq. (3.75)
1	$9.3 \cdot 10^{-4}$	67.1	0.5	$1.94 \cdot 10^{-3}$	$2.00 \cdot 10^{-3}$
2	$9.6 \cdot 10^{-4}$	84.4	0.58	$2.74 \cdot 10^{-3}$	$2.24 \cdot 10^{-3}$
3	$10 \cdot 10^{-4}$	94.8	1.35	$0.946 \cdot 10^{-3}$	$1.12 \cdot 10^{-3}$
4	$10.3 \cdot 10^{-4}$	113.4	0.84	$2.82 \cdot 10^{-3}$	$2.22 \cdot 10^{-3}$
5	$7.9 \cdot 10^{-4}$	114.0	1.86	$0.942 \cdot 10^{-3}$	$0.77 \cdot 10^{-3}$
6	$9.0 \cdot 10^{-4}$	126.2	0.72	$3.32 \cdot 10^{-3}$	$2.52 \cdot 10^{-3}$
7	$8.4 \cdot 10^{-4}$	153.6	1.7	$1.66 \cdot 10^{-3}$	$1.21 \cdot 10^{-3}$
8	$11.4 \cdot 10^{-4}$	154.2	1.67	$5.98 \cdot 10^{-3}$	$1.68 \cdot 10^{-3}$
9	$8.8 \cdot 10^{-4}$	164.6	0.94	$2.84 \cdot 10^{-3}$	$2.46 \cdot 10^{-3}$
10	$8.6 \cdot 10^{-4}$	165.5	1.77	$1.57 \cdot 10^{-3}$	$1.29 \cdot 10^{-3}$
11	$7.5 \cdot 10^{-4}$	174.7	1.45	$1.75 \cdot 10^{-3}$	$1.44 \cdot 10^{-3}$
12	$8.6 \cdot 10^{-4}$	181.4	2.21	$2.36 \cdot 10^{-3}$	$1.13 \cdot 10^{-3}$
13	$11.5 \cdot 10^{-4}$	185.9	1.53	$2.08 \cdot 10^{-3}$	$2.23 \cdot 10^{-3}$
14	$12 \cdot 10^{-4}$	191.1	2.87	$3.48 \cdot 10^{-3}$	$1.28 \cdot 10^{-3}$
15	$12 \cdot 10^{-4}$	198.1	2.77	$5.41 \cdot 10^{-3}$	$1.37 \cdot 10^{-3}$
16	$9.6 \cdot 10^{-4}$	198.7	1.38	$3.71 \cdot 10^{-3}$	$2.21 \cdot 10^{-3}$
17	$12.7 \cdot 10^{-4}$	209.7	1.21	$4.74 \cdot 10^{-3}$	$3.52 \cdot 10^{-3}$
18	$12.7 \cdot 10^{-4}$	210.6	1.16	$4.42 \cdot 10^{-3}$	$3.69 \cdot 10^{-3}$
19	$11.5 \cdot 10^{-4}$	210.9	1.64	$2.58 \cdot 10^{-3}$	$2.37 \cdot 10^{-3}$
20	$12 \cdot 10^{-4}$	235.0	1.39	$3.04 \cdot 10^{-3}$	$3.25 \cdot 10^{-3}$

Observations of the suspended load of streams leads to an empirical law in the form

$$C_s = a\, U^b \qquad (3.79)$$

which seems to correspond to equation (3.75). The factor a should depend on the sediment supply, and thus significantly on climatic influences and other factors related to intensity of weathering. b depends primarily on gradient and, therefore, should be constant for a given stream.

A comparison of equations (3.75) and (3.79) is difficult, since in the latter only the mean current velocity appears as a variable. In equation (3.75) the stream gradient and settling velocity, that is, the grain size of suspended particles, are likewise stated variables. It is now certain that grain size influences the amount of suspension load. Under essentially comparable conditions coarse-grained sediment is transported less than fine-grained sediment. On the other hand stream gradient is exponentially related to current velocity (see equation 3.8). For this reason, and also based on the conditions expressed in equation (3.75), the suspension load concentration increases non-linearly with current velocity.

3.16 Marine transport processes

3.161 General

The sediment carried by the rivers into the ocean does not come to rest immediately, but is further transported by marine transport processes. It is sorted and in some cases also mechanically altered. The agents are waves induced by the winds and tides, as well as ocean currents. In addition submarine land slides and suspension currents occur along the continental slopes, acting under the influence of gravity. These different kinds of marine transport will be considered in the following sections.

3.162 Transport by ocean currents

Ocean currents can transport the suspension load out into the open sea. According to the principles dealt with above, current velocity and turbulence, water depth, and particle settling velocities determine transport distances. In the large ocean currents huge masses of water are moved which are many hundreds of times greater than the large rivers of the continents. For example, the Florida current, the beginning of the Gulf Stream, flows along the Florida coast in the amount of $26 \cdot 10^6$ $m^3 sec^{-1}$. After the influx of additional component currents, the approximately 50 km wide Gulf Stream south of the Newfoundland Bank has a water capacity of about $55 \cdot 10^6$ $m^3 sec^{-1}$. As an immense "river in the sea", it bends eastward at that point, divides into several branches and reaches European waters still at the rate of $10 \cdot 10^6$ $m^3 sec^{-1}$. Within the laterally sharply bounded Gulf Stream, high velocities are established which extend to considerable depth. For example, at the latitude of Chesapeake Bay, at a depth of 100 m velocities up to 2 m sec^{-1} have been measured; at 500 m, about 1 m sec^{-1}; at 700 m, about 0.5 m sec^{-1}; and at 1000 m, still about 0.1 m sec^{-1} (DIETRICH & KALLE 1957). The Blake Plateau off the coast of Florida and Georgia lies between 750 and 1000 m below sea level. Here all fine sediment has been winnowed away by the Gulf Stream. Ocean currents of this sort are in a position to carry fine clay particles with low settling velocities over long distances. This is especially true of those currents extending extraordinary distances, such as the Gulf Stream or the large North and South Pacific currents which traverse the entire ocean from continent to continent. It is remarkable, however, that actually only an extraordinary small amount of sediment is transported by ocean currents. We conclude this from the low concentrations of suspended mineral matter encountered in sea water of the open ocean (for example, in the north Pacific on the average only $23 \cdot 10^{-6}$ g/l, ARRHENIUS 1963) and also from the very slow deep sea sedimentation rates ($5 \cdot 10^{-5}$ to $5 \cdot 10^{-4}$ cm/yr, ARRHENIUS 1963). The basis for the low suspension load of ocean currents is probably the flocculation of terrigenous clay by the ions of sea water, which takes place in the region of coastal waters. There by coagulation the finest particles form coarser particles, whose settling velocities are so great, that they can be carried no longer by the current turbulence and sink to the bottom.

That deep sea currents can move a bed load also must be concluded from the occurrence of ripple-marked sands, encountered also at great depths up to 2000 m (SHEPARD 1963 b).

3.163 Wave transport

In coastal waters sedimentary material is transported, sorted, and mechanically affected by wave action. These processes are of considerable importance for all sediments which form not far from the coast line. In order to understand them, the mechanics of water waves, especially those in shallow water, must be dealt with (see DIETRICH & KALLE 1957; STOKER 1957; SCHULEIKIN 1959; SCHEIDEGGER 1961; INMAN 1963; BAGNOLD 1963).

Fig. 3-14 shows the idealized form of a water wave, emphasizing the orbital motion of water particles. This shall be considered in more detail in the following section. First consider the most important wave parameters: the wave length L is the distance between two wave crests or wave troughs. The wave height or amplitude H is the height of a wave crest above a trough. The wave period T is the time

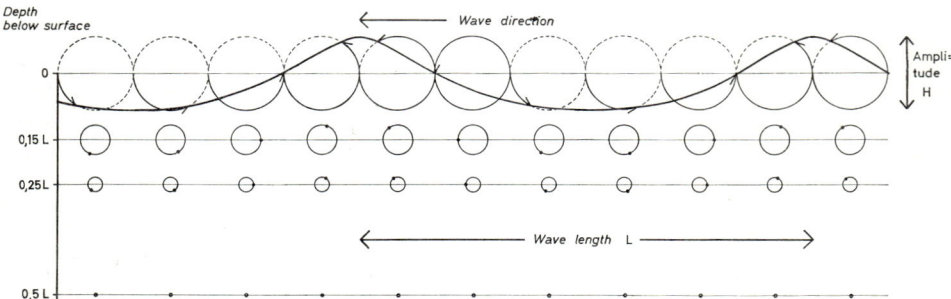

Fig. 3-14. The form of water waves and orbital motions of water particles as a function of depth beneath water surface (explanation in text).

a complete wave (crest to crest) takes to pass a given point. The wave velocity C is then

$$C = L/T \qquad (3.80)$$

Waves form when wind blows over a water surface. As a result a wide spectrum of different periods or wave lengths forms, which interfere with one another leading to complex and irregular wave forms. If these waves spread out over the open sea from their points of origin without further influence by the wind, the short and irregular waves are lost, the spectrum becomes narrower, and certain frequencies clearly predominate. Thus the quite periodic waves such as we observe at the coasts finally result.

Providing the wave amplitudes are not too large, the use of principles of hydrodynamics leads to the following relation between wave velocity (C), wave length (L), and water depth (h):

$$C = \sqrt{\frac{gL}{2\pi} \tanh \frac{2\pi h}{L}} \qquad (3.81)$$

g is the acceleration of gravity. In deep water, far from the coast, h/L is large, so that $\tanh \frac{2\pi h}{L} \simeq 1$, and, therefore

Deep water
$$C_t = \sqrt{\frac{gL}{2\pi}} = \frac{gT}{2\pi} = 156 \, T \; [\text{cm sec}^{-1}] \qquad (3.82)$$

$$L_d = \frac{gT^2}{2\pi} = 156 \, T^2 \; [\text{cm}] \qquad (3.83)$$

In deep water, for a given wave period, wave velocity and wave length are fixed. The longer the period the greater the velocity and wave length. Long waves are propagated more rapidly than short ones. The relations are shown in fig. 3-15.

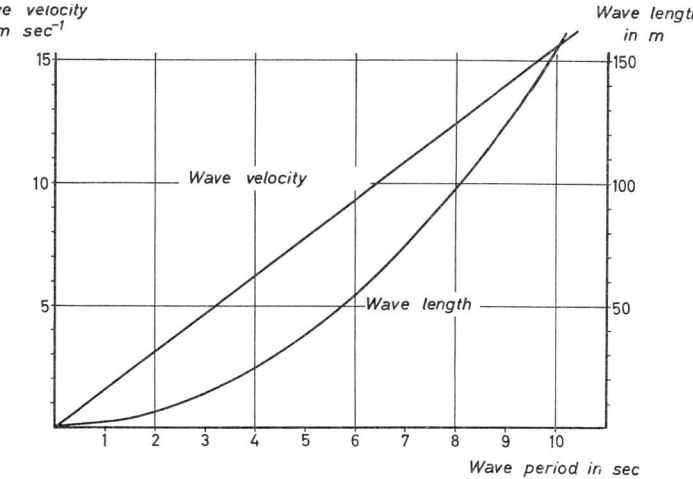

Fig. 3-15. Wave velocities and wave lengths in deep water.

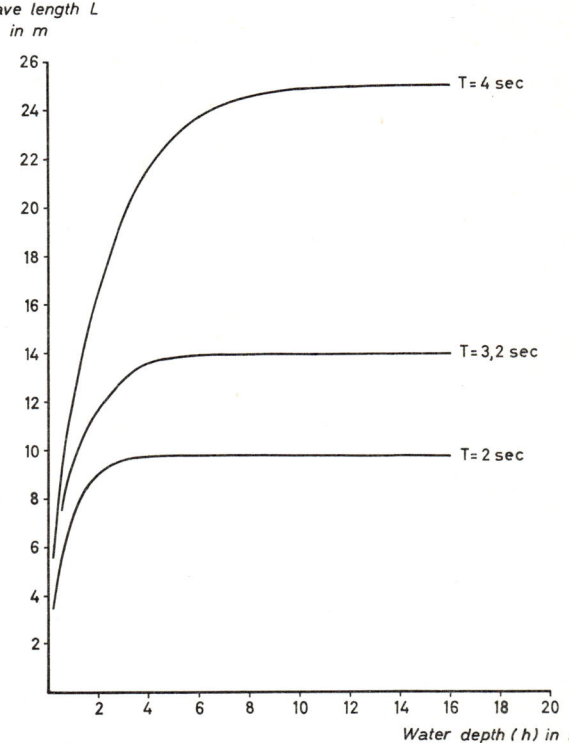

Fig. 3-16. Wave lengths of water waves of given frequency (T) in deep water.

Upon approaching the coast the water depth (h) becomes shallower. Therewith, according to equation (3.81), the wave velocity decreases. Since the wave period T remains constant, according to equation (3.80), the wave length must become less also with decreasing water depth. In addition the wave amplitude increases. To calculate the wave length for a given water depth h, one can use the following approximation (after ECKART 1952) (L_d = wave length in deep water):

$$L = L_d \sqrt{\tanh \frac{2\pi}{L_d} h} \qquad (3.84)$$

This relation for several wave periods is plotted in fig. 3-16. One can read off the wave lengths for given water depths. In addition one can determine the wave velocity according to equation (3.80).

If water depth is very low, $\tan h \frac{2\pi h}{L}$ approaches $\frac{2\pi h}{L}$, leading to another simplification of the equations for wave velocity and wave length:

Shallow water: $\quad C_s = \sqrt{gh}$ \hfill (3.85)

$\qquad\qquad\qquad L_s = T \sqrt{gh}$ \hfill (3.86)

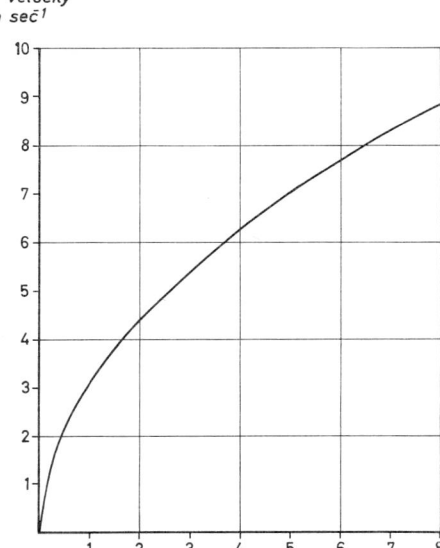

Fig. 3-17. Wave velocities in shallow water.

Therefore, near shore the wave velocity depends only on the water depth, the wave length or the period and water depth. The initial portion of the curves shown in fig. 3-15 (to a water depth of h = 0.05 L) corresponds to equation (3.86). In fig. 3-17 equation 3.85 is plotted, relating wave velocity to depth in shallow water. One obtains the corresponding wave length by multiplying the given velocity by the wave period.

If one allows a variation of 5% between calculated and actual values, we obtain the following limits in the use of equations (3.82) to (3.86) (INMAN 1963):

$$\left.\begin{array}{l} h > 0.25\ L \\ h > 39\ T^2 \end{array}\right\} \ldots \text{deep water} \ldots \begin{array}{l} C_d = 156\ T \\ L_d = 156\ T^2 \end{array}$$

$$\left.\begin{array}{l} 0.25\ L > h > 0.05\ L \\ 39\ T^2 > h > 2.5\ T^2 \end{array}\right\} \ldots \begin{array}{l} \text{intermediate} \\ \text{depth} \end{array} \ldots L = L_d \sqrt{\tanh \frac{6 \cdot 28\ h}{L_d}}$$

$$\left.\begin{array}{l} h < 0.05\ L \\ h < 2.5\ T^2 \end{array}\right\} \ldots \text{shallow water} \ldots \begin{array}{l} C_s = \sqrt{981\ h} \\ L_s = T \sqrt{981\ h} \end{array}$$

If T is expressed in sec and h in cm, all velocities are expressed in cmsec^{-1} and wave lengths in cm.

The use of equations (3.80-3.86) and fig. 3-15-3.17 is illustrated by an example. A wave of period 4 sec has in deep water a wave length of 25 m and a velocity of

6.25 msec^{-1} (fig. 3-15; equations 3.82, 3.83). If the water becomes shallower, at a depth h = 39 T^2 = 6.2 m, an appreciable decrease in wave length and velocity occurs, which is shown in fig. 3-16. At a water depth of 2 m the wave length is only 16.5 m and the velocity 4.1 msec^{-1} (fig. 3-16; equation 3.84). Shallow water begins at a depth h = 2.5 T^2 = 40 cm. At this depth the wave velocity is 2 msec^{-1} and the wave length 8 m (fig. 3-17; equations 3.85, 3.86).

The change in wave length, velocity, and amplitude upon approaching the coast is caused by the interaction of the moving water mass and the bottom. Therefore, we can conclude that a strong influence of the bottom on transport by water waves is expected to be significant only in the intermediate and shallow water zones, that is, when h < 0.25 L or h < 39 T^2. In these zones the most important transport processes and mechanical effects take place.

In order to understand the effects of water waves on the sea bottom, it is necessary to consider the detailed movement of water particles in the wave. The wave motion occurs as a result of the circular motion of individual water particles. When the amplitude is not too great and in deep water, the water particles move in closed circular orbits or paths. The mode of flow as it appears to a fixed observer is shown in fig. 3-14. Shown are the orbital paths of water particles at the surface and at depth, as is the mode of movement of the water surface. At the surface water particles on the wave crests are moving in the direction of the wave; in the wave troughs movement is in the opposite direction. The diameter of the circular orbit at the surface is equal to the wave amplitude H, as is evident from fig. 3-14. Downward the amplitude decreases markedly, according to the equation:

$$d = H \frac{\cosh \frac{2\pi z}{L}}{\sinh \frac{2\pi h}{L}} \tag{3.87}$$

In deep water both h and z are large. Thus equation (3.87) can be replaced by an exponential function, which is more convenient to handle:

Deep water:
$$d = H e^{-\frac{2\pi}{L}(h-z)} \tag{3.88}$$

Based on this expression, the diameters of orbital paths in deep water have been computed and plotted in fig. 3-18. It is seen that the orbital paths diminish markedly with distance below the water surface. At a depth corresponding to one-half wave length, the orbital diameter is only 4% of the wave amplitude. Action on the bottom sediment is to be expected only when the orbital diameter there still has some finite value. Substituting z = 0 in equation (3.87), the following expression is obtained for the orbital diameter at the bottom:

$$d_o = \frac{H}{\sinh \frac{2\pi h}{L}} \tag{3.89}$$

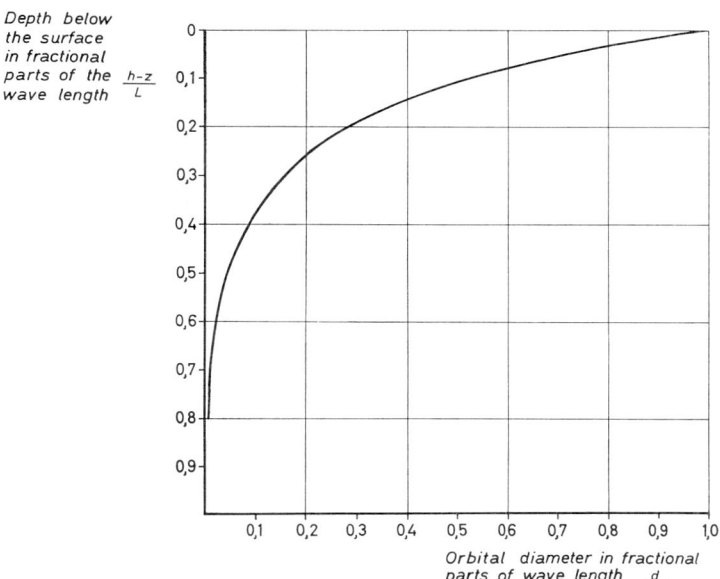

Fig. 3-18. Orbital diameters in a deep water wave in relation to depth below the surface.

The circumference of the orbital path at the bottom is πd_0. The maximum flow velocity for the water moving back and forth across the bottom is obtained as follows:

$$u_0 = \frac{\pi d_0}{T} = \frac{\pi H}{T \sinh \frac{2\pi h}{L}} \qquad (3.90)$$

Using equations (3.89) and (3.90), values for orbital diameters and maximum flow velocity at the bottom are plotted in fig. 3-19 for different water depth/wave length ratios, amplitudes, and wave periods.

If the waves are propagated into shallow water, as defined above, a simplification of equations (3.89) and (3.90) is permitted. The term $\sin 2\pi h/L$, for small values of h/L, approaches $2\pi h/L$, so that the orbital diameter d_0 and maximum flow velocity u_0 at the bottom are obtained as follows:

Shallow water
$(h < 0.005\ L)$

$$d_0 = \frac{HL}{2\pi h} = \frac{HT}{2\pi} \sqrt{g/h} \qquad (3.91)$$

$$u_0 = \frac{HL}{2hT} = \frac{HC_s}{2h} = \frac{H}{2} \sqrt{g/h} \qquad (3.92)$$

Therefore, in shallow water both the orbital diameters and maximum flow velocities are constant from the water surface to the bottom.

88 3.1 Transport and deposition in water

The action of wave motion on loose bottom sediment, according to equations (3.89—3.92) is determined by wave length and wave amplitude.

The above-mentioned increase in wave amplitude, which accompanies the decrease in wave length, is important for wave action when waves approach the shoreline and pass into increasingly shallower water. That the wave height must increase in passing into shallower water can be deduced qualitatively from energetic considerations. The energy of a wave generated in deep water is half potential and half kinetic energy. The potential energy is expressed in the height, the kinetic in the orbital movement of water particles. The advance of the wave corresponds to a transport of energy. Since wave velocity decreases directly with water depth, more energy is released per unit time in a wave approaching the shoreline from the open ocean and thus slowing down, than is disipated at the same time in the direction of propagation. The result is an increase in wave height. The process is difficult to

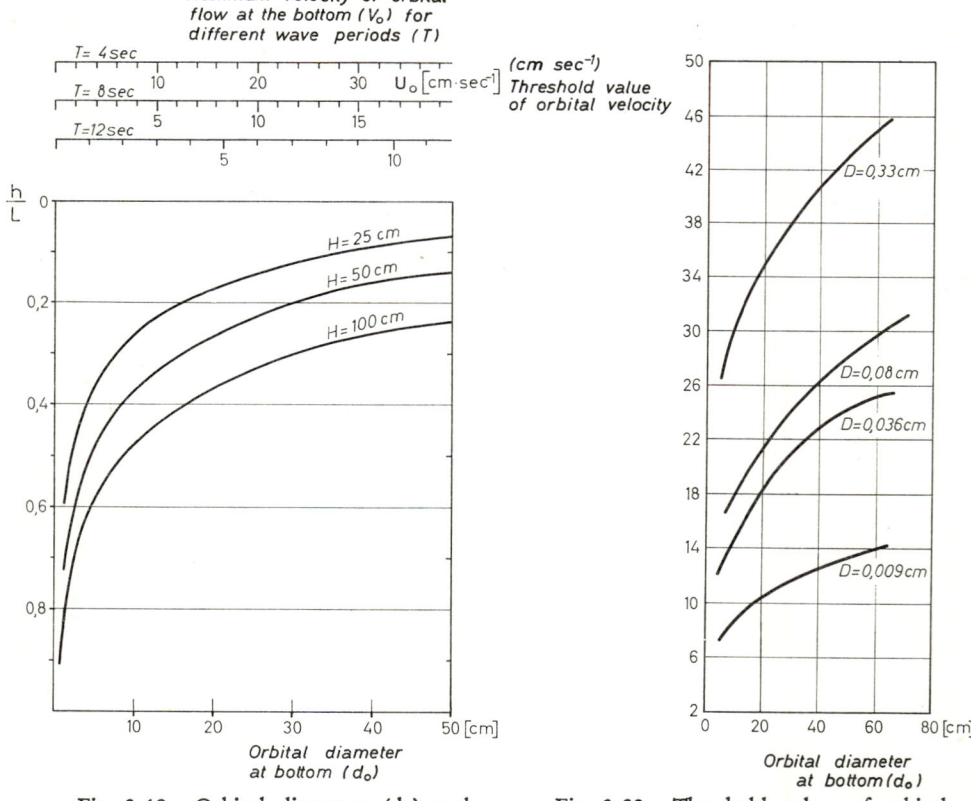

Fig. 3-19. Orbital diameters (d_o) and maximum velocity of orbital flow (u_o) at the bottom in relation to water depth (h) and wave length (L) for different wave periods (T) and amplitudes (H).

Fig. 3-20. Threshold values of orbital velocity at the bottom necessary for movement of quartz grains of different diameters D, in relation to orbital diameters (BAGNOLD 1963).

assess quantitatively, especially since the theory of water waves as presented here rests on the assumption that amplitudes are not too high. Thus is cannot be applied to the large wave heights of the breaker zone.[3] Of the different formulae presented by INMAN (1963), only the equation derived by MUNK will be cited here, which relates wave height in the breaker zone (H_b) to the wave height (H_d) and wave length (L_d) of the same wave in deep water:

$$\frac{H_b}{H_d} = 0.30 \left(\frac{L_d}{H_d} \right)^{1/3} \tag{3.93}$$

According to this equation the wave height decreases less if the deep water wave height is originally high, than if it were low.

From equation (3.92) it is apparent that the velocity of water flow finally equals the wave velocity, when wave amplitude becomes double the water depth. At this depth the wave collapses or breaks. From experience the wave amplitude in the breaker zone is about equal to the water depth. Each breaker throws a mass of water as swash on to the shore. This then returns in response to gravity as backwash.[4]

In considering transport processes along a coast, one must distinguish between the shallow sea outside the breaker zone, and the narrow strip between breakers and shoreline. Outside the breaker zone waves are active; between the breakers and the shore, swash and backwash are active. In the zone of wave activity, no water transport and thus little or no sediment transport can take place, if the orbital paths of water particles are circular or elliptical. This we can assume with no great error, as long as the H/L ratio is small. In such cases, when the wave motion extends to the bottom, loose sediment is moved back and forth along the bottom, but is not transported. The resulting oscillatory flow is different from simple flow in that full turbulence can not build up as a result of the very frequent directional changes. The principles established for simple flow cannot be applied in this case. Nevertheless there are in this case also certain threshold values of sand grain size necessary for the initiation of motion. Decisive is not only the maximum value of the orbital velocity at the bottom, but also the acceleration at the point of reversal of the motion. If waves are passing over sands with ripples, one can observe from time to time, at the end of the back and forth water motion, small clouds of sand rising up from the ripple crests. Since the acceleration at constant orbital velocity decreases with increasing orbital diameter, greater velocities are necessary for the motion of grains of given weight with increasing orbital diameter. Fig. 3-20 shows the results of some experiments by BAGNOLD (1963) involving quartz grains.

---

[3] In shallow water, especially in the breaker zone, waves can be described using the model of so-called "solitary waves". These are waves which consist of isolated high wave crests separated by relatively shallow troughs (see INMAN 1963).

[4] The impression that in a breaking wave an undertow exists which is directed seaward beneath the surface rests on an illusion. Within the breaking wave all motion is toward the shore. In the wave trough following the breaker the water swings again seaward against the next breaker. Discussion follows later concerning the strong seaward directed rip currents of the breaker zone.

Fig. 3-20 can be used in connection with fig. 3-19 to evaluate the action of given waves on sandy bottoms. We can elaborate by means of an example: Let us assume a wave motion of period T = 4 sec., and a wave height of 50 cm. From fig. 3-15 or equation (3.83) the corresponding wave length is 25 m. At a depth of $h = 0.3$ $L = 7.50$ m the orbital diameter at the bottom $d_o = 16$ m and the orbital velocity $u_o = 12.5$ cmsec^{-1}, according to fig. 3-19. It is concluded from fig. 3-20 that as a result of this wave motion, sand grains of about 0.01 cm diameter would just begin to move at a depth of 7.50 m. At the somewhat greater depth $h = 0.4$ $L = 10$ m, according to fig. 3-20, these sand grains could no longer be moved, for here (according to fig. 3-19) the orbital diameter would be 8 cm and the orbital velocity only 6.5 cmsec^{-1}. If, however, the wave height is raised from H = 50 cm to H = 100 cm, then, according to fig. 3-19, at 7.50 m depth $d_o = 30$ cm and $u_o = 23.5$ cmsec^{-1}. At this depth, according to fig. 3-20 even quartz grains of 0.08 cm diameter would be moved.

Oscillation ripples form on a sandy bottom as a result of the water oscillating in closed orbital motion. This happens as soon as the threshold value for movement of sand grains is exceeded. If the orbital velocity becomes too great, the ripples disappear and the sandy bottom becomes flat again, analogous to the case of current ripples. In contrast to the latter, oscillation ripples are symmetrical. The distance from crest to crest corresponds, up to a maximum ripple separation, to the orbital diameter at the bottom (INMAN & BAGNOLD 1963). The maximum ripple size depends on grain size. For a fine sand a maximum value of 15 cm has been observed; for coarse sand 125 cm. Onward from a water depth corresponding to that depth at which the amplitude of surface waves can act on the bottom, a region of oscillation ripples extends along sandy coasts as far as the breaker zone. According to INMAN & BAGNOLD (1963), divers have observed such ripples at depths down to over 52 m. In the breaker zone the ripples are absent and the bottom becomes smooth, since the water velocity at the bottom is so great.

The assumption that orbital paths are closed is a valid approximation as long as the wave amplitudes are small. On approaching the breaker zone the amplitudes grow as a result of interaction with the bottom. The assumption of very small amplitudes is no longer valid. For waves of greater amplitude the theory is modified in representing the orbital paths as open curves. Thus the oscillating water particles progress forward a bit with each wave in the direction of propagation of wave motion. Therefore a small amount of water transport is associated with each wave, which becomes more perceptible, the higher the amplitude. For waves in deep water the modified theory leads to the following expression for transported water volume per unit time in a wave of amplitude H and period T per unit length of wave crest:

$$q = \frac{\pi H^2}{4T} \qquad (3.94)$$

This equation is useful only for water depths exceeding 1/4 of the wave length. In shallow water wave flow at the surface and at the bottom has the same direction as the wave; in an average zone the water flows in the opposite direction (LONGUET

-HIGGINS 1953). Bottom wave flow which affects sediment transport is obtained as follows (LONGUET-HIGGINS 1953):

$$u_{0s} = \frac{5 u_0^2}{4C} \qquad (3.95)$$

u_0, the bottom orbital velocity, can be taken from equation (3.92) or from fig. 3-19. C, the wave velocity, is obtained from equation (3.82) or fig. 3-15. As an example, calculation of the bottom wave flow from equation (3.94) is carried out as follows:

Assume:	Wave period	T =	4 sec
	Water depth	h =	500 cm
	Wave amplitude	H =	50 cm
From fig. 3-15:	Wave length	L =	2500 cm
	Wave velocity	C =	630 cmsec^{-1}
From fig. 3.19:		u_0 =	23.5 cmsec^{-1}
Calculating from equation (3.95):		u_{0s} =	1.1 cmsec^{-1}

The total water movement at the bottom of the shallow zone outside the breakers results from the superimposition of the wave flow considered above (velocity u_{os}) and the oscillatory back and forth movement (amplitude d_o; velocity u_o). Sediment transport at the bottom of the zone comes about, in that the sediment already in suspension as a result of the oscillatory water movement is seized by the wave flow and carried forward without further work expenditure. The direction of wave propagation is, therefore, at the same time always a direction of sediment transport. Insofar as sufficient coarse particles (sand, gravel) are available, they are transported on the bottom toward the shore, while at the same time the finer sediment is washed out repeatedly and put into suspension. This fine material can be sedimented only at great depths, where wave motion no longer extends. The coarse sediment moved to the shore builds up the shallow sea bottom, sloping seaward.

Sediment transport in coastal regions is strongly influenced by changes in wave direction, which occur as a result of reflection, deflection, and refraction. Like light waves, water waves are also reflected, bent, and refracted, when they approach the coasts from the open sea.

For wave reflection at the coast the angle of incidence equals the angle of reflection, and that portion of wave energy reflected increases with the bottom slope angle. A vertical wall reflects most completely. Where obstacles (islands, cliffs) or where narrow passages connect broad basins (lagoons) with the open sea, waves are bent or deflected.

Refraction of waves results from the fact that in shallow water wave velocity decreases with decreasing water depth. Waves running toward a coast, whose crest line initially forms an angle to the depth contours, have their direction changed by refraction: that portion of the wave proceeding in shallow water is slowed down relative to the part in deeper water, so that the wave crest line is bent around such that its angle to the depth contours is diminished. If α_d is the angle between wave normal and the normal to the depth contour and C_d the wave velocity in deep water, and if α and C are the corresponding values at any depth outside the breaker zone, then, analogous to the law of refraction for light rays:

$$\sin \alpha : \sin \alpha_d = C : C_d \qquad (3.96)$$

Sediment transport in the zone between breakers and the shoreline is affected by swash and backwash. In the breaking wave the flow velocity of water has attained its maximum value, which, according to equation (3.92) is determined by water depth and amplitude. Since on the other hand the wave breaks where the water depth is approximately equal to the amplitude, the flow velocity in the wave, immediately before its collapse is:

$$u_B \simeq \sqrt{g\,H/4} \qquad (3.97)$$

With this high velocity (for an amplitude $H = 1$ m, $u_B \simeq 1.6$ msec^{-1}) water from the breaking wave is thrown against the shoreline. This surge has the direction of the incoming wave, which generally does not strike exactly perpendicular to the shoreline. On the other hand the water runs down the beach slope as backwash, under the influence of gravity, withdrawing in a direction perpendicular to the shoreline. In this way, due to the difference in direction of movement of swash and backwash, a current is generated running parallel to the coast. This is of considerable importance in the transport of beach sediments. Its direction is determined by the prevailing wind direction. Water dammed up against the shoreline at certain places is likely to break through the surf in a concentrated rip current. In these current canals high velocities are encountered at certain places, which under certain circumstances extend several hundred meters out to sea.

Tides are waves with periods of about 25 or about 12.5 hours, which act like waves in shallow water. The orbital velocity of the water, therefore, should be constant from the surface to the bottom. Actually the water is subject to friction and thus a lowering of orbital diameter and velocity at the base. This interaction leads in this instance also to mobilization of loose sediment lying on the bottom. The current velocity is zero at the instant of low and high tide, that is in the wave crest and trough. Maximum flow velocity is attained between wave crest and trough at the moment when the water level is changing most rapidly. If both velocities were equal, the sedimentary material would not be transported, but only moved back and forth. Actually the flood tide velocity is generally clearly greater that that of the ebb tide (SEIBOLD 1964, p. 294). Therefore, sediment can be transported shoreward by tidal waves. This is especially effective in narrow bays, at river mouths, fiords and the like, since at such restrictions the height between high and low tides is very much exaggerated.

3.164 Sub-marine landslides and suspension currents

If the suspended and bed-load material transported by streams, currents, and wave action are deposited in the horizontal parts of marine and continental basins, it permanently comes to rest there as sediment. However, when masses of loose sediment are deposited on inclined slopes, they can be set into motion again later as a result of their weight. This can happen in lakes, but especially in the ocean. The continental shelf regions are inclined on the average only about 7'. At the edge of the continental shelf (on the average 68 km wide; ahead of mountainous terrane like the California coast, < 15 km) the continental slope falls off with an average slope of 4° to the flatter continental foot which passes continuously into the deep sea or abyssal plain. On the continental slope very many steeper slopes occur locally. Off the coast of Georgia and Florida the slope is 45°. Into the continental slope are cut the steep channels of submarine canyons (on the morphology of the ocean bottom see, for example: SHEPARD 1963 a, 1963 b; HEEZEN & MENARD 1963; GUILCHER 1963; SEIBOLD 1964; MENARD 1964).

On inclined slopes unconsolidated sediments can be set into motion, if the components of gravity exceed the frictional or plastic flow limits of the sediment. This results in subaquatic landslides which transport the sediment down over the slope. As a result primary sedimentary structures can be more or less deformed or even completely destroyed. It can lead eventually to the occurrence of so-called suspension currents (turbidity currents), if the slope angle and the mass of material set in motion are sufficiently large. In these currents all components of the primary sediment are reworked and mixed and then thrust beneath the stationary water layer at the ocean bottom far out onto the abyssal plain.

Detailed investigation of the continental slope and deep sea floor has led to the realization that subaquatic landslides and suspension currents are among the most important processes of marine sedimentation. Phenomena, such as the occurrence of shallow marine or even continental sands and organic remains on the deep sea bottom, and the great sedimentary fans at the mouths of submarine canyons, attest to this fact. Also only by the activity of suspension currents is it possible to explain the building up of the planar surface of the deep sea bottoms and trenches adjacent to the shelf areas. From the geologic past, the very thick clayey, sandy geosynclinal flysch-type sediments, known from many formations, attest to the sediment forming activity of landslides and suspension currents (see KUENEN & MIGLIORINI 1950 and KUENEN 1951 as well as more recent reviews by HEEZEN 1963 and BOUMA & BROUVER 1964, which contains of special importance the article of KUENEN 1964 and the literature summary of KUENEN & HUMBERT 1964).

Since submarine landslides and suspension currents occur only in very deep water, and then only relatively seldom, their direct observations is even more difficult than for the other modes of sediment transport in water. Also because laboratory studies are especially difficult in this case, current theory can not offer more than a very general and preliminary view of the mechanisms of submarine landsliding and suspension currents.

We should first consider under what conditions on an inclined slope layered, unconsolidated sediments can be suddenly mobilized, after they have remained at rest for a long time. The rheologic properties of unconsolidated sediments need to be considered. As two extreme cases we can consider on the one hand the properties of a coarse sand, and on the other, a mud consisting of fine-grained clay particles. If a packed layer of sand of thickness s is to be moved over a stationary layer of the same sort, a certain shear stress is necessary, which is proportional to the pressure P, exerted by the weight of the mobilized sand layer on its sandy substrate. The ratio of the shear stress τ and pressure P is the frictional coefficient α that we have used previously in considering bottom transport of sand:

$$\alpha = \frac{\tau}{P} \tag{3.98}$$

For densely packed sand grains $\alpha = 0.65$. Accordingly a pile of sand grains will assume an angle of repose of about 33° (tan 33° = 0.65).

On the other hand a clay mud behaves like a plastic body under the influence of a shear stress. This is illustrated in fig. 3-21. Shown is the dependence of the exerted shear stress on the velocity gradient in a flowing substance. Fig. 3-21 shows that in normal fluids (Newtonian fluids) all shear stresses, even the smallest, cause flow, that is, a certain velocity gradient. The flow curve is a straight line, whose slope $d\tau/d\,(dv/dx) = \mu$ is called the viscosity. Plastic substances behave on the other hand like solid bodies at shear stresses below a certain threshold value — the flow

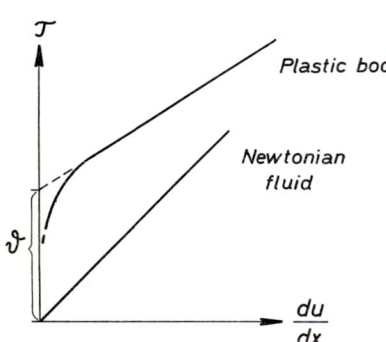

Fig. 3-21. Relation between shear stress (τ) and velocity gradient (du/dx) for Newtonian fluids and plastic bodies.

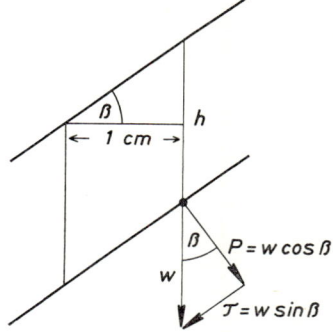

Fig. 3-22. A plastic sediment at equilibrium on an inclined surface (explanation in text).

limit ϑ. They begin to flow only at shear stresses $\tau > \vartheta$. The slope of the flow curve $d\tau/d\,(dv/dx) = \mu'$ is called the plastic viscosity. The flow behavior of suspensions of small particles is further complicated in that in many such systems the flow limit of the moving suspension is very much smaller than when it is at rest. The dependence of flow limit on the state of motion and time is called thixotropy. Solidified thixotropic suspensions at rest can be "liquified" by gentle shaking.

The forces effective in a sediment deposited on an inclined slope are shown in fig. 3-22. The weight w of an oblique prism of height h (layer thickness $s = h \cos \beta$) and of width = depth = 1 cm (base = 1/cos β) in water is equal to $(\sigma - \varrho_1) g \cdot h$, if

σ is the volume weight of the sediment and ϱ_1 the specific gravity of the water. At the base of the prism a normal pressure P and a shear stress τ is in effect:

$$\left. \begin{array}{l} P = w \cos \beta \\ \tau = w \sin \beta \end{array} \right\} \tag{3.99}$$

The movement downslope acts in opposition to a frictional force $R = \alpha P$ and the flow limit $T = \vartheta/\cos \beta$ at the base of the prism. At equilibrium:

$$\begin{array}{l} \tau = R + T \\ w \sin \beta = \vartheta/\cos \beta + \alpha w \cos \beta \end{array} \tag{3.100}$$

For a slope angle β and a layer thickness s, we obtain:

$$\operatorname{tg} \beta = \alpha + \frac{\vartheta}{\cos^2 \beta (\sigma - \varrho_1) g h} \tag{3.101}$$

$$s = h \cdot \cos \beta = \frac{\vartheta}{\cos \beta (\sigma - \varrho_1)(\operatorname{tg} \beta - \alpha) g} \tag{3.102}$$

For the limiting case represented by coarse sand, the flow limit vanishes so that $\tan \beta = \alpha$. Equation (3.102) loses its meaning. As long as the slope angle is smaller than the frictional coefficient, sand layers of any thickness are stable. If the slope is greater than the frictional coefficient no sand at all can be deposited.

However, if the sediment is plastic and the frictional coefficient very small, if the flow limit is known, the maximum layer thickness can be calculated for any slope angle from equation (3-102). A plastic sediment can, therefore, be deposited initially on an inclined slope. If the layer thickness finally exceeds a critical value, a layer of this thickness will slide down the slope.

Natural sediments are for the most part mixtures of sand and clay. Depending on the grain size of their constituents, each will likewise have a frictional coefficient or flow limit which dominates the system. Accordingly its behavior will be better represented either by the model for coarse sand or by that for clays. For the activation of submarine landslides and the occurrence of suspension currents, plastic behavior, that is, a certain content of fine-grained components may be the necessary prerequisite.

The flow limit of a clay mud is higher the smaller the particle size and the lower the water content. It is especially high for montmorillonite clays (bentonites). For recent pelitic ocean sediments flow limits measured cover a wide range between a few g/m² and about 50 g/cm² (1-50 10³ dyn/cm²) (EINSELE 1967). Corresponding to these values and to different slope inclinations, maximum sediment thicknesses calculated from equation (3.102) are plotted in fig. 3-23. The density difference $(\sigma - \varrho_1)$ was taken as equal to 1.0 and the frictional coefficient disregarded ($\alpha = 0$). If the frictional coefficient has a value significantly greater than zero, the maximum sediment thickness becomes greater, especially at small slope angles.

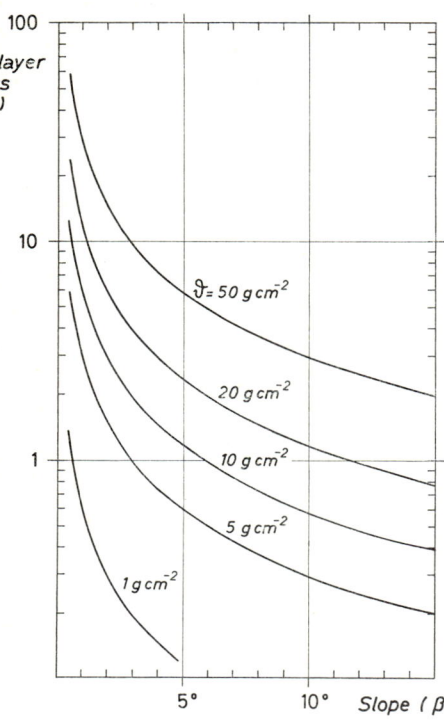

Fig. 3-23. Maximum layer thickness for plastic sediments with different flow limits (ϑ) at different slope inclinations.

Therefore, plastic sediments deposited on a continental slope are set into motion and slide down slope, if the critical thickness is exceeded. The accumulation of sediment by supply of new material must of course take place rapidly enough that no significant diagenetic consolidation can occur.

Rapid sedimentation of fine-grained material on a not too gentle slope will be a prerequisite of submarine landslides. Since unconsolidated fine-grained sediments are more or less thixotropic, sliding can take place before the critical thickness for the static flow limit is reached, if earthquake shocks suddenly decrease the flow limit and the sediment becomes "fluidized" as a result. Actually to some considerable extent related to earthquakes, different submarine landslides have been observed. They are listed in table 3-7 after a summary by MORGENSTERN (1967).

Table 3-7. Submarine landslides triggered by earthquakes (after MORGENSTERN 1967).

Location	Slope angle	Literature
Grand Banks 1929	3.5°	HEEZEN & EWING 1952
Orléansville 1954	4°—20°	HEEZEN & EWING 1955
Straits of Messina 1908	4°	RYAN & HEEZEN 1965
Suva (Fiji Islands) 1953	3°	HOUTZ 1962
Chile 1922	6°	GUTENBERG 1939
Valdez, Alaska 1964	6°	COULTER & MIGLIACCIO 1966
Aegean Sea, 9.7.1956	10°	AMBRASEYS 1960

3.16 Marine transport processes

The transatlantic cable break associated with the quake in the area of the Grand Banks, south of Newfoundland and observed in 1929 (HEEZEN & EWING 1952), just like the cable disturbance off the African coast after the quake of Orleansville in 1954 (HEEZEN & EWING 1955) both have been traced back to suspension currents which developed from large landslides. It is assumed that suspension currents form to a great extent from landslides which occur on long, steep slopes, and that these currents slide down the continental slope beneath the mass of static ocean water, spreading out extensively on the abyssal plain. From the characteristics of sediments (turbidites) which have been ascribed to such currents, a turbulent mixing of moving material takes place. For lack of any precise data on this sort of suspension current, there is still no complete theoretical basis for this phenomenon. It is especially difficult to understand how and over what distances a suspension current may flow beneath stationary water without its energy being dissipated in the surrounding water and without loosing its sharp boundary.

In any case the transport process in suspension currents must be essentially different from normal suspension transport in streams und currents. In normal streams and currents the amount of suspended matter is so low that the density of the flowing suspension is practically equal to the density of water. Under this supposition the amount which a given current is able to carry is limited and the maximum suspended load of a stream was calculated in section 3-15. The submarine suspension current flows, however, because its volume weight is much higher than that of the surrounding and overlying water. Here, therefore, we are not concerned with the flow of water that is able to carry a certain amount of material in suspension. The concentration of suspended particles is so high that its mass constitutes an essential part of the flowing suspension and constitutes itself the carrying power of the current. BAGNOLD (1962) has shown that this situation prevails when the suspension current velocity, the slope inclination, and the mean settling velocity of the suspended particles stand in a certain relationship to one another. In such a case, as the following considerations show, the solid content rises without limit. This is the essential difference between suspension currents and normal streams laden with suspended load.

A suspension contains particles with settling velocity v and specific gravity ϱ_2. Its volume concentration is C. For the work produced per unit time and volume from the swirling of the turbulent flow, to hold the particles in suspension:

$$A_1 = (\varrho_2 - \varrho_1) C g v \qquad (3.103)$$

This work must be supplied by the flowing suspension, which consists of water and its suspended particles. For normal streams that amount of energy which originates from the mass of suspended particles can be neglected. In the case of suspension currents, however, the amount of work derived from the suspended mud predominates. Thus per unit time and volume:

$$A_2 = (\varrho_2 - \varrho_1) C g U \sin \beta \qquad (3.104)$$

U is the mean current velocity (it is assumed that there is no velocity difference between particles and water), β is the slope angle. For the difference between both quantities of energy, we obtain:

$$\Delta A = A_1 - A_2 = (\varrho_2 - \varrho_1) C g U \left(\frac{v}{U} - \sin \beta \right) \quad (3.105)$$

If ΔA is positive, for maintaining the suspension, energy must be taken away from the energy supply of the flowing water. Then, as in the case of normal streams, the amount of suspended particles is limited. However, if ΔA is zero or negative, we have the case of the suspension current: From the energy of the flowing particles (A_2) there is just as much or more energy available as is needed to maintain the suspension. Then the suspended solid concentration can rise without limit and a self-supporting suspension current of high energy occurs. The condition leading to the occurrence of a suspended current can be expressed as follows:

$$U \geq \frac{v}{\sin \beta} \quad (3.106)$$

A suspension current of particles of a certain settling velocity v can form only when a current velocity U is achieved which, according to this equation, must be higher the lower the slope angle. On the other hand for a given slope angle the necessary current velocity is the higher, the greater the settling velocity, or particle size. Some appropriate values are listed in table 3-8.

Table 3-8. Minimum velocities U of suspension currents containing particles with settling velocities (v) at different slope angles (β). Calculated from equation (3.106).

β	$v = 0.1$ cm sec^{-1} U [cm sec^{-1}]	$v = 0.001$ cm sec^{-1} U [cm sec^{-1}]
0.5°	11.5	0.11
1°	5.7	0.057
2°	2.87	0.029
4°	1.43	0.014
8°	0.72	0.007

A suspension current can occur, therefore, only when a suspension is brought to a certain minimum flow velocity. In the case of marine suspension currents this happens apparently if fine-grained sedimentary layers deposited on the continental slope, crumble under the shock of earthquakes or because an minimum layer thickness is exceeded, and by sliding down over the steep continental slope and into the channels of submarine canyons achieve the necessary acceleration and turbulent mixing.

Nothing certain is known about the actual velocities, thicknesses and dimensions of the suspension currents occurring on the continental slopes. The high velocities of 6 to 20 msec^{-1} which HEEZEN & EWING (1952) have calculated from the successive cable breaks during the incident south of Newfoundland (1929) often have been questioned. A theory offered by PLAPP & MITCHELL (1960), based on the use of hydrodynamic turbulence theory, arrived at similarly high velocities, but this theory contains a number of arbitrary assumptions.

In laboratory experiments KUENEN and co-workers (KUENEN & MIGLIORINI 1950; KUENEN 1951; KUENEN & MENARD 1952) have produced suspension currents and studied them in a 30 m long flume. With sandy clay suspensions of densities between 1.5 and 1.8, initial velocities of about 80 cmsec^{-1} were attained, which in the 24 m long horizontal portion of the flume diminished gradually to about 10 cmsec^{-1}. As a result vertical as well as horizontal graded sediments were deposited, which were similar to natural turbidities.

3.17 Changes during transport and deposition

3.171 Mechanical effects of water transport

3.171.1 General

Particles produced by the destruction of primary rocks in the course of weathering experience further changes in form during transport as they strike and abrade one another. Particles transported in suspension impact one another only occasionally and then only with slight mechanical effect. Therefore, the sizes and shapes only of bottom transported sedimentary particles are altered. According to fig. 3-12, particles with the specific gravity of quartz are carried primarily in suspension by water currents when their equivalent diameters are less than 0.2 mm. The phenomena we shall consider in this section of the mechanical effects of transport will concentrate particularly on particles of greater than 0.2 mm diameter.

The attrition of particles moving on the bottom leads first of all to a decrease in size and secondly to a change in the geometric state of the particle. Thereby two different geometric aspects must be considered, the form or shape of the particle and the quality of its surface. At the particle surface the distribution of various radii of curvature is especially important. This is measured as the roundness of the particle. The degree of roundness is low when portions of the surface have especially small radii of curvature (edges, corners). Also the surface microstructure can be important. For example one can distinguish smooth, polished, dull, or rough surfaces. Before we consider the mechanical effects of transport in detail, the definitions of the parameters with whose help shape and roundness of sedimentary particles are measured, will be considered in the next section (see also Part I of this series, p. 104-107 and p. 16 f.).

3.171.2 Definition of form and roundness[5]

The methods for describing geometric particle shapes consist of measuring the three mutually perpendicular major diameters (a $>$ b $>$ c). For describing the shape,

---

[5] See also the comprehensive descriptions by PETTIJOHN (1957) and KÖSTER (1964).

one can select two of six possible ratios formed from the three quantities. As an example, one uses the ratio of intermediate to largest diameter and smallest to intermediate diameter:

$$\alpha = b/a \qquad \gamma = c/b$$

Those two ratios can assume all values between 0 and 1. All particle shapes so far as they can be defined by the major diameters alone, can be plotted in a square diagram with the coordinates α and γ, as shown in fig. 3-24. In the diagram are given the various shape classifications as defined by ZINGG (1935) (table 3-9).

Table 3-9. Shape classification of sedimentary particles (after ZINGG 1955).

	$\alpha = b/a$	$\gamma = c/b$
Disks (oblate)	> 0.67	< 0.67
Spherical (equiaxial)	> 0.67	> 0.67
Rods (prolate)	< 0.67	> 0.67
Blades (triaxial)	< 0.67	< 0.67

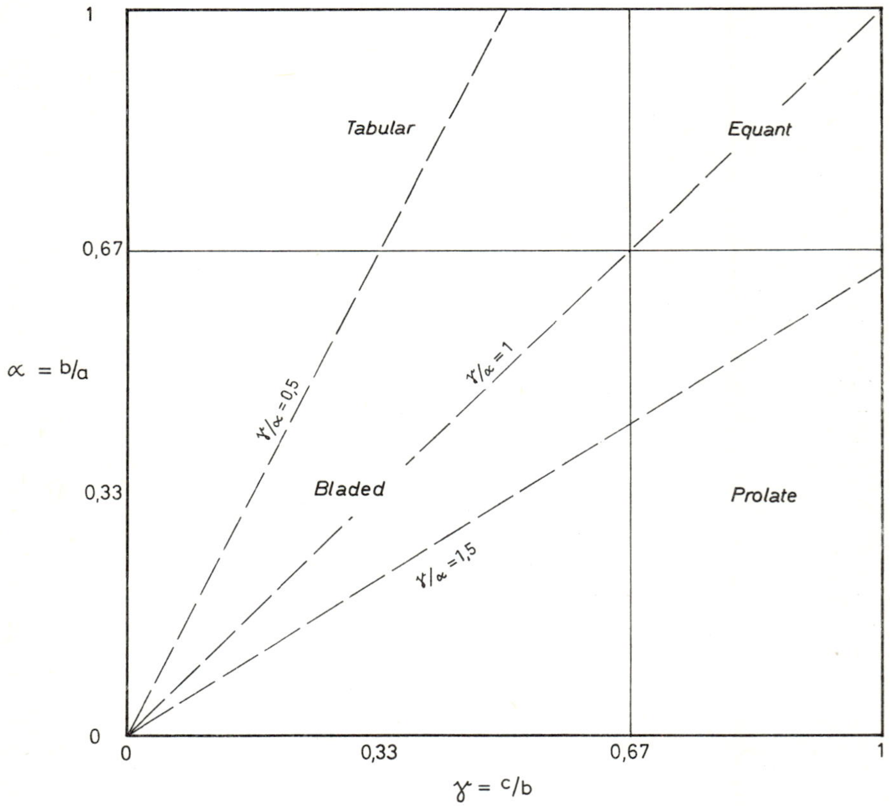

Fig. 3-24. Shapes of sedimentary fragments (after ZINGG 1935).

Naturally one can use the ratios b/a and c/b, or for that matter others, to describe particle shape. SNEED & FOLK (1958) proposed the ratios c/a and $(a-b)/(a-c)$, with whose help the classes given in table 3-10 are distinguished.

Table 3-10. Shape classification of sedimentary particles (after SNEED & FOLK 1958).

	c/a	$(a-b)/(a-c)$
Compact	0.7—1	0 —1
Compact-platy	0.7—0.5	0 —0.33
Compact-bladed	0.7—0.5	0.33—0.67
Compact-elongated	0.7—0.5	0.67—1
Platy	0.3—0.5	0 —0.33
Bladed	0.3—0.5	0.33—0.67
Elongated	0.3—0.5	0.67—1
Very platy	0 —0.3	0 —0.33
Very bladed	0 —0.3	0.33—0.67
Very elongated	0 —0.3	0.67—1

While the three major axes can be determined for large pebbles, this is not possible for sand grains. In this case one must be satisfied to measure those two diameters which appear in cross section when the grain is viewed in a microscopic image. This cross section gives a larger (a') and a smaller diameter (c'). Loose sand grains generally arrange themselves on a flat surface so that the greatest cross section is parallel to the substrate. For randomly scattered grains a' corresponds approximately to a and c' to b. For characterizing the geometric shapes of such sand grains, one can use the ratio c'/a' which corresponds approximately to the quotient α. Special problems arise in thin section, since the largest and smallest diameters measured in random grain cross sections are not necessarily the major grain diameters (SCHNEIDER-HÖHN 1954).

Different suggestions have been made to establish a single parameter for characterizing particle shape. With respect to all of these it is important to note that the geometric form of a particle can never be fully described by a single parameter. From the quotients α and γ one can devise the form factor (ZINGG 1935):

$$F = \gamma/\alpha = a \cdot c/b^2 \qquad (3.107)$$

The straight lines drawn in fig. 3-24 indicate equal form factors. It is clear from this diagram that very differently shaped particles can have the same form factor.

Sphericity was introduced as a parameter to describe the deviation of particle shape from that of a sphere. This was probably because it was felt that the shapes of transported particles tend in the long run toward a spherical shape. This, as will be shown, is not a generally valid assumption. According to the original definition of WADELL (1932), sphericity is the ratio of the surface area of a sphere of the same volume as a particle in question (s) to the actual surface area of the object (S):

$$\Psi = s/S \qquad (3.108)$$

Surface areas and volumes of particles are difficult to measure. Therefore, KRUMBEIN (1941 a) suggested a modified definition of sphericity, concluding that the form and volume of a particle can be represented by a triaxial ellipsoid with a, b, and c axes:

$$\Phi = \sqrt[3]{\frac{\text{volume of the ellipsoid}}{\text{volume of the circumscribed sphere}}} = \sqrt[3]{\frac{abc}{a^3}} = \sqrt[3]{\frac{bc}{a^2}} \quad (3.109)$$

If one introduces the parameters α and γ to describe the particle shape, then:

$$\Phi = \sqrt[3]{\alpha^2 \gamma} \quad (3.110)$$

Another definition of sphericity was proposed by ASCHENBRENNER (1956) which introduced a distorted cubo-octahedron as a model for pebble shapes.

The concept of flattening or oblateness (aplatissement) was introduced by CAILLEUX (1952):

$$A = \frac{1}{2}\left(\frac{a+b}{c}\right) = \frac{1}{2}\left(\frac{1}{\alpha \gamma} + \frac{1}{\gamma}\right) \quad (3.111)$$

Since sand grains, because of their smallness can be observed only in cross section (microscopic procedures), comparison to a circle of equal area must take the place of the comparison with a sphere. According to WADELL (1935) the two dimensional or projection sphericity is defined as the ratio of the diameter of a circle with the same area as the particle projection (d) to the diameter of the circumscribing circle (D):

$$\Phi' = \frac{d}{D} = 2 \frac{\sqrt{A/\pi}}{a'} \quad (3.112)$$

A is the planimetrically determined cross sectional area of the grain and a' the longest diameter. If the particle cross section is an ellipse with diameters a' and c', a clear relation exists between the ratio c'/a' and the projected sphericity:

$$\Phi' = \sqrt{c'/a'} \quad (3.113)$$

A summary of other definitions of projected sphericity is provided by RILEY (1941).

As a measure of particle roundness WENTWORTH (1922) first introduced the relation:

$$\text{roundness index} = r_1/R \quad (3.114)$$

r_1 is the smallest radius of curvature of the particle, R is the average radius of the particle, and approximately equal to $1/6 \, (a+b+c)$. Roundness as defined by CAILLEUX (1952),

$$\text{roundness index} = 2 \, r_1/a \quad (3.115)$$

is a modification of the WENTWORTH definition. Instead of a mean radius, the largest particle diameter is used. In order to take into account not only the sharpest corners or edges, but rather the whole surface, WADELL (1932) proposed measuring the radii (r) of all convex parts of the periphery of a particle cross section and the radius (R) of the largest inscribed circle.

The roundness (P) is given as the mean of N measurements:

$$P = 1/N \, \Sigma \, r/R \tag{3.116}$$

This roundness is determined, therefore, from a cross section of the particle. To characterize the particle, the values must be averaged properly over several differently oriented sections. Accordingly the roundness P can be determined also on the grain cross sections in a thin section.

In order to avoid the tedious measurement of radii of curvature on the circumference of each grain, various authors have published pictures of particle cross sections, corresponding to the degrees of roundness of WADELL: RUSSELL & TAYLOR 1937, KRUMBEIN 1941 a, POWERS 1953, PETTIJOHN 1949, 1957, SCHNEIDERHÖHN 1954. Roundness determinations of given particles can be made by usual comparison with these standard pictures. Individual scales differ in their intervals:

RUSSELL & TAYLOR: 0 — 0.15 — 0.30 — 0.50 — 0.70 — 1
KRUMBEIN: 0 — 0.1 — 0.2 — 0.3 — 0.4 — 0.5 — 0.6 — 0.7 — 0.8 — 0.9 — 1
POWERS: 0 — 0.12 — 0.17 — 0.25 — 0.35 — 0.49 — 0.70 — 1
PETTIJOHN: 0 — 0.15 — 0.25 — 0.40 — 0.60 — 1
SCHNEIDERHÖHN: like PETTIJOHN

A method offered by SZADECZKY-KARDOSS (1933) likewise concerns itself with the entire surface: The even (P), concave (C) and convex (V) portions of the circumference of a particle section are measured and reported as percent of the total circumference. A so-called rolling index is defined by the three values P, C, and V and can be plotted on a three-component triangular diagram. This method can be used also with thin sections.

For evaluation of experimental findings and natural observations, it is to be noted that degrees of roundness determined by different methods cannot be compared directly with one another. Rounding as measured by WENTWORTH and CAILLEUX considers only the least rounded places on the particle and emphasizes low degrees of roundness. The methods of WADELL and SZADECZKY-KARDOSS both take into account the entire surface, but in different ways, in that the former leads to a single number, the latter to two defining parameters (for a comparison of methods see SCHNEIDERHÖHN 1954).

3.171.3 Experiments on the mechanical effects of water transport

Although it is difficult to reproduce natural conditions, experimental studies of the mechanical action of transport are especially important. The change in sedimentary particle size and form, observed in the course of a stream or along the coast, cannot always be traced back alone to attrition during transport.

Selective transport and the addition of material by tributaries can have a disturbing influence on the relations. Only by experiment can the attrition due to transport be studied in isolation.

First studies of this kind were published as early as 1879 (DAUBRÉE). His simple experimental procedure has served as a model for many later investigations. Water with rock and mineral fragments is allowed to rotate in a horizontal cylinder. After given times, or distances of travel, the size and shapes of the sample introduced are determined. Using a very similar scheme WENTWORTH (1919), MARSHAL (1927), SCHOKLITSCH (1933), KRUMBEIN (1941 b), RAYLEIGH (1943, 1944) and BERTHOIS & PORTIER (1956, 1957 a) later investigated the behavior of large rock and mineral fragments. The samples in these studies moved at velocities of about 0.5 to 1 msec^{-1}. Instead of using a rotating drum, ERDMANN (1894) used a flume to simulate better natural conditions of stream flow. Similarly KUENEN (1956) let pebbles rotate at the bottom of a cylindrical vessel through a circulating water current. The results of these studies can be summarized as follows: If the material consists of rock and mineral fragments of uniform size, the mean diameter decreases rapidly at first with distance travelled and then gradually slower and slower. Pebble sphericity increases, as does roundness, with distance travelled. The increase in these quantities per unit distance travelled diminishes gradually, and finally approaches a constant roundness and sphericity, although the mean particle size continues to decrease. In fig. 3-25 some typical results are reproduced, which KRUMBEIN (1941 b)

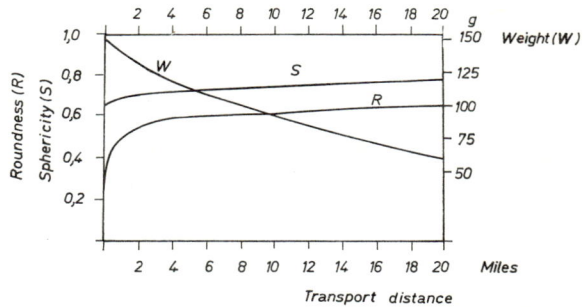

Fig. 3-25. Experiments by KRUMBEIN (1941) on the mechanical effects of water transport on limestone pebbles.

obtained by rotating limestone fragments of original 5 cm diameter in a drum. Only at the beginning, as long as the pebbles still had sharp corners and edges, were large pieces broken off. Later only small and the smallest of the rounded particles were separated. Therefore, the decrease in particle weight is greater at the beginning than later. After a certain transport distance smooth surfaces with a very uniform radius of curvature develop.

3.17 Changes during transport and deposition

Attempts to express the experimentally established decrease in particle weight or size by a mathematical function are not lacking. Repeatedly the so-called STERNBERG law (STERNBERG 1875) is cited, which rests on the simple premise that a small decrease in pebble weight (W) or diameter (D) over a small transport distance (s) is proportional to the initial weight or diameter:

$$dW = -c \cdot W \cdot ds, \qquad dD = -k \cdot D \cdot ds \tag{3.117}$$

Following integration the particle weight or diameter, following transport over distance s, are given as follows:

$$W = W_0 e^{-cs}, \qquad D = D_0 e^{-ks} \tag{3.118}$$

W_0 and D_0 are the initial weight and diameter.

KRUMBEIN (1941 b) found that in his studies with limestone pebbles a linear relationship developed, when the log of the pebble weight or diameter was plotted against transport distance. Only at the very beginning of transport was the curve somewhat bowed. KRUMBEIN concluded, therefore, that the pebble reduction follows equation (3.118).

If an object is ground on a rough surface or by means of an abrasive, studies have indicated that the volume of material abraided per unit of transport is proportional to P/F, where P is the force with which the body presses against the substrate and F is the grinding surface area (see, for example, v. ENGELHARDT & HAUSSÜHL 1965). Over a narrow range of abrasion velocities these factors have no influence on the amount abraded. The STERNBERG proportionality between weight loss and original weight conforms to the abrasion process, if one assumes the pebble weight to be the effective force and that the abrading surface area is independent of pebble size. The latter, however, is unlikely. One would presume rather that the abrading surface area becomes smaller with decreasing pebble size and is at any given time proportional to the square of the pebble diameter. Therefore, for a weight loss (dW) over a small distance (ds):

$$dW = -\text{const.} \frac{D^3}{D^2} ds = -\text{const.} D \cdot ds \tag{3.119}$$

and for reduction in the diameter (dD):

$$dD = -\text{const.} \frac{1}{D} ds \tag{3.120}$$

Pebble weight and diameter, therefore, would change more slowly with transport, than prescribed by the STERNBERG relationship. By integration one derives, instead of equation (3.118), the following equation for the diameter following transport:

$$D = \sqrt{D_0^2 - \text{const.} s} \tag{3.121}$$

Which of the two statements, equation (3.118) or (3.121), represents the correct relationship, that is, whether particle diameter decreases as a parabolic or as an exponential function of transport distance, is not readily ascertained from the afore-mentioned experiments. If one focuses on the initial rapid abrasion, the experimental results of KRUMBEIN (1941 b) can be represented just as well by either function.

Experiments by KUENEN (1956) have led to somewhat different relations between pebble size and transport distance. He allowed cube-shaped samples of different rocks to roll around at the bottom of a cylinder in a circulating stream of water. He determined changes in size and shape, investigating especially the influence of velocity. The resulting weight loss over a given travel distance, at different flow velocities (v), can be expressed by the following equation:

$$(W_0 - W)_s = cD_0 + lv^2 \tag{3.122}$$

W_0 is the initial weight; c and l are constants. KUENEN interpreted this finding by proposing two processes of attrition: the one is an abrasion process and proceeds proportional to the particle diameter; the other occurs as the result of impact, and is proportional to the square of the velocity. With small particles KUENEN found that the second process predominated, with large ones the first.

KUENEN's so-called abrasion process does not correspond exactly to the relation expected from equation (3.119):

$$\left(W_0 - \frac{D_0}{D} W\right)_s = c D_0 \tag{3.123}$$

The weight loss for a given travel path should be somewhat smaller than predicted by the first term of the KUENEN formula. Still that part of the particle reduction observed by KUENEN to be independent of velocity simulates better the abrasion mechanism than does the STERNBERG statement.

The abraded volume depends, per unit travel path at given pebble size, also on the material. The studies of SCHOKLITSCH (1933) and KUENEN (1956) produced data on the relative abrasion resistance of different rocks. Flint and quartzite abrade the slowest, friable sandstone the most rapidly. Other rocks such as limestones, crystalline schists, and igneous rocks exhibit a broad spectrum of durabilities, lying between these two extremes. This property depends not only on the mineral composition, but also significantly on the rock fabric. As a result no generalized order of resistance to abrasion can be indicated.

Exact values for the abradability of pure minerals can be given. Table 3-11 gives values for the abrasion resistance of minerals as measured by ROSIVAL (after TERTSCH 1949) and v. ENGELHARDT & HAUSSÜHL (1965). In both cases the volume loss suffered by the mineral, when polished by an abrasive (corundum powder) on a flat surface, was determined (by v. ENGELHARDT & HAUSSÜHL in xylol). ROSIVAL fixed the time (8 minutes) and abrasion surface area (4 cm²). v. ENGELHARDT & HAUSSÜHL determined the abrasion for a given travel distance and a given pressure. The abrasion restistance is proportional to the reciprocal of the volume loss and is set equal to 1000 for quartz.

There is not much we can infer about changes in pebble shape during transport from all of those studies in which only indices of sphericity and roundness were determined. These express very little of the actual shape. In the studies by KRUM-

Table 3-11. Relative abrasion resistance for different minerals.

Mineral	after ROSIVAL (TERTSCH 1949)	after v. ENGELHARDT & HAUSSÜHL (1965)
Beryl (0001)	—	2300
Topaz (0001)	1210	860
Quarz (0001)	1000	1000
K-feldspar	257 (010)	240 (110)
Fluorite (111)	28.3	96
Calcite (10$\bar{1}$1)	19.1	25
Dolomite (10$\bar{1}$1)	—	71
Apatite (0001)	5.30	—
Gypsum (010)	3.6	—
Talc (compact)	5.9	—

BEIN (1941 b) and KUENEN (1956) it was determined that sphericity and roundness initially change rapidly, but later on very slowly, appearing to approach asymptotically maximum values. When the limestone pebbles in KRUMBEIN's experiments were reduced to one third of their original volume, the sphericity and roundness indices had the values 0.77 and 0.64 (initially 0.65 and 0.13). With further transport these values barely changed. Spherical pebbles could never form, because previously they were completely abraded before they could achieve the shape of a sphere.

Only RAYLEIGH (1943, 1944) has made detailed observations of the changes in pebble shape during abrasion in a rotating drum. He found that flattened and elongate ellipsoids formed from chalk as a result of abrasion, leading to equalization of their major diameters; yet a spherical shape was never reached before the pebble disappeared. Original spheres showed a gradual flattening at two poles. Marble prisms assumed the following sequence of shapes with increasing abrasion: prism → cylinder with rounded edges → somewhat spherical form → flattened form → disc with flattened edges. With marble cut into other starting forms, after a certain transport distance, pebbles formed which were similar to flattened or elongate ellipsoids. The main cross section (through longest and shortest axes) of these pebbles deviated in charcaterisitic ways from an ellipse. The outline of the pebble cross section always laid outside the perimeter of an ideal ellipse with the same major axes. RAYLEIGH's experiments allow us to surmise that the spherical form is not the endpoint of pebble abrasion, but rather a favored flatter shape develops, whose roundness and sphericity is less than unity.

SCHUBERT (1964) studied the grain size distribution of particles produced by the rotation of quartz and orthoclase pebbles in polyethylene flasks. He found an arithmetic normal distribution with a small excess of fine material (< 0.3 mm). The principal grain size fraction produced by this process had a diameter around 0.78 mm.

Sand grains are abraded during transport only when they are rolled around on the bottom. Small grains suffer very little change in size and shape. With increasing current velocity the grain size limit below which abrasion practically ceases, is displaced to greater diameters. With his studies in a rotating cylinder DAUBREE

(1879) found no abrasion of sand grains below 0.1 mm diameter for a velocity of 1 msec^{-1}; grains over 0.5 mm experienced a 0.01 percent weight loss per kilometer transport. Later investigations have confirmed that grains of less than 1 mm diameter experience very little abrasion. BERTHOIS & PORTIER (1957 b), experimenting with quartz fragments between 1.5 and 2 mm, found, after 11 km transport distance in a rotating cylinder, no visible sign of abrasion. Only after 1125 km were 22% and after 2040 km 65% of the grains visibly abraded (usé). The velocity in these experiments was 0.4 msec^{-1}. In separate experiments SCHUBERT (1964) let quartz and orthoclase fragments of sand size (0.2—1.6 mm; median at 1.0 mm) rotate in distilled water in polyethylene flasks. After 912 hours, corresponding to a distance of 1160 km, the cumulative grain size distribution curve had not changed. Therefore, essentially no measurable abrasion had taken place. In his previously described experiments in which sand grains were propelled around at the bottom of a rotating cylinder, KUENEN (1958, 1959) found the abrasion of small limestone and quartz cubes, as well as feldspar grains below 0.4 mm to be imperceptably low. For particles above 1 mm a discernable increase in roundness and a corresponding weight loss could be confirmed, especially during the initial part of the experiment. The abrasion of feldspar was about double, and of limestone 5 to 20 times greater than that of quartz. The increase in roundness and decrease in weight was significant only during the first kilometer of transport and then fell off strongly. The weight loss of quartz particles was very much less than that of limestone fragments. Some of these results are reproduced in table 3-12.

Table 3-12. Weight loss of limestone and quartz cubes during bottom transport; after experiments of KUENEN (1958).

Size mm	Velocity cm sec^{-1}	Transport distance km	Weight loss in % limestone	quartz
2.5	35	32	4.05	0.13
1.0	40	64	1.85	0.17

THIEL (1940) investigated the change in size and sphericity (projection sphericity of WADELL) of grains of different minerals. The experiments were made in a rotating drum at a velocity of 0.8 msec^{-1}. The results attained for grains of 1—2 mm diameter over a distance traveled of 8000 km are summarized in table 3-13.

Table 3-13. Weight loss and change in shape for mineral grains (1—2 mm diameter) after transport of 8 000 km in a rotating drum (after THIEL 1940).

	Weight loss %	Sphericity initial	final
Quartz	24.2	0.72	0.79
Garnet	41.7	0.77	0.80
Hornblende	82.2	0.65	0.78
Apatite	84.3	0.74	0.91
Microcline	—	0.67	0.76
Tourmaline	—	0.74	0.79

3.171.4 Observations of the mechanical effects of water transport[6]

The STERNBERG (1875) rule (see p. 105) has been confirmed by a number of investigators (see, for example, BARREL 1925; KRUMBEIN 1942; BLISSENBACH 1952). According in this relation, pebble weight and downstream transport are related as follows:

$$D = D_0 e^{-cs} \qquad W = W_0 e^{-ks} \qquad (3.118)$$

where D_0 and W_0 are initial diameter and weight, s is transport distance, and c and k are constants. Since the gradient of a stream decreases downstream, frequently exponentially, it is not certain whether an observed decrease in pebble size is related to abrasion during transport or merely reflects the decrease in transport power of the stream. Therefore, one must evaluate the effects of selective transport and abrasion. In a study of terrace gravels from three rivers in South Dakota, PLUMLEY (1948) attempted such an evaluation. Over a distance of 47 km a reduction in pebble size averaging 24 to 5 mm was observed. At the same time he observed a concentration, due to selective abrasion, of the more resistant rocks: from an original 40% to a final 90% of the total pebble count. As interpreted by PLUMLEY 80 to 85% of the reduction in pebble size can be attributed to the downstream decrease in transport power resulting from the diminished gradient. Therefore, one may not regard to so-called STERNBERG law as a model for abrasion during transport.

Shape and surface characteristics are more informative than pebble size in evaluating the mechanical effects of transport. As pointed out previously, data on the three major axes is necessary to characterize shape. Such complete descriptions are not found frequently in the literature. Most authors have confined their shape characterizations to a single index (sphericity, flattening).

An exception is the study by ZINGG (1935) of the distribution of pebble shapes in gravels of an alpine stream. ZINGG described these pebbles according to the scheme reproduced in fig. 3-24. This study showed a distinct dependence on pebble size. With increasing size the proportion of tabular forms increased over equant ones. For grains of a few mm equant forms predominate; over 20 mm the flat ones. Flat and rounded rod-like forms show no definite size dependence. In the case of these gravels investigated by ZINGG, there is little dependence of shape on rock type: crystalline rocks show some tendency toward equant forms; most of the flattened shapes consist of limestones. SNEED & FOLK (1958) also found that in the Colorado River (Texas) the three most abundant rocks, granite, limestone, and chert, have the tendency to form flattened pebbles (platy, see table 3-10). Quartz pebbles had rather elongated shapes. Also on the shore of Lake Michigan LANDON (1930) found a gradual increase in flatter pebble shapes in the direction of the coastal current: Pebbles eroded from a boulder clay cliff become, during coastal transport, less angular and increasingly equant in shape, but then more flattened, so that finally (36 km from the source) disk-shaped pebbles with sharp edges prevail. On the sea coast of New Jersey, in the

---

[6] See also FÜCHTBAUER in volume II of this series (p. 43 ff).

case of pebbles over 10 mm diameter, platy and bladed forms (SNEED & FOLK 1958 classification) predominate (BLATT 1959). Along lake and seacoastal shores selection according to shape, in addition to specific abrasion, plays an important role. VAN ANDEL, WIGGERS & MAARLEVELD (1954) found on a beachwall on the North Sea a concentration of flat pebbles, while at the foot, equant shaped pebbles were concentrated. Here, therefore, flattened pebbles predominantly accumulate in the beach deposits during transport along the coast.

That pebble shape is determined also by rock type is indicated by VALETON's (1955 a) measurements of Main River gravels: Cherts form approximately equant pebbles, graywackes slightly flattened cylinders, and slates flattened forms.

The fact that especially the large pebbles are observed to form preferentially flattened shapes may result in part from the fact that large pebbles are rarely able to roll and thus remain stationary on the bottom for a long time. During this stationary period they are abraded by finer material passing over them. Moreover we can recall the investigations of RAYLEIGH (1943, 1944). He determined that in a rotating cylinder flattened shapes were formed preferentially and found also that natural flints, which appear equant or ellipsoidal, in reality have flattened "ends".

Sphericity as determined by WADELL, according to present observations, is not especially well suited to determine changes in pebble shape during transport. Thus KRUMBEIN (1940, 1942) when studying the gravels of the San Gabriel Canyon (California) found over a transport distance of 10 km a small increase in mean sphericity from 0.66 to 0.71; in the gravels of Arroyo Seco Canyon (California) essentially no change in sphericity was noted over a 16 km stretch. The gravels in the South Dakota rivers investigated by PLUMLEY (1948) showed essentially no change in pebble sphericity over 48 km.

The roundness of stream pebbles does change in a characteristic manner during transport, which corresponds also to the results of laboratory experiments (KRUMBEIN 1940, 1942; PLUMLEY 1948; SNEED & FOLK 1958). During the initial period of transport, roundness of primary angular pebbles increases very rapidly. Subsequently roundness changes very slowly, apparently approaching an endpoint asymptotically. This endpoint and the travel distance for the initial rapid rounding depends under like conditions on the rock type. In the stream deposits studied by PLUMLEY (1948) the roundness of limestone pebbles increased from an initial approximate value of 0.2 to 0.5—0.6 within the first 15 km and then changed very slowly, so that after a stretch of about 50 km the roundness was still only about 0.65. In the Colorado River (Texas) SNEED & FOLK (1958) found about the same roundness values (0.60—0.65) over a 400 km stretch in the case of limestone pebbles. Apparently this survey did not include the initial stretch along which the limestone pebbles achieved their ultimate roundness. In contrast quartz pebbles in the same stream attained roundness values of about 0.65 not earlier than after 150 km, and maintained that value subsequently. At the same time the roundness of chert pebbles reached only 0.55 over a 400 km stretch, increasing further during additional transport. It would appear that in the case of this stream all pebbles finally approach a roundness maximum of about 0.65. This happens in the case of limestone after a few kilometers transport;

for quartz 150 km and for chert several hundred kilometers are necessary.

Ultimate rounding probably depends also on gradient and other particulars of flow. For pebbles in two California Canyons, which were deposited by a single flood, KRUMBEIN (1940, 1942) found an increase in roundness in the direction of transport (11 or 18 km), but observed maximum values of only about 0.4. INNERBRITZEN (1959) found in the recent deposits of Alameda Creek (California) an irregular increase in roundness from about 0.2 to about 0.5 over a 30 km stretch. These final values are clearly less than those observed for large streams.

Also in the case of coastal transport, roundness increases rapidly initially and then the rate declines. Thus VAN ANDEL, WIGGERS & MAARLEVELD (1954), studying the heterogeneous gravels from a bar on a North Sea island, found a rapid increase in roundness from 0.3 to about 0.5 over the first kilometer, followed by a further increase to only 0.6 for the remaining 2 kilometers. Also observations by GROGAN (1945) on the roundness of rhyolitic pebbles along Lake Superior show that coastal transport is more effective than stream transport in producing strongly rounded pebbles. In this case roundness increased from about 0.25 to about 0.60 over a distance of 700 m.

While the previously cited roundness values were estimated visually, using the KRUMBEIN scale (1941 a), additional measurements have been made using the roundness index of CAILLEUX. From the results of individual studies CAILLEUX (1952) estimated the mean values for limestone pebbles, as reproduced in table 3-14.

Table 3-14. Roundness indices for limestone pebbles (after CAILLEUX 1952).

	$2 r_1/a$
Streams in temperate-warm climates (Appenines, Alps)	0.100—0.540
Streams in periglacial climates	0.100—0.200
Fluvioglacial	0.240—0.300
Lake shores (Lake Geneva)	0.300—0.370
Sea shores	0.170—0.610

Also CAILLEUX (1952) found that coastal transport is a more effective rounding agent than stream transport. Using CAILLEUX's methods, POSER & HÖVERMANN (1952) investigated the roundness of fluviatile gravels from the Harz Mountains and the Alps. Values between 0.100 and 0.300 were noted, with a definite increase downstream. VALETON (1955) found that Main River gravels, consisting of slates, graywackes, and cherts, had roundness indices between 0.2 and 0.3, the lower values being for cherts, and the higher values for slates.

The results of laboratory studies have confirmed the fact that pebble abrasion is reduced the smaller the pebble size. Therefore, one would expect that sand grains transported by streams would experience only slight size and shape alteration. In the river bed of the 80 km long Tessin (Alps) BURRI (1929) found coarser sands at points of steeper stream gradient. Therefore, river bed sands in the upper course of the stream are on the average coarser than near the mouth in Lake Maggiore. In addition coarser sands were found to prevail where tributaries discharge into the main stream, supplying fresh material. The degree of sorting increases on the average downstream. Indications of rounding could not be confirmed in this stream. The green hornblende content of the sands decreases downstream as does the mean grain size of this mineral. Since hornblende stems from rocks occurring only in the upper course of the stream, it must be concluded that here on the one hand dilution by hornblende-free sediment and on the other probably size reduction during transport both play a role.

Investigation of sands over the stretch of the Mississippi River from Cairo to the mouth (1730 km) indicated (RUSSELL & TAYLOR 1937, RUSSELL 1937) that roundness and sphericity decrease with decreasing grain size, so that grains below about 0.2 mm show very little sign of abrasion. An increase in roundness of sand grains in the direction of transport is not discernable. Roundness of grains between 0.6 and 0.2 mm diameter even appears to decrease somewhat downstream, so that the authors assume that these grains breakup during transport. Change in mineral composition along the course of the river is slight. Feldspar content decreases from about 25 % at Cairo to about 20 % near the mouth. Also pyroxene decreases somewhat, whereas hornblende, zircon, sphene, rutile, and monazite remain constant. The total heavy mineral content varies and appears to increase a bit in the direction of transport. From these observations it follows that selective mineral abrasion plays a minor role in this river.

A detailed study of the river bed sands of the south Canadian River in Mexico, Texas, and Oklahoma, carried out by POLLACK (1961) is of special interest, because this river receives essentially no tributary flow over a stretch of about 1000 km. Over this entire distance median grain size and sorting of the sands do not change. The skewness of the grain size distribution remains constant. Only the kurtosis[7] increases from about 0.9 in the upper course to about 1.1 at the mouth of the river. The heavy mineral content is higher in the fine grain size classes than in the coarser. The amounts vary irregularly and exhibit no systematic discernable change during the course of transport. No systematic variation was established for the distribution of different heavy mineral contents. The quartz/feldspar ratio for the bulk sand likewise does not change. Feldspar content of the coarse fraction decreases, however, from the upper stretch to the mouth of the river. This indicates that during transport some feldspar grains are broken up by virtue of their good cleavage. For zircon, sphene, pyroxene, hornblende and feldspar, roundness, as estimated visually by the KRUMBEIN (1941 a) method, is greater for large grain diameters than for smaller ones. The roundness of some heavy minerals (sphene, ilmenite, hematite, pyroxene, hornblende, tourmaline) and small feldspar grains increases somewhat downstream (from values less than 0.6 to values between 0.6 and 0.7). In contrast roundness of the coarser quartz (above 0.5 mm) and feldspar grains (above 0.25 mm) decreases downstream from values between 0.5 and 0.6 to values around and below 0.4. These observations

---

[7] The kurtosis used here is defined as $(Q_3 - Q_1) : 2 (P_{90} - P_{10})$. Q_1 and Q_3 are the first and third distribution quartiles; P_{90} and P_{10} the grain diameters below which 90 % and 10 % of the total falls.

are in agreement with the findings for Mississippi River sands, and must be interpreted to conclude that during transport of sand, not only is the finest material abraded, but also the larger grains are fragmented. Sphericity (after Aschenbrenner 1956) of the smaller grains of pyroxene, hornblende, quartz and feldspar is greater than for the larger grains. Only in the case of tourmaline do the larger grains have greater sphericity than the smaller ones. The sphericity of hematite and magnetite increases downstream from below 80 to a value between 80 and 90. The form factor (after Zingg 1935) is higher for small grains than for large ones for all minerals studied. Downstream the form factor for coarser feldspar and quartz grains increases somewhat; for coarse-grained pyroxene and hornblende it decreases. The lack of or slight change in size, roundness and shape of sand grains in the South Canadian River over a long distance of 1000 km, shows very expressively, how slight are the mechanical effects which sand grains experience during fluviatile transport.

In the case of sand transport along the coasts of lakes or the oceans, a decrease in mean grain size and roundness in the transport direction has been established repeatedly (Atlantic coast between Delaware and Chesapeake Bay: MacCarthy 1933; between Long Island and Georgetown: MacCarthy 1933; Lake Erie: Pettijohn & Lundahl 1943). This phenomenon results perhaps from a selective transport of suspended load, in that angular and smaller grains are sorted out by the surf and are transported further than the large, well-rounded ones.

A statistical study of the shape of quartz grains of different sands, carried out by Schumann (1941), indicated that in many sands quartz particles are elongated preferentially in a direction, inclined about 40—70° to the c-axis. In an alpine sand the grains were elongated preferentially along the c-axis. While this last phenomenon can be traced back to the c-elongated grains in the primary rock, Schumann maintains that elongation along a 40—70° inclined axis is the result of the anisotropic abrasion resistance of quartz. According to previous experiments this should be minimal parallel to the rhombohedron $(10\bar{1}1)$.

3.171.5 Summary

In summary the following picture emerges of the mechanical effects of water transport, based on experiment and natural observations.

The effect on bottom transported particles decreases markedly with decreasing size. Therefore, significant and marked effects are exerted essentially only on pebbles, while sand grains are changed very little or very slowly.

The change in pebble size, shape and roundness diminishes rapidly during the course of transport. Change is much more prevalent during the initial part of travel than later. Abradability depends on the rock type: limestone pebbles are very rapidly rounded during fluviatile transport. In contrast well-rounded quartz or chert pebbles probably always require several hundred kilometers transport in streams to reach their ultimate roundness. Independent of rock type, and especially with larger pebbles, there is a tendency toward forming flattened forms.

In the case of transport along coasts of lakes and the ocean a relatively more rapid and stronger abrasion takes place than in streams, because pebbles are moved repeatedly back and forth by swash and backwash, covering a much greater transport distance than measured parallel to the shore.

Quartz grains of less than 1 mm diameter experience, even after stream transport of over many hundreds of kilometers, almost no perceptible rounding. The origin of well rounded sand grains, such as occur in some sands and sandstones is not easy to understand. In part this may occur at the coast through wind action. Fluviatile transport can produce marked effects only under special conditions (very long transport distances and not too high velocities). In many cases well-rounded sand grains are relics of multiple cycles of sedimentation.

Since roundness of pebbles and coarse sand grains moving along the bottom as a rule increases during transport, the material abraded from the moving particles is of much smaller grain size. It may most likely belong chiefly to the silt and clay fraction. By the mechanical action of transport, the number of particles greater than about 1 mm decreases and silt and clay, rather than sand, results. The repeatedly established relative rarity of sediments in the fine sand-coarse silt range may be due in part to the mechanical action of transport producing essentially only material whose grain size lies below this size range (ROGERS, KRUEGER & KROG 1963; SCHUBERT 1964).

3.172 Chemical changes during transport and deposition: subaquatic weathering

3.172.1 General

Materials which have been transported and deposited in water, before they are buried within the growing sediments, undergo chemical changes which are analogous to weathering processes in soils. These involve solution processes, the transformation of primary and the formation of new minerals. Depending on the composition of the transport and depositional medium, these processes can act either as a continuation of soil weathering, or take different directions. In contrast to subaerial soil weathering, these changes take place completely under water and without direct contact with the atmosphere. All chemical processes acting during transport and deposition under water can be included under the therm subaquatic weathering (NIGGLI 1952).

Subaquatic weathering can be subdivided according to chemical conditions which prevail: s u b f l u v i a l   w e a t h e r i n g during transport and deposition in fresh river water leads to products different from s u b m a r i n e   w e a t h e r i n g (CORRENS 1939) in sea water. The term halmyrolysis (HUMMEL 1922) is used also for the latter. Not only solution processes take place in the sea, but in fact neoformation and reconstitution predominate. Therefore, the more general term submarine weathering is preferable. Finally it is possible to differentiate s u b l a c u s t r i n e   w e a t h e r i n g also (CORRENS 1939; thololysis, according to WETZEL 1923) which can be further characterized according to the special characteristics of the lakes involved (eutrophic, oligotrophic lakes, soda lakes, sulfate lakes, etc.). Thus we arrive at the following classification:

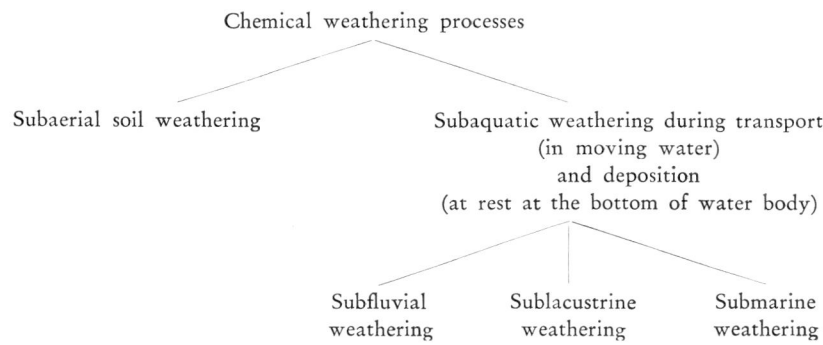

Subaquatic weathering must be differentiated distinctly from diagenesis, which includes all processes active within the sediment. Especially in the case of submarine weathering, the boundary between diagenetic and pre-diagenetic processes is defined quite differently by various authors.

From the kinds of reactions between water and solid phases, one can draw ideally a boundary between subaquatic weathering, which involves only the sediment surface, and diagenesis, which takes place within the sediment: All weathering processes, including subaquatic weathering, take place in open systems, that is, they involve uninhibited material exchange between the solid phases and a practically unlimited amount of solution. The course of the reactions (dissolution, neoformation) is not arrested, in that solution compositions are not significantly changed by reaction, since either, as in soils, new solvent is supplied repeatedly, or because as in the ocean, the supply of solvent is infinitely great.

In contrast all diagenetic processes involve reactions in more or less closed systems, since they take place between minerals and a given pore solution whose amount is limited and can be resupplied only relatively slowly. Diagenetic processes, therefore, alter considerably the amount and composition of pore solutions. Diagenetic reaction cannot continue for any great length of time in one given direction.

3.172.2 Subfluvial weathering

The chemical conditions of subfluvial weathering are not essentially different from those of soil weathering. In comparison to sea water almost all stream waters are low in dissolved salts. The hydrogen-ion concentration (pH) of streams waters lies, according to ALEKIN (1962), between 6.0 and 9.0. A summary of recent measurements from 20 large rivers (most from the North American continent) gives a range of values between 5.9 and 8.0 (DURUM, HEIDEL & TISON 1960). The hydrogen-ion concentration of stream waters is, therefore, on the average less than in most soils.

Since this is dependent essentially on CO_2-content, river water is usually more acid in winter than in summer.

Dissolution of feldspars and other primary minerals, begun within the soil, is continued in the streams. Since the hydrogen-ion concentration usually lies near neutrality, dissolution is slower than in the soil. On the other hand, at least as far as bed load transport is concerned, dissolution is accelerated as the protective surface layer formed from residual components, is repeatedly abraded away. Also phyllosilicates carried in suspension are, during stream transport, altered in ways similar to their alteration in soils. However, at the higher pH, degradation processes may proceed more slowly. With increasing pH aggradation processes are to be reckoned with also.

The oxygen contents of stream waters are quite variable. They depend on temperature, biological processes, and on depth of water. In general the oxidation potential of stream water is higher than in soils.

In addition to the chemistry of stream waters, the subfluvial weathering activity depends on the kind and rapidity of transport. Material that migrates slowly downstream as a result of low gradient, and is repeatedly deposited and transported, experiences more significant alteration on the way to the sea than does detritus, which along a steep gradient is washed rapidly and without pause into the ocean.

Investigations of subfluvial weathering appear to be lacking. Because of admixing of new material it might be difficult to establish unequivocably the chemical changes experienced by the load of a stream over its entire course. On account of the low hydrogen-ion concentrations of most stream waters, subfluvial weathering certainly is not very intensive. Rivers transport material altered in the soil into basins of deposition without significant further chemical alteration.

3.172.3 Submarine weathering

Whether submarine weathering of sediments and primary rocks at the bottom of the ocean really plays an important role, often has been doubted. Various facts suggest that submarine weathering takes place at significantly slower rates than subaerial weathering in soils (CORRENS 1939). Because of the high pH and high salt concentration of sea water the dissolution rate of minerals is, in any case, less than in acid soil solutions. Also the missing comminutive processes and the absence of active humic substances such as occur in soil diminish the effects of all decomposing reactions.

Study of recent sea bottoms will reveal products of submarine weathering only in those places where little sediment is being deposited so that the bottom surface is in contact with sea water for a long time and any weathering products are not contaminated with clastic material. The effects of submarine weathering are thus especially evident and unequivocable in the deep sea.

It is a different situation in ocean basins adjacent to the continents, where eventual submarine weathering products become mixed with much larger amounts of clays supplied by streams and are thus difficult to recognize. While some authors have drawn conclusions concerning reconstitution and neoformation from the distribution of clay minerals over the ocean floor, others (especially WEAVER 1958) have offered the opinion that all clay minerals, as we observe them on the sea bottom, originate by continental weathering and experience at most trivial alteration on the sea floor.

That submarine weathering processes do occur is confirmed with certainty by comparing the chemical composition of river and sea water (table 3-15). From the ^{18}O-, Sr-, and Mg-contents of recent and fossil brachiopod shells, it can be concluded that the composition of sea water has not changed noticeably since the late Paleozoic (200—250 million years) (LOWENSTAM 1961). If the composition of river waters is now different from sea water, and if we at the same time note that the composition of sea water has remained constant for a long time, although large amounts of river water have been supplied to the ocean, chemical reactions must constantly take place to maintain the equilibrium. These must include precipitation from sea water as well as reaction with deposited material.

Table 3-15. Average composition of the dissolved substances in river and sea water.

	Rivers of the World (LIVINGSTONE 1963)	Sea water (CULKIN 1965)
	ppm	ppm
HCO_3	58.3	142
SO_4	11.2	2712
Cl	7.8	19353
NO_3	1.0	0.0043—3.5*
Ca	15	413
Mg	4.1	1294
Na	6.3	10760
K	2.3	387
Fe	0.67	0.02**
SiO_2	13.1	1.1—4.3 Surface
		5.4—10.8 Deep waters***

* RANKAMA & SAHAMA 1950
** RÖSLER & LANGE 1965
*** WEDEPOHL 1967

Bicarbonate is the most important anion in river water. The ocean would take on the character of a soda lake if it were continually supplied by rivers without chemical reactions taking place (MACKENZIE & GARRELS 1966). For the maintenance of equilibrium, besides reduction of part of the dissolved sulfate to insoluble sulfide and precipitation of $MgCO_3$, $CaCO_3$ and NaCl, reactions are necessary which remove from solution the major part of the bicarbonate and SiO_2 supplied by the rivers. In addition reactions must occur which increase the Na:K-ratio from 2.75 in fresh water to 28 in sea water. This can occur as a result of the aggradation of phyllosilicates, which gives rise to micas by K-uptake and to montmorin minerals by uptake of Na, Mg, and Ca. As a result hydrogen-ions are liberated, which combine with bicarbonate liberating CO_2 and H_2O. MACKENZIE & GARRELS (1966) propose that as a part of this process SiO_2 is also combined and used up, as more or less amorphous phases of kaolinitic composition form three-layer type clay minerals (micas, montmorins). SiO_2 can be removed in other ways also as will be mentioned later (HARDER experiments, 1965).

Experimental studies of the dissolution of feldspars and other primary minerals by sea water have not yet been made. Undoubtedly though the same principles apply as in dilute soil solutions. The high electrolyte content presumably slows down the dissolution rate. Based on experiments in pure solutions, dissolution by sea water (pH = 8.2) should proceed somewhat more rapidly than in neutral solutions. Feldspars are dissolved in sea water as they are in soil solutions, if they remain long enough in contact with sea water. This is confirmed by observations of granodiorites obtained from the walls of submarine canyons off the California coast (Carmel and Monterey Canyons). The K-feldspar of the rock is intermixed with neoformed minerals (muscovite, kaolinite, halloysite), which have formed at the expense of the dissolved feldspar (REX & MARTIN 1966).

Clay minerals react very little and slowly with sea water of normal composition. When montmorillonite, illite, kaolinite, halloysite and a mica-montmorillonite mixed-layer mineral were treated with sea water, CARROL & STARKEY (1959) found that SiO_2 and lesser amounts of Al_2O_3 went into solution. MACKENZIE, GARRELS and co-workers (1965, 1967) treated kaolinite, chlorite, muscovite, and montmorillonite with sea water, that contained both less and more SiO_2 than normal sea water. In SiO_2-deficient solutions (0.03 ppm SiO_2) these minerals dissolved incongruently, and SiO_2 increased in solution. In contrast in the solutions with higher SiO_2-content (25 ppm SiO_2) a take-up of SiO_2 by the clay minerals from the solution was observed. The authors attributed this to the formation of ill-defined SiO_2-rich silicates. Both this clay mineral dissolution and SiO_2-uptake are slowly acting processes. In closed systems these processes had not yet gone to completion after several months. The dissolution rate was greatest for montmorillonite, followed by the mica minerals, and finally the most stable minerals of the kaolinite group. Since the SiO_2-concentration produced by the artificial dissolution corresponded approximately to that of ocean water (6 ppm SiO_2), it is possible that the concentration in normal sea water corresponds to the maximum stability of the clay minerals. At lower SiO_2-content they dissolve to form an aluminium-rich residue; at higher SiO_2-content SiO_2-rich phases form.

The first reactions which clay minerals experience when they pass from the fresh water environment into the sea are changes in cation exchange capacity and composition. According to experiments by CARROL & STARKEY (1959) with montmorillonite, illite, halloysite, kaolinite, and a mica-montmorillonite mixed-layer mineral, and by POTTS (after KELLER 1963) with a predominantly montmorillonitic Missouri River clay, treatment with sea water results in replacement of a substantial part of the exchangeable Ca by Mg and much smaller amounts of Na and K. Such behavior is to be expected since Mg occurs in higher concentration than Ca in sea water and because the binding energy of Mg on clays is higher than for Ca and much higher than for univalent cations (see GRIM 1953; MARSHALL 1954).

In addition to exchange of Mg for Ca sea water causes a slight increase in exchange capacity of the two-layer minerals and a reduction in the case of three-layer minerals. Reduction in the number of exchangeable cations can be explained only by introduction of ions into fixed positions in the structure. It is known that K-ions are fixed quite readily by illites and montmorillonites. Apparently the reduction in exchange capacity noted can be attributed in part to K-fixation by the three-layer silicates. Evidently bonding of potassium in the sediment is involved, reflecting the geochemical balance between river and sea water. Submarine weathering in this case acts as an aggradation process, leading to more or less complete mica minerals, acting in opposition to the degrading effects of subaerial weathering, which produces deficient micas and expandable three-layer minerals from micas.

It will be difficult always to recognize these aggraded micas in marine sediments, that is, to distinguish them from unaltered detrital micas.

Very often more or less complete micas are the predominant clay minerals in sediments, and many authors have attributed the distribution of mica, montmorin and mixed-layer minerals in different basins to the aggradation of previously degraded micas and montmorin minerals (see MILLOT 1964, p. 211 ff and 367 ff).

The experimentally determined uptake of magnesium from sea water by clay minerals leads not only to a change in the exchangeable cation population, but as a result of interlayering of $Mg(OH)_2$ sheets in alkali-free three-layer minerals, to the formation of chlorite. GRIM & JOHNS (1954) found the following relations at the mouth of the Guadelupe River in the Gulf of Mexico: The river carries a Ca-montmorillonite into the sea. The new sediment in the Gulf contains with increasing distance from the mouth decreasing amounts of montmorillonite an increasing illite and chlorite contents, which apparently result from reaction of montmorillonite with sea water. From phases transitional between montmorillonite and chlorite one can conclude that transformation into chlorite proceeds first with the interlayering of individual islands of $Mg(OH)_2$ between layers. followed by gradual lateral growth of brucite layers, until all interlayers are filled. Therefore intermediate chlorite-montmorillonite mixed-layer structures form (see section 2.332). As a result of the dioctahedral character of the montmorillonite this chlorite has a b-parameter characteristic of dioctahedral phyllosilicates, in spite of its brucite layers. In confirmation of this interpretation JOHNS (1963) established that during the transition from fresh water into the marine milieu, simultaneously with the increase in chlorite content of these sediments, the content of water-insoluble chloride also increased significantly. Chloride ions apparently are taken into some of the hydroxyl positions in the brucite layers, when these are built into the clay minerals.

JOHNS & GRIM (1958) found less significant indications of montmorillonite-chlorite alteration in Mississippi delta sediments. POWERS (1954, 1957) found, in the region of Chesapeake Bay on the North American Atlantic coast, a gradual transformation of a weathered illite into a chlorite mineral by reaction with sea water. In this case also intermediate phases occur, which this author described as illite-vermiculite-chlorite mixed-layer structures.

Green sedimentary grains, called glauconite, are a widely distributed product of submarine weathering: In many of today's oceans glauconite occurs in areas of the shelf zone and upper continental shelf, especially where sedimentation is minimal.

Since observations have indicated that glauconite originates exclusively in the marine environment, it serves as a criterion for the marine origin of a sediment. Therefore, this mineral has been of special interest to geologists, and an extensive literature exists about it. (A complete bibliography up to 1958 was prepared by VALETON, 1958: later bibliographies are found in the works cited below by BURST 1958; BELL & GOODELL 1967; EHLMANN & SAND 1959; as well as by MILLOT 1964). Utilization of the older literature is difficult because numerous substances have been named glauconite which have a similar external appearance, but which through the use of modern X-ray methods, are shown to be quite different from one another.

According to BURST (1958) and HOWER (1961) the following forms of "glauconite" can be differentiated.

(1) Glauconite, narrowly defined = dioctahedral iron-illite, ordered, with high K-content.
(2) Disordered glauconite with low K-content.
(3) Glauconite-montmorillonite mixed-layer minerals.
(4) Mixtures of two or more minerals.

In addition green grains consisting of chamosite occur, which cannot be distinguished from glauconite without X-ray characterization.

Glauconite in the strictest sense is an illite in which about half of the octahedral sites are occupied by iron. This differs from the common aluminum-illite, in which only about one fifth of the octahedral positions are occupied by iron. The $Fe^{+3}:Fe^{+2}$ ratio varies in the case of the ordered and disordered glauconites (BURST 1958) between 1.7 and 7.4. To illustrate the chemical differences the following two structural formulae are reproduced:

Ordered glauconite (Fe-illite)[8]: $\quad K_{1.58}Ca_{0.10}(Al_{0.70}Fe^3_{2.11}Fe^2_{0.49}Mg_{0.82})$
$(Si_{7.22}Al_{0.78})O_{20}(OH)_4$

Illite (Al-illite)[9]: $\quad K_{1.16}(Al_{2.76}Fe^3_{0.74}Fe^2_{0.08}Mg_{0.68})$
$(Si_{6.82}Al_{1.18})O_{20}(OH)_4$

Types (1) to (3) apparently form a continuous series, terminating in montmorillonite. In this series the K-content decreases continuously. Glauconites contain less potassium, the more disordered and the more expandable they are (HOWER 1961).

The more recent observers agree that glauconite forms on the sea floor by a recrystallization process involving detrital three-layer minerals (BURST 1958; HOWER

---

[8] Bonne Terre, Missouri, Cambrian, after BURST (1958).
[9] Aver. composition of 5 illites, after JASMUND (1955).

1961; EHLMANN, HULINGS & GLOVER 1963; BELL & GOODELL 1967). Since the parent material is dioctahedral and aluminum-rich and contains little potassium, aluminum must be replaced by divalent or trivalent iron in the octahedral sheet and also take up potassium. For the latter process substitution of aluminum for tetrahedral silicon is also necessary. The necessary amounts of iron and potassium are taken up from sea water. Since the mineral contains both divalent and trivalent iron, its formation can take place only in an environment in which the oxidation potential permits both types of iron ions. Such conditions seem to prevail where organic remains or excrements are lying on the sea floor. Thus glauconite is found frequently in the empty tests of foraminifera (HOWER 1961; BELL & GOODELL 1967) and as pseudomorphs after fecal pellets.

BELL & GOODELL (1967) investigated six recent occurrences of glauconite. On the Scotia-Ridge east of the Falkland Islands (water depth between 880 and 3260 m), and on the southeast Atlantic Shelf off the North American coast (water depth between 274 and 365 m), the correspondence in mineral composition of the detrital clays and associated glauconite grains indicates that the glauconite forms as a result of aggradation of the detrital clay minerals. In sediments off the California coast and off Morocco similar comparisons indicate that the glauconite is probably a detrital constituent of the sediment. Sediments from the Pacific east of New Zealand and off the Atlantic coast of Guinea contain glauconites that probably are in part detrital and in part newly formed.

Neoformations are especially notable on the ocean bottom where the sedimentation rate is low; therefore, chiefly in the deep sea. Such neoformations form either by the interaction of sea water and detritus or by direct precipitation from sea water.

From the difference in SiO_2-concentration of river and sea water (table 3.15) it follows that somewhere in the marine domain SiO_2 must be incorporated into the sediment. A part of the SiO_2 is removed from sea water by living organisms, forming siliceous tests (diatoms, radiolaria, silicoflagellates, siliceous sponges). The remains of these organisms become incorporated into sediments and can form silica-rich rocks (radiolarites, cherts).

HARDER (1965) was able to show that SiO_2 can be precipitated quantitatively by inorganic means from very dilute solutions (3 ppm), if hydroxides of aluminum, iron, manganese, magnesium, etc. are precipitated also. SiO_2 is adsorbed on the precipitating hydroxide, forming X-ray amorphous precipitates, whose SiO_2-content depends on a number of factors, especially the SiO_2:hydroxide ratio of the solution. When the initial solution contains approximately equal amounts of SiO_2 and Al_2O_3, it forms a precipitate containing 43% SiO_2. If SiO_2 is present in excess, a precipitate with 80% SiO_2 is produced. It can be presumed, therefore, that the increase in pH in passing from river water into the sea will produce precipitation of hydroxides of the sesquioxides, in which is bound most of the SiO_2 originally in solution in the river water. The finest, still X-ray amorphous particles are carried far out to sea and finally deposited. During this time different alteration processes can take place: In siliceous Fe- and Mn-hydroxide precipitates HARDER & FLEHMIG (1967) and

HARDER & MENSCHEL (1967) found quartz and cristobalite after only a month. From siliceous Al-hydroxide precipitates quartz and an ill-defined clay mineral formed. Ageing these precipitates at 60° gave kaolinite, chlorite, illite, and feldspar, in addition to quartz.

It is assumed that in a similar manner clay minerals and quartz are formed in the sea and on the sea bottom. It would be difficult of course to identify these neoformations. The occurrence of up to 30 μ idiomorphic quartz crystals in deep sea manganese nodules is explained by this process by HARDER & MENSCHEL (1967).

From weathered soils of the tropics, diaspore, boehmite, gibbsite, and kaolinite are carried into the ocean basins. Thus appreciable amounts of aluminum hydroxy oxides are found in Atlantic sediments east of the Brazilian coast, west of Central Africa and south and east of Madagascar (BISCAYE 1965). Since, on the other hand, gibbsite, boehmite or diaspore are absent in ancient marine sediments and are contained only in continental deposits (bauxites), it must be assumed that these minerals are unstable and disappear under conditions of marine deposition or diagenesis. In fact, according to the thermodynamic arguments of GARRELS & CHRIST (1965, p. 361) gibbsite is not stable in solutions containing more than about 10^{-5} mol/l SiO_2. The low SiO_2-content of sea water of 10^{-4} mol/l still lies above the stability limit of gibbsite. Accordingly SWINDALE & FAN (1967) have observed in sediments of Waimea Bay off Kanai Island of the Hawaiian group, that stream-borne gibbsite particles are altered by the action of sea water into pseudomorphs of a phyllosilicate, which, based on its optical properties (anomalous interference colors, $\eta_\beta = 1.58 \ldots$ 1.60, pleochroism between colorless or green and dark brown) was described as chlorite. Since the newly formed mineral was found even at the mouth of a stream coming from the island's interior, it is evident that the gibbsite dissolution and chlorite formation are fast acting processes. Since this chlorite forms essentially at the expense of gibbsite, we may be dealing with a dioctahedral mineral like sudoite. The necessary confirming X-ray study is lacking, however.

BONATTI & ARRHENIUS (1965) reported on the formation of an iron-rich chlorite. In a sediment core from the floor of the Pacific near Guadelupe Island off the Mexican coast (water depth 3530 m) quartz grains were found with a thick coating of small, pale green crystal platelets of a 14 Å-chlorite of likely intermediate iron content. Quartz grains of the recent desert sands of the neighboring mainland are coated with layers of iron and aluminum hydroxides. The authors propose that such grains were brought by the wind into the ocean, and chlorite formed by reaction of sea water with the hydroxide layer. Since these quartz grains were obtained from a sediment layer about 5 m below the present sediment surface, it is certainly possible, that this chloritization already involves an early diagenetic change.

Recently chamosite, a trioctahedral two-layer mineral, with 7 Å basal thickness (therefore a member of the berthierine or septachlorite group), was described as a neoformed mineral in recent marine sediments. v. GAERTNER & SCHELLMAN (1965) found grains of a black to dark green aggregate of goethite and a iron-rich 7 Å layer silicate in sands off the coast of Guinea. The authors presume that this mineral formed by reaction of lateritic goethite with sea water. PORRENGA (1967) found chamosite in the form of fecal pellets and shell fillings in shelf sediments off the mouth of the Orinoco and Niger Rivers, as well as off the coasts of Sarawak, NW-Borneo. GIRESSE (1965) found chamosite off the coast of Gabon. All these occurrences are in the tropics. The iron necessary for the formation of chamosite, may, like the aluminum for submarine formation of aluminum-rich chlorites, stem from tropical lateritic weathering.

Chamosite formation appears to take place preferably in the shallowest regions of the continental shelves. It occurs on the shelf off the Niger delta and off Borneo to a water depth of about 60 m; off the Orinoco delta it occurs down to about 150 m. In contrast newly formed glauconite is found at greater depths (off the Niger delta, between 100 and 400 m). Glauconite and chamosite grains look very much alike and can be differentiated only by X-ray methods. In older descriptions it is possible that they often may have been mistaken for one another. The green grains found in shallow water sediments of the Gulf of Paria, presumed by VAN ANDEL & POSTMA (1954) and KOLDEWIJN (1958) to be glauconite, are in fact chamosite (PORRENGA 1967).

Newly formed minerals are found in especially high concentrations on the sea bottom where sedimentation of detrital material is low. Such accumulations are important components of pelagic sediments (ARRHENIUS 1963). The most important of these are manganese nodules and zeolites.

Manganese nodules are deep sea concretions, consisting of iron and manganese oxides and hydrated oxides, admixed with detrital material and other neoformed minerals such as titanium oxides, barite, nontronite, quartz and opal. The iron is tied up mostly as goethite; manganese occurs as at least three different oxides in the tetravalent state (BUSER & GRÜTTER 1956). Manganese nodules contain a series of heavy metals. Especially noteworthy are the high contents (1—2%) of Cu, Co, and Ni.

The concentric layered structure of these nodules shows that they grow slowly on the sea floor. That they occur just in the deep sea is compatable with the high oxidation potential of the oxygen-rich deep water. It is still not clear how the iron, manganese, and other enriched elements in the nodules originate. WEDEPOHL (1960) showed that there is an interrelation between the chemistry of deep sea clays and the composition of manganese nodules. In comparison to shallow sea clays, a series of elements is enriched in deep sea clays, which include Mn, Fe, Zn, Ni, Pb, Cu, Co, and Mo. Also calculations of the geochemical balance for Mn indicate an excess in deep sea deposits (RUBEY 1951; HORN & ADAMS 1966). These differences cannot be explained if it is assumed that shallow sea and deep sea clays both originate exclusively as continental detritus. It must be assumed rather that there is a source in the ocean itself for these excess elements, especially for the constituents of the manganese nodules. The oldest explanation derived manganese and iron from the submarine weathering of basalts (MURRAY & RENARD 1891). To form nodules, however, very large amounts of basalt would have to be altered, which is incompatable with the very slow decomposition of submarine basalts and lavas. WEDEPOHL (1960) proposed a supply of Mn, Fe, and the other elements mentioned, from volcanic exhalations. BOSTRÖM (1967, also earlier literature) assumed the action of ascending solutions, coming from volcanic or metamorphic sources. As a result of their reducing character these solutions should take on Mn and Fe in the divalent state from the perfused rocks and sediments and transport them upward. In the uppermost sediment layer or on the sea floor precipitation of Fe and Mn in the tri- or tetravalent state would occur, as a result of the high oxidation potential there.

Phillipsite and members of the phillipsite-harmotome series are widely distributed in pelagic sediments (ARRHENIUS 1963). These zeolites occur as idiomorphic crystals enclosing other mineral grains, so that there is no doubt about their neoformation. Very often they make up to over 50 % of the total sediment. Presumably submarine weathering of basaltic lavas and glasses supplies at least a part of the constituents for zeolite formation. In the Gulf of Naples MÜLLER (1961) confirmed the neoformation of analcite, quartz, and clay minerals with the simultaneous submarine weathering of volcanic glasses.

3.18 Deposition in water

3.181 Grain size distribution

When sediment, transported as suspension or bed load, is deposited, it becomes sorted according to the settling velocities and resistance to bottom transport of its constituent particles. The resulting sedimentary deposits represent populations of characteristic particle types, differentiated by specific gravity, size, and shape. The population frequencies can be described by distribution functions. Preliminary to dealing with the most important types of water-lain clastic sediments, the basic concepts for characterization and representation of such distribution functions will be summarized. For further details and literature, the reader is referred to the treatment of grain size analysis in the first volume of this series (MÜLLER 1964) and the mathematical basis by MARSAL (1967), as well as the summary presentations of PETTIJOHN (1957), INMAN (1952), and FRIEDMAN (1962, 1967).

If one considers first of all only spherical grains of equal specific gravity, the distribution functions of grain diameters can be considered. Such distributions can be determined by methods described in the first volume of this series. Only when particles are larger than a few microns does the determined grain size distribution give direct information on the state of sedimentary material during transport and deposition. Special methods (for example, ultracentrifuge) must be used for size analysis of clay particles. These do not always give results that are simply interpreted. In addition it is questionable if one can reproduce, by manipulating clays in the laboratory, the natural size distribution of the clay. Particle sizes in clay suspensions depend on the degree of flocculation of secondary particles, which is influenced strongly by electrolytes.

To graphically represent the grain size distribution of a sediment, one plots the frequency (weight percent) of the individual grain size classes against grain size, so that the area over each grain size interval is proportional to the frequency of this size class (see fig. 3-29). By reducing the class intervals one goes from a step-like histogram to a grain size distribution curve. Since the investigations of UDDEN (1914), it has been confirmed repeatedly, that the grain size distribution curves of many silts, sands, and gravels are bell-shaped, and approximately symmetrical. This is true if the grain size is represented in geometrical progression, by using a logarithmic scale. Since the logarithm to the base 10 gives inconveniently large intervals, the logarithm to the base 2 has been adopted, giving the so-called Φ-scale. Φ is the negative logarithm to the base 2 of the diameter D measured in mm:

$$\Phi = - \log_2 D \tag{3.124}$$

For D = 1 mm, $\Phi = 0$; diameters greater than 1 mm give negative Φ values; less than 1 mm positive values. A table for converting D to Φ is found in the appendix of the first volume of this series (MÜLLER 1964).

In addition to the distribution curve, the cumulative curve is used also to represent the grain size of a sediment (see fig. 3-30). It has, relative to the former, the disadvantage of being less lucid, but has the advantage that certain quantities for characterizing grain size distribution can be read directly from it. In the cumulative curve the cumulative frequency F is plotted as a function of grain size. The cumulative frequency corresponding to a given grain size is the relative amount (weight percent) of all grains of smaller grain size.

For characterizing grain size distribution, the quantities used, which can be taken from the graphic representations, are enumerated in table 3-16. Percentiles, quartiles, medians, and modal values are given in normal units (mm, cm), unless specified otherwise. The modal value appears on the distribution curve as the maximum, on the cumulative curve as the inflection point. For distribution functions which are symmetrical when plotted on a logarithmic size scale, skewness (Sk) is 1. If the finer constituents predominate, Sk is less than 1, and for coarser particles greater than 1. The quantity S_0 is smaller, the better the sorting.

Table 3-16. Parameters for characterizing grain size distribution and cumulative curves. (F = cumulative frequency).

Term	Symbol and definition
Percentile	P_n, those grains for which F = n %
Quartile	$Q_1 = P_{25}$; $Q_2 = P_{50}$; $Q_3 = P_{75}$
Median	$Md = Q_2$
Mode	M, diameter of most frequent grain size
Sorting	$So = \sqrt{Q_3/Q_1}$
Skewness	$Sk = \dfrac{Q_1 \cdot Q_3}{(Md)^2}$

Grain size distributions are characterized more exactly by the use of the statistical quantities summarized in the left column of table 3-17. Measurement of grain size is not based on diameter D as in table 3-16, but on the corresponding Φ-value

($-\log_2 \cdot D$). f_n is the frequency (weight percent) of the grain size class Φ_n. Mean grain size in Φ units is obtained as the arithmetic mean from the individual grain size classes and their frequency. The standard deviation (σ) calculated in the usual way, serves as a measure of sorting. Asymmetry of monomodal distribution curve is measured by the skewness α or so-called 2nd moment. For a symmetrical distribution α is zero. With predominantly coarse particles α is negative, with fine ones it is positive (note the difference from the skewness calculated from the two quartiles and the median, table 3-16). The peakedness of a distribution curve is measured by the kurtosis β or the 4th moment. The greater the kurtosis, the sharper the peak.

Examples of grain size distribution and cumulative curves and the statistical quantities taken from them, as defined in tables 3-16 and 3-17, are given in fig. 3-29 and 3-30.

Table 3-17. Statistical parameters for characterizing grain size distributions, according to the Φ scale.

Term	Statistical Quantity	Proposed by INMAN (1952)
Median	— — —	$Md_\varphi = \Phi_{50}$
Mean Grain Size	$\overline{\Phi} = \dfrac{1}{100} \sum_n f_n \Phi_n$	$\overline{\Phi}^* = \tfrac{1}{2}(\Phi_{16} + \Phi_{84})$
Deviation (sorting)	$\sigma = \left\{ \dfrac{1}{100} \sum_n f_n (\Phi_n - \overline{\Phi})^2 \right\}^{1/2}$	$\sigma^* = \tfrac{1}{2}(\Phi_{84} - \Phi_{16})$
Skewness	$\alpha = \dfrac{1}{100} \sigma^{-3} \sum_n f_n (\Phi_n - \overline{\Phi})^3$	$\alpha^* = \dfrac{\overline{\Phi}^* - \Phi_{50}}{\sigma^*}$
Kurtosis	$\beta = \dfrac{1}{100} \sigma^{-4} \sum_n f_n (\Phi_n - \overline{\Phi})^4$	$\beta^* = \dfrac{\tfrac{1}{2}(\Phi_{95} - \Phi_5) - \sigma^*}{\sigma^*}$

f_n: Frequency (weight percent) of the grain class with mean diameter Φ_n.
Φ_5, Φ_{16}, Φ_{50}, Φ_{84}; Φ values which correspond to the cumulative frequencies $F = 5\%$, 16%, 50%, 84% (Percentile in Φ units).

The grain size distribution of many sediments corresponds approximately to a normal or Gaussian distribution when a logarithmic scale is used. That is, they give a single modal, symmetrical distribution function for which median, modal and mean values are superimposed. There is no theoretical basis as to why sediment distribution curves should have normal distributions only with logarithmic and not with arithmetic size scales. Since, however, such distributions are wide spread in nature and their treatment quite familiar it would seem appropriate to describe grain size distributions using parameters derived from normal distributions and to measure deviations from this ideal model.

The normal distribution for grain sizes in Φ-units corresponds to the following function:

$$f = \dfrac{1}{\sigma \sqrt{2\pi}} \, e^{-\dfrac{(\Phi - \overline{\Phi})^2}{2\sigma^2}} \tag{3.125}$$

f (ordinate) is the frequency (weight percent) corresponding to the grain size Φ (abscissa). Each given normal distribution is defined by the two parameters Φ (mean grain size in Φ units) and σ (standard deviation or sorting), which are defined as shown in the left column of table 3-17. The skewness α is equal to zero for a normal distribution. Each skewness of an actual distribution that deviates from zero is a first measure of the deviation of the real distribution from the ideal.

Kurtosis can serve as the second such measure, having the value 3.0 for a normal distribution. Curves with kurtosis values above 3 have sharp peaks, below 3 broad peaks.

Furthermore a normal distribution has the property that there are certain Φ-values of cumulative frequencies which are related to the mean value Φ. These are given in table 3-18. As an example, in a normal distribution, between the grain sizes $(\Phi-\sigma)$ and $(\Phi+\sigma)$, $84.135-15.86 = 27\%$ of the total grains fall. These regularities can be used for near-normal distributions to designate the quantities for sorting, skewness, and kurtosis, extracting the exact statistical quantities from the graphical cumulative curve, instead of by tedious calculation. Such quantities as defined by INMAN (1952) are given in the right-hand column of table 3-17. The mean grain size Φ, for example, is defined as the mean of the Φ-values with cumulative frequencies 16 and 84 % (Φ_{16} and Φ_{64}), since their positions for a normal distribution are exactly $-\sigma$ and $+\sigma$ from the mean grain size. Correspondingly sorting, skewness, and kurtosis can be determined from data taken directly from the cumulative curve. The difference between quantities also derived from the cumulative curve and calculated from the quartiles (table 3-16), is that a broader range of the distribution curve is considered. The percentiles used are more susceptible to theoretical statistical interpretation than are the quartiles.

Calculation of the quantities for sorting, skewness and kurtosis according to the definitions of INMAN (1952) considers only that portion of the cumulative curve between Φ_{16} and Φ_{84}. Redefinition, based on the range between Φ_5 and Φ_{95} was suggested by FOLK & WARD (1957). This has the advantage that deviations from log normal distribution at the fine and coarse ends are better shown, but it requires greater calculation.

Table 3-18. Cumulative frequencies for a lognormal distribution.

Φ	$\overline{\Phi}-3\sigma$	$\overline{\Phi}-2\sigma$	$\overline{\Phi}-\sigma$	$\overline{\Phi}$	$\overline{\Phi}+\sigma$	$\overline{\Phi}+2\sigma$	$\overline{\Phi}+3\sigma$
F in %	0.135	2.275	15.865	50	84.135	97.275	99.865

Whether a given grain size distribution corresponds to a normal distribution is shown best by plotting the cumulative curve on probability paper. The ordinate scale is subdivided so that a normal distribution plots as a straight line. The slope of the line is proportional to the sorting σ. The steeper the slope, the better sorted is the sediment.

In the discussion so far it has been presumed that the grain size distribution is monomodal. However, bimodal or polymodal distributions are not uncommon.

These can be described usually as a combination of several log normal distributions. On probability paper these appear as several contiguous straight line segments.

If sediment grains do not have the same specific gravity, different size grains are transported and deposited together. The size relations of equivalent grains of different specific gravity are generally different in the case of suspension and bottom transport.

In the bed load all grains which require the same shear stress for rolling or saltating motion behave alike. For the ratio of diameters D_{ar} and D_{br} of spherical equivalent particles of specific gravities ϱ_a and ϱ_b, we obtain, considering equations (3.57) and (3.60):

$$D_{ar} : D_{br} = \Theta_b(\varrho_b - 1) : \Theta_a(\varrho_a - 1) \tag{3.126}$$

Θ_a and Θ_b are constants related to rolling and saltating transport, which are dependent on grain size. For grain sizes above the Hjulström minimum (that is, for sands above about 0.2 mm diameter), it is probably justified to neglect the size dependence for these constants. We then obtain as an approximation:

$$D_{ar} : D_{br} = (\varrho_b - 1) : (\varrho_a - 1) \tag{3.127}$$

During transport and deposition of the suspended load, all grains behave alike that have the same settling velocities. For Reynolds numbers less than 1 (quartz grains <0.2 mm) the STOKES formula can be used. For the ratio of diameters D_{as} and D_{bs} of spherical grains of equal settling velocity and specific gravities ϱ_a and ϱ_b:

$$D_{as} : D_{bs} = \sqrt{(\varrho_b - 1) : (\varrho_a - 1)} \quad (\text{STOKES}) \tag{3.128}$$

For higher Reynolds numbers, that is for larger grains, the settling velocity is calculated from the OSEEN formula or some other approximation. A general expression for the size ratios of grains of equal settling velocity cannot be formulated in this transitional size range. In any case the ratio changes with increasing grain diameter in the sense that the difference between D_{as} and D_{bs} increases, until finally in the applicable range for the NEWTON formulae (Reynolds numbers >1000; quartz diameters >2.6 mm) the following relationship applies:

$$D_{as} : D_{bs} = (\varrho_b - 1) : (\varrho_a - 1) \quad (\text{NEWTON}) \tag{3.129}$$

Note that for the largest grains the equivalent expressions are identical for both bottom and suspension transport (equations 3.127 and 3.129). For small grains the size difference of equivalent particles of different specific gravity is greater in the case of bottom transport than for transport in suspension.

In fig. 3-26 the relations for equations (3.127) and (3.128) for different specific gravities are plotted. In sands which have been transported and deposited in water, the size differences between light and heavy mineral constituents have been shown to lie indeed within the range predicted by the above-mentioned equations (v. ENGELHARDT 1940a; RITTENHOUSE 1943; WALGER 1966).

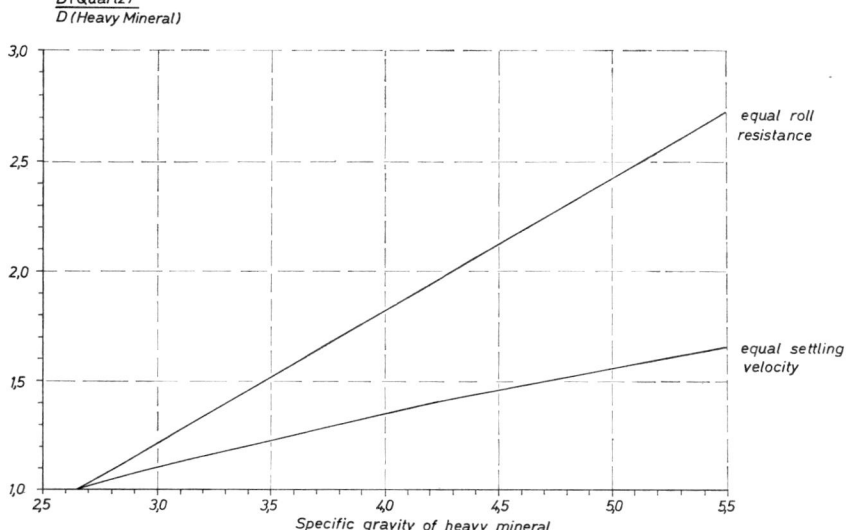

Fig. 3-26. Ratios of diameters of spherical quartz and heavy mineral grains of equal settling velocity or roll resistance in water.

3.182 Streams[10]

The discharge (m³/sec) of most streams transporting sedimentary material into lakes or into the sea increases from the headwaters to the mouth. The so-called mean stream velocity (cm/sec) is obtained by dividing the discharge by the appropriate charge sectional area. Although the stream cross section increases as a rule toward the mouth, so does its mean velocity and discharge, at least in the cases investigated (LEOPOLD 1953). However, it is not this mean velocity which is important in stream transport and deposition, but rather it is the shear force at the stream bottom that is decisive. This is proportional to the product of the stream gradient (tan β) and the hydraulic radius (m) or stream depth (h) (see section 3.12):

$$\tau_0 = m\, g\, \mathrm{tg}\, \beta \simeq h\, g\, \mathrm{tg}\, \beta \qquad (3.130)$$

The way in which the most important transport and depositional factors change along the course of a stream is shown as an example in fig. 3-27. These data come from studies by LEOPOLD & MADDOCK (1953) of numerous rivers of the mid-western United States. The relations noted between gradient, depth, mean velocity and transport power (h · tan β) on the one hand, and discharge on the other have been averaged from observations of many individual streams and illustrate the relations in a typical model stream. Increasing discharge corresponds to a progression from the source region to mouth. In this direction gradient decreases, while simultaneously depth increases. Since the decrease in gradient is more marked than the increase in

---

[10] See also LEOPOLD, WOLMAN & MILLER (1964), and their bibliography.

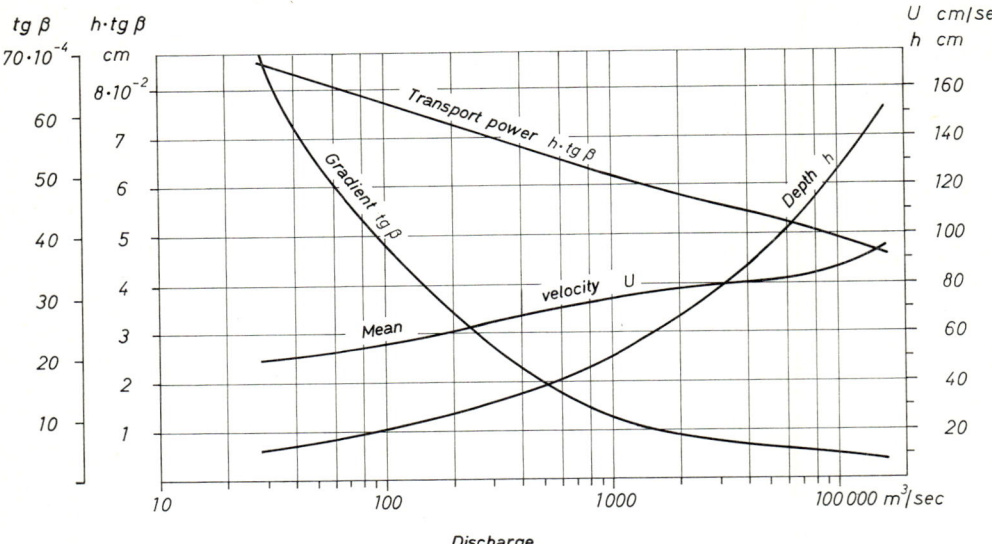

Fig. 3-27. Mean values for gradient, velocity, transport power and depth for rivers in the mid-west United States as a function of increasing discharge from source to mouth (after LEOPOLD & MADDOCK 1953).

depth, the transport power for this stream model decreases toward the mouth, although the mean velocity increases.

The decrease in transport power or shear stress at the base of the stream must result in deposition of transported material, if the river is carrying with it material of the appropriate grain size. This deposition is accomplished in different ways for bed and suspended load.

As was indicated in section 3.14, two critical shear stresses can be distinguished for bottom transport. The first critical shear stress τ_{01} signifies initiation of rolling motions of a grain; the second critical shear stress, the initiation of saltation. The appropriate formulae derived in section 3.14 are repeated once again:

$$\tau_{01} = \Theta_1 (\varrho_2 - 1) g D \quad \text{(rolling transport)} \qquad (3.131)$$
$$\tau_{02} = \Theta_2 (\varrho_2 - 1) g D \quad \text{(saltating transport)} \qquad (3.132)$$

Θ_1 and Θ_2 are functions of the diameter and ϱ_2 is the specific gravity of a grain. All non-suspended grains, for which the effective bottom shear stress is greater than the threshold value for rolling or saltating transport, are moved as bed load. The bed load consists of grains, for which the product $(\varrho_2 - 1) D$ varies from a maximum value to small values. The lower limit is determined by the carrying capacity for suspension load of the corresponding current. Grains for which $(\varrho_2 - 1) D$ lies between the maximum values for rolling transport and saltation form current ripples at the stream bottom as transport bodies.

3.18 Deposition in water

Deposition of bed load takes place when the transport power of the stream decreases, so that at the bottom a diminished shear stress prevails. Then certain grains which were previously saltating can no longer do so and can at most only roll along the bottom. Other grains which previously were rolling along the bottom can no longer do so. Material moving in current ripples can move only very slowly and can be considered as deposited sediment. The grain size range of the deposited material becomes narrower, the smaller the amount the shear stress is diminished.

As already noted in section 3.14, there is a minimum critical shear stress for bottom transport (fig. 3-11), applicable to quartz grains of approximately 0.2 mm diameter. The shear stress just necessary to mobilize these grains is the smallest shear stress that permits bottom transport. For even smaller grains the shear stress increases with decreasing diameter, and for larger ones with increasing diameter. The existance of this minimum is probably the reason that sands with grain size around 0.2 mm are often especially well sorted (INMAN 1949). For shear stresses somewhat above the minimum the bed load consists of coarser grains down to about 0.2 mm. Finer ones are transported in suspension. If the transport power decreases the coarser grains are gradually sorted out until only those grains remain which correspond to the minimum shear stress. These would be grains of 0.2 mm diameter (quartz) without admixed finer or coarser material.

In suspension fine material can be transported over long distances because all natural water currents are turbulent (section 3.12). Settling particles are lifted up repeatedly by the turbulent mixing motions. As long as no deposition takes place, the number of suspended particles remains constant, that is, a steady state is reached. At each elevation above the bottom, in a unit of time the number of settling particles is equal to the number of particles ascending as a result of the turbulence. As was developed earlier, an exponential particle distribution develops over the cross section during this steady state condition, when the exchange is constant. More complex distributions develop if the exchange changes with distance from the bottom. The latter condition applies in the case of natural streams.

Deposition occurs when this steady state is disturbed. Disturbances can arise in different ways. In streams the steady state is tied in to the condition that all material falling on the bottom from the lower-most suspended layer is mobilized immediately and repeatedly and put into suspension again. If this is not the case and suspended matter is removed continuously, a steady particle distribution cannot develop. As long as coarser material is being transported as bed load on the bottom, conditions may remain favorable for the greater part of the falling suspended load to be remobilized. This is because the movement of suspended grains on the bottom will stir up the finer material. If the shear stress at the bottom drops so that a portion of the bed load is immobilized and forms slowly migrating ripples, then the possibility exists for falling particles of the suspension load to become included between the coarser grains. As a result a bed load sediment forms which includes a small admixture of the suspended load. If the bottom shear stress drops below the minimum value corresponding to the movement of quartz grains of 0.2 mm diameter, bottom transport ceases. Then material settling out of suspension can no longer be mobilized and the suspension load is deposited continuously.

Essentially complete steady suspension transport is possible, therefore, only when the bottom shear stress is greater than the minimum value of about 15 dyn/cm^2, corresponding for quartz particles to a transport power of m tan $\beta = 1.5 \cdot 10^{-2}$ cm. At lower shear stresses suspended load material will deposit on the bottom and remain there.

The distribution of suspended particles over the cross section of the stream depends on the ration of particle settling velocities to exchange integrated over the entire cross section of the stream (equation 3.50). The quotient $p = v : 0.4\sqrt{\tau_0}$ determines approximately the suspended particle distribution in a stream (section 3.12). v is the particle settling velocity and τ_0 the bottom shear stress. Therefore the bottom shear stress of a stream is crucial for transport and deposition of the suspended load. If shear stress is diminished the quotient p becomes smaller. All particles transported in suspension experience displacement such that they are concentrated more in the lowermost layers of the stream. This particle concentration is most significant for those particles that settle most rapidly. The material settling out of the stream has a grain size distribution similar to that of the suspension load of the layers nearest the bottom. In contrast to the bed load sediment, the sediment deposited from the suspended load contains all grain sizes, but with a greater prevalence of rapidly settling particles. Therefore, sedimentation of the suspended load, in contrast to bed load, does not result in good sorting of particles.

Although transport power decreases downstream because of the diminishing gradient, the maximum stream transport capacity, that is, the amount of material transported per second past a fixed point, does not slacken. In fact, according to equations (3.73) and (3.74), the maximum transport capacity for a model stream (fig. 3-27) increases downstream for both the suspended and bed loads. Such a stream would be in a position, in spite of decreasing transport power, in the direction of flow to transport increasing amounts of bed and suspended loads, if progressively finer material also is supplied toward the mouth. It is presumed naturally that sufficient numbers of particles are supplied, that can be moved by the available transport power at the bottom or in suspension.

Stream valley sediments can be classified into two genetic groups: stream bed sediments and flood sediments (DOEGLAS 1950).

S t r e a m   b e d   s e d i m e n t s consist mainly of bed load. They occur where a decrease in transport power causes previously rolling and saltating grains to come to rest. At the bottom of a stream channel sediments are found that increase in fineness from source to mouth. Fig. 3-28 indicates this relationship for the Mississippi River between Cairo, Illinois and the Gulf of Mexico. While the mean grain size decreases continuously, but only slightly, the disappearance of gravel and coarse sand from the bed load is apparent.

Deposition of sand and gravel in a stream channel takes the form of sand or gravel bars. Such bars of coarse material alternate in the stream channel with bedload in which fine sediment is deposited at lower points. Meandering streams deposit bed

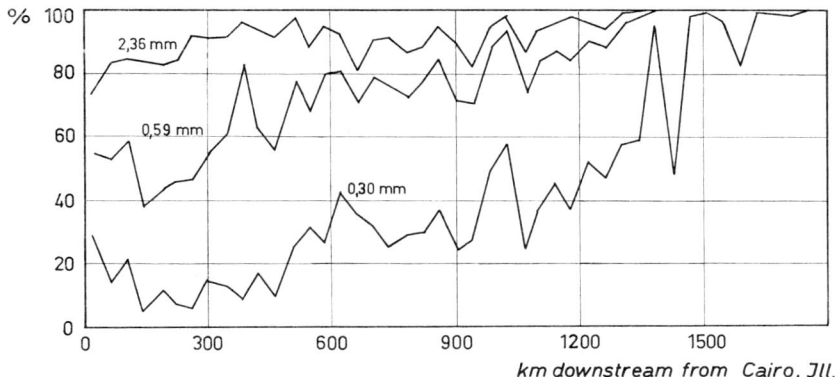

Fig. 3-28. Grain size distribution of stream bed sands of the Mississippi River between Cairo (Illinois) and the mouth (after LEOPOLD, WOLMAN & MILLER 1964).

load at sand bars (point bars) which form along concave banks of the stream. To a very great extent deposition of stream bed sediment takes place wherever a sudden widening of the stream channel or a sudden decrease in gradient causes a marked diminution in transport power. At such places delta-like accumulations of sand bars form, between which the stream is subdivided into several smaller, frequently changing channels (braided stream).

Stream bed sediments are deposited always from more or less strong currents. The resulting sedimentary structures generally can be recognized clearly. Inclined or cross-bedding is observed in current ripples and sand bars. The locally usually rapidly changing flow relationships produce a fine layering of deposits of different grain size. Each individual layer is generated from a single regime of certain transport power. Since the layers are very thin, in certain instances, they are difficult to sample. The grain size distribution of a sample collected without special care can consist frequently of a mixture of several individual distributions, each one corresponding to a single layer. If the individual sand layers each have log normal distributions, the cumulative curve for the mixture, when plotted on probability paper, is composed of several straight line segments (WALGER 1961).

There appears to be a regular interrelation between the mean grain size and thickness of individual layers — the thickness increases with mean grain size (SCHEIDEGGER & POTTER 1967). This probably is related to the fact that the amplitudes of ripples and sand waves increase with increasing transport power and that the thickness of layers deposited on the downstream side of these structures grows with amplitude.

The sorting of sands and gravels from stream beds is generally relatively good. For a large number of river sands FRIEDMAN (1967) found deviations in Φ-units between $\sigma = 0.50$ and 1.50. In many cases river sands exhibit a characteristic intermixing of finer material, which is greater than for an ideal log normal distribution. Positive skewness (α or α^*) often is noted. For 180 river sands studied by FRIEDMAN

(1967), 134 have positive skewness to values up to $\alpha = +2.3$ (most between 0 and $+1.5$). Presumably clay and silt is admixed, having been deposited from the suspended load and intermixed among the coarser grains so that it could not be mobilized again.

F l o o d s e d i m e n t s form when a stream flows over the banks of its narrow channel flooding small or large portions of the valley floor. In the flood region, because of the sudden reduction of depth and gradient, a significant diminution in transport power occurs. All of the sediment carried along by the floodwaters, especially the fine suspended load, is deposited all at once. The deposits contain all grain sizes with a good share of fine components. The sorting is exceptionally poor and criteria for current flow, such as finebedding and crossbedding, are absent. The resulting strata are either homogeneous or exhibit a graded texture, in which the coarse components are concentrated in the lower layers with the fine constituents increasing upward.

To exemplify the differing characters of stream bed sands and flood sediments for the same stream, cumulative curves and grain size distributions, and the statistical parameters for two sediments are given in fig. 3-29 and 3-30. These represent sediments from the upper Euphrates River (SCHNEIDERHÖHN 1957). Sediment A is a typical poorly sorted flood sediment from the floodplain of the river. Sediment B comes from a several meter high terrace deposit. It is a bed load sediment whose mean grain size corresponds approximately to the minimum shear stress for bottom transport. This sediment is much better sorted, but has a slightly positive skewness, that is, a slight excess of fine material.

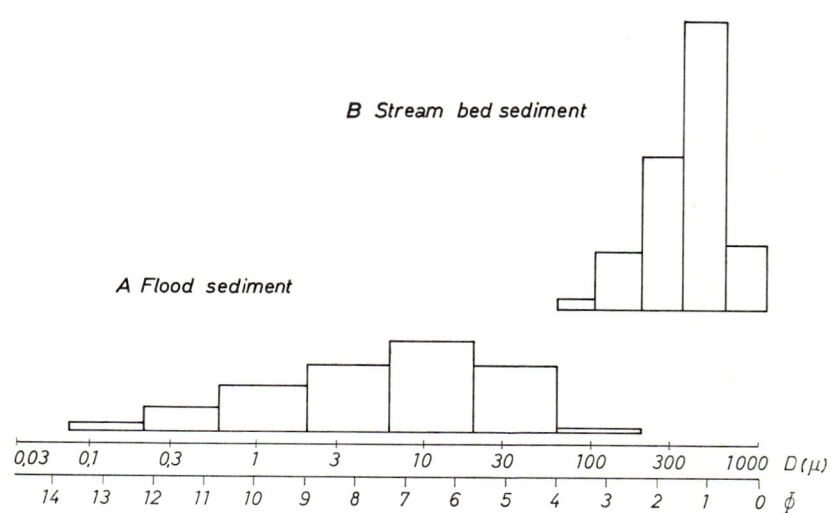

Fig. 3-29. Grain size distribution of a flood sediment (A) and stream bed sediment (B) of the upper Euphrates River (after SCHNEIDERHÖHN 1957).

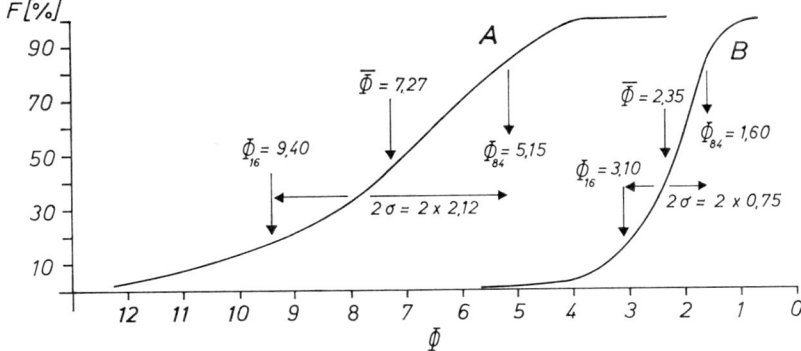

Fig. 3-30. Cumulative curves for sediments shown in fig. 3-29, giving the following statistical parameters:

	A	B
Md = Φ_{50}	0.008 mm	0.22 mm
Q_1	0.0024 mm	0.15 mm
Q_3	0.020 mm	0.30 mm
So	2.86 mm	1.44 mm
Sk	0.75 mm	0.88 mm
Φ^*	7.27 mm	2.35 mm
σ^*	2.12 mm	0.75 mm
α^*	+0.127 mm	+0.267 mm

River valley sediments are subject to repeated redeposition, because stream discharge varies greatly. At any particular place in the stream bed the transport power varies within wide limits. At low water levels, sediments deposit in the channel, which are mobilized again at higher water levels and augmented transport power. If the water rises so high that the stream flows over its banks, chiefly suspended sediment is deposited on the inundated valley floor. Even these sediments are not deposited permanently. The meandering stream will erode later sediments which had been deposited previously on the flood plain. They are mobilized again and transported further downstream. As evidence of erosion of fine-grained flood sediments, clay pebbles are frequently found in stream bed sands. Therefore, sediments deposited in stream valleys are to a great extent temporary deposits. They consist of material involved in a repeatedly interrupted and slow migration which leads from the region of denudation toward the river mouth. For all large streams the time during which a sedimentary particle lies at rest in a deposit is probably much longer than the time spent in actual transport. The grain deposited in a flood plain remains there a very long time unitl the meandering stream reworks the earlier deposited material. During this time the particle is subject to the processes of chemical weathering. It is apparent that during the entire time span involved in transport to the mouth of the stream, the sedimentary material is exposed to chemical influences for a much longer time than to mechanical transport abrasion.

Whether in the long run accumulation of sediment occurs in the stream valley, whether the valley develops only a transport channel as a result of equilibrium between transport and deposition, or finally whether erosion predominates, causing the stream to be incised, these events depend on large scale crustal movements.

3.183 Deltas

At the points where rivers discharge into lakes or into the ocean, their transport power and capacities suddenly decrease and rapid deposition of transported material takes place. Thus deltas form as accumulations of clastic sediment. Many recent deltas of the great rivers of the world are well known (Nile, Niger, Mississippi, Rhone, Rhine, Danube, Orinoco, Mekong, Yangtse, and others; see SHIRLEY & RABSDALE 1966). Thick deltal deposits can be formed especially where continued subsidence enables the accumulation of large quantities of sediment. A thick wedge of sediment develops whose thicker end abuts against the continent.

In general along the course of a stream, as shown above, the mean current velocity increases while the transport power decreases. As a result grain size and mean settling velocity of suspended particles are diminished. According to equation (3.78) the ratio of suspended to bed load increases. Delta sediments, therefore, consist primarily of fine sediments (silt and clay). Smaller quantities of sand are embedded in and among the finer sediment. These sands are deposited in the numerous distributary channels that dissect the delta platform and in which the streams' waters seek their way to the ocean through their own deposits.

Stream channel sands of delta deposits are usually relatively fine-grained. In the Mississippi delta, for example, the median size of these sands lies between 0.1 and 0.2 mm; at most at 0.3 mm. They exhibit admixtures of finer constituents typical of fluvial sands, giving a positive skewness in Φ-units (FISK, MCFARLAN, KOLB & WILBERT 1954). Also the stream channel sands of the Rhone delta are quite fine-grained (the median for the coarsest sediment is at 0.2 mm), containing more fine material than would a normal distribution (KRUIT 1955).

The stream channels are bound by natural levees consisting of finer material (silt and clay) that was laid down during higher water levels. The delta regions lying between distributaries are only occasionally flooded, at which times poorly sorted suspension load is deposited as flood sediment, whose grain sizes range from fine sand to clay. Fig. 3-31 illustrates some examples of typical sediments of the Rhone delta (after KRUIT 1955).

Stream channels often extend from the actual delta far out into the ocean, since the natural levees which laterally bound the stream are built up repeatedly. These arms of the river in the Mississippi delta extend 25 to 30 km into the ocean forming the so-called Bird Foot Delta. At the mouth of each distributary alluvial cones form from bed and suspension load. Adjacent to the mouth chiefly sands are deposited, which, with the advance of the channel protrude as finger-like levees into the basin. In the Mississippi delta fine-grained sands are noted to water depths of about 15 m. At greater water depths only silt and clay arrive as suspended material. Water

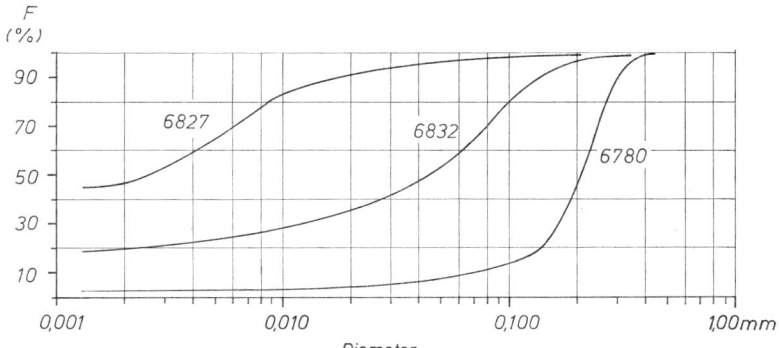

Fig. 3-31. Examples of typical delta sediments from the Rhone River (after KRUIT 1955). 6780: Fine stream channel sand with admixed silt and clay. Md = 0.21 mm.
6832, 6827: Poorly sorted silty-clay deposits typical of flood sediments. Md = 0.046 and 0.0026 mm.

which is visibly turbid from the suspension load is observed up to 100 km off the mouth of the Mississippi (FISK 1955). Also at the Rhone mouth, at the front of the delta, rapid deposition of material transported in distributary channels takes place. Off the mouth of the main stream the median size of recent sediments is still 0.21 mm in 3 m deep water, 0.05 mm at 8 m, and only 0.01 mm at 30 m depth (KRUIT 1955).

At the outer edge of the delta pile, beyond the mouths of the intermittantly active stream channels, reworking of deposited material by incoming waves takes place. Here material transport parallel to the coast can occur, as can aeolian and aquatic sorting processes typical of the shallow sea coast. In the outer zone of most deltas deposits of marine coastal sands and dune formation occur. They differ characteristically in their grain size distributions from the primary delta deposits.

3.184 Sea coasts

Along a coast with little tidal influence, transport and deposition is determined exclusively by the action of heavy seas. As was indicated in some detail in section 3.163, waves approaching the coast from the open sea interact with the sea bottom beyond a certain depth, causing the waves to change their form and break in the surf zone and to move loose material on the bottom within a coastal strip parallel to the shore. In this zone in which the wave motion extends to the bottom, silt and clay cannot be deposited and retained permanently. Sand which rolls and saltates on the bottom is freed of all material which can be carried in suspension by the wave currents. Also within this coastal strip in the region farthest from shore, where only oscillation ripples form, and the bed load is essentially stationary, simply rolling back and forth, fine particles are washed away continuously. Sands occurring along open coasts sufficiently removed from river mouths or other sources of fine material contain almost no grains below about 0.15 mm diameter (INGLE 1966, p. 174).

Since the draught of wave motion depends on wave length and intensity of orbital motion at the bottom as well as on wave height (see section 3.163), the width of the sandy strip off a shallow coastline is strongly dependent on the local hydrographic and climatic conditions. Along the German North Sea coast, from the East Frisian Islands to the mouth of the Elbe, the sea bottom has collected to a water depth of about 20 mm fine sand (coarser than 0.18 mm) (REINECK 1963). Also on the Baltic coast between Kiel and Mecklenburg sands with oscillation ripples occur to water depths of about 20 m; in Dantzig Bay sands extend even to depths of 70 to 80 m (SEIBOLD 1964). Off the mouth of the Orinoco only sediments occur in water depths over 20 m which are finer than 0.05 mm (NOTA 1958).

Seaward from the breaker zone, bottom transport related to wave action, neglecting rip currents, is directed toward the shore. In the breaker zone an especially intensive interaction affects the bottom, resulting in considerable stirring up of loose material. Between the breakers and the shore line sedimentary material is moved toward the beach by incoming swash, but transported seaward again by the outgoing backwash. Along stationary shorelines neither accumulation nor erosion take place. Rather a dynamic equilibrium prevails, during which supply and removal of material are held just in balance. Along a given section of the seaward beach slope swash and backwash each perform a given amount of work, in that sand is transported first toward the shore and then seaward. As a result of various energy losses the energy of the outgoing water is always less than the incoming. On a horizontal base chiefly the swash deposits more material than the backwash can carry away. A seaward inclined beach slope is formed whose slope is so disposed that at any point just as much material is removed by the backwash as was brought in by the swash. The slope is steeper the greater the energy difference between swash and backwash currents. An essential factor in causing this energy loss is the percolation of water into the beach sand. This is more important, the greater the permeability, that is, the coarser the sand. Therefore, the slope of a beach consisting of coarse constituents is steeper than one consisting of fine sand. Since fine sand is transported farther up the beach by the swash, as a rule a sorting of sand results on the beach. The mean grain size increases from the shoreline seaward. Correspondingly the slope of the beach changes also. The slope is least at the shoreline and increases out to sea, until it passes over into the flatter base with a change in slope. For the quantitative calculation of the inclination of the beach slope reference should be made to the formulae and examples cited by INMAN & BAGNOLD (1963).

Within the coastal strip between the breaker zone and the strand line sand transport parallel to the coast takes place, since waves generally impact the coast at a certain angle, and also because coastal currents occur. Rip currents, breaking through the surf, repeatedly carry out to sea material moving parallel to the coast. Fine suspension load can be deposited only far outside the coastal strip. The coarser constituents are moved along toward the shore and as a whole finally in a direction parallel to the shore, which corresponds to the prevailing wind and/or coastal current direction. This results in the formation of elongated sand ridges or sand bars, which are interrupted by stream channels through which rip currents transport seaward again water thrown as surf onto the beach.

3.18 Deposition in water

If the coastline is moving continuously landward in slow transgression all of the previously formed sand bodies in the beach zone can be preserved and a widespread sand layer can be developed in continuous stages, because it is buried by the transgressing silt and clay sediments of the deeper shallow sea.

If, however, the coastline is stable along a sea coast not broken up by inlets, these processes lead at best to temporary accumulation of sandy sediments, which are in constant motion along the coast. If a bay or inlet somewhere breaks the smooth course of the coastline, at the point where the coast suddenly veers inland, sand must remain. It forms a spit which continues to grow in the direction of coastal transport and has a tendency to close off the inlet, finally forming a bay or lagoon.

All underwater deposits in the shore region exhibit the typical characteristics of bottom transport, such as finebedding, crossbedding and current ripples, as well as symmetrical oscillation ripples resulting from wave action.

Pebbles and sand grains, moved a long distance along the coast, back and forth between the breaker zone and the shoreline, are more strongly mechanically stressed before final deposition than by any other kind of water transport. Therefore, as noted earlier, sand grains and pebbles of beach deposits generally are better rounded than those from fluvial environments.

In addition the long and intensive coastal transport has a significant sorting influence. Sorting according to size as well as specific gravity takes place.

Emery (1955) compared the properties of gravels of different origin and determined the values given in table 3-19. These data show the better sorting of beach gravels.

Table 3-19. Comparison of different gravels (after EMERY 1955).

Origin	Number	Median (mm)		Sorting $(Q_1:Q_3)^{1/2}$	
		average	range	average	range
Sea coasts	32	56.0	(10.8—750)	1.25	(1.13—2.14)
Lake shores	37	47.0	(13.0—125)	1.15	(1.09—1.21)
Streams	35	19.0	(10.4—355)	3.18	(1.34—5.49)
Alluvial fans	58	16.5	(10.0—64.0)	5.33	(2.50—8.95)

Beach sands on the average are better sorted than river sands. For a large number of beach sands FRIEDMAN (1967) found a deviation of 0.8 Φ at most, while river sands give values between 0.5 and 1.5 Φ. The admixing of fine material common to river sands, leading to a positive skewness (in Φ-units) seldom occurs in beach sands. Among 155 beach sands FRIEDMAN (1967) noted only 54 with positive skewness. For most beach sands wave action has caused all of the admixed fine suspension load to be washed out.

Frequently one observes on the beach slope, formed as a result of the interaction with swash and backwash, the occurrence of quite relatively ephemeral heavy mineral placers. For their formation simple sorting due to roll resistance (bed load) or settling velocity (suspension load) would never suffice, because in both cases the heavy mineral grains would be diluted by a large number of rapidly settling or equally mobile quartz grains and could not be separated from them. However, a combination of both sorting processes can produce such a selection. If at a given point on the beach slope, grains of equal settling velocity are deposited by the incoming swash, transported there in suspension, using STOKES law, and according to equation (3.128), the following relation applies for the diameters of specific light (specific gravity ϱ_a, diameter D_a) and heavy mineral grains (sp. gr. ϱ_b, diameter D_{bs}):

$$D_{bs} = D_a \sqrt{\frac{\varrho_a - 1}{\varrho_b - 1}} \qquad (3.133)$$

Quartz grains ($\varrho_a = 2.65$) of 0.200 mm diameter would be deposited, for example, with garnet grains ($\varrho_b = 4.00$) of 0.148 mm diameter. If the outgoing backwash at the same point has a velocity such that it can just barely mobilize the lighter quartz grains, the heavy mineral grains of equal settling velocity would remain behind, since a greater force is necessary to move them. Equation (3.127) gives us approximately the diameter D_{br} for those heavy minerals which are mobilized by the equivalent shear stress for the lighter grains of diameter D_a:

$$D_{br} = D_a \frac{\varrho_a - 1}{\varrho_b - 1} \qquad (3.134)$$

Therefore, D_{br} is smaller than D_{bs}. Those garnet grains which offer the same resistance to movement on the bottom as the 0.200 mm quartz grains, would have a diameter of 0.108 mm. Therefore, if the backwash can just barely remove the quartz grains, garnet grains of 0.148 mm diameter would remain behind.

In heavy mineral concentrates formed in this way, the diameters of mineral grains of different specific gravities should have ratios to one another as defined by equation (3.134). This was confirmed for a concentrate from the beach of the Baltic at Warnemünde, which consisted chiefly of quartz, garnet, and ore minerals (magnetite and ilmenite) (v. ENGELHARDT 1937 b). The following grain size distribution maxima were established: quartz 0.480 mm, garnet 0.256 mm, ore minerals 0.224 mm. Assuming specific gravities of 4.0 for garnet, and 5.0 for ores, according to equation (3.134) the diameters should be as follows: quartz 0.480 mm, garnet 0.264 mm, ores 0.200 mm.

Relative to further details on the origin of beach placers, especially with regard to the influence of wave action and coastal development, the studies by BÜLOW (1951) and of LUDWIG & VOLLBRECHT (1957) are recommended.

Equation (3.133) applies strictly only to spherical quartz particles of less than 0.100 mm diameter. The settling velocity of larger quartz grains must be calculated by means of the OSEEN or other approximations. Then in place of the term $\sqrt{(\varrho_a - 1) : (\varrho_b - 1)}$ (section 3.13) a smaller factor applies; that is, the size difference between a quartz grain and an equally rapidly settling heavy mineral grain becomes larger than predicted by the STOKES formula. Finally in the range of applicability of the NEWTON formula (for quartz, grains larger than 2.6 mm) the ratios of diameters of grains of like settling velocity can be determined from equation (3.134). Then the described mechanism of placer formation is no longer possible. Based on these considerations, we can conclude that beach placers of this kind most readily will occur in the fine sand range and that their formation becomes increasingly unlikely when quartz grains are larger than 1 mm.

While along a level coast without strong tidal influence, sandy sediments are found along the shore and finer deposits seaward, coasts subjected to strong tides exhibit just the opposite relations. At low water level the effects of waves and currents are felt most strongly. At this time conditions prevail which are similar to those where tidal influences are absent. Fine-grained material cannot be deposited and thus pure sandy deposits are formed. On the other hand the flood tide carries predominantly suspension load landward, and this is deposited between the low- and high-water lines. The receeding ebb tide transports only a portion of this material seaward again. A more or less broad marshy strip occurs, which is traversed by a system of ebb tide-cut stream channels. The marsh sediments are extraordinarily multiform. Fine clay and silt deposits prevail. Poorly sorted sediments, stemming from the suspended load of the flood tide, correspond to stream valley and delta flood sediments. In between in multiple succession sands with fine- and crossbedding and ripples are interbedded. These were deposited by the stronger currents, for example, in the ebb stream channels. Examples of such sediments and sedimentary structures were described from the German North Sea coast by REINECK (1963, 1967).

3.185 Lakes and sea

As a river enters a lake or ocean basin, conditions suitable for continuing suspension transport no longer prevail. The sudden diminution in current velocity causes a reduction in turbulence and exchange. The sediment particle distribution in the stream is altered correspondingly and sedimentation of the suspended load results, wherein chiefly the coarser constituents are deposited near the mouth of the stream. They are accompanied always by a certain portion of finer particles. Also apart from the decrease in carrying capacity of the stream as a result of decrease in exchange, after entrance of the stream into the basin, suspension load is sedimented. This happens to a large extent, because that portion of the suspended load continuously settling from the stream can be mobilized no longer since movement is lacking at the bottom of the basin.

The spatial distribution in the basin of sediments of differing grain size which results from these processes depends on the special localized stream relationships. MÜLLER's (1966) studies in Lake Constance serve as an example of lake sedimentation. Material is supplied there by the Rhine River. At the east end of the lake it forms a delta. Sediments lying on its seaward slope consist of about 60% silt, 30% sand and 10% clay. On the lake bottom in front of the delta only silt and clay occur, which are distributed by currents active in the lake. In regions of high current velocity and higher turbulence along the northern shore of the lake, the clay components remain largely in suspension and silt is deposited. In still water areas apart from the main current, bottom sediments consist of clay.

If water flow and transport power of a stream supplying sediment changes seasonally (for example, glacial streams which flow more in summer than in winter), regularly banded sedimentary sequences form in the lake basin. Such banded glacial clays are called varves and consist of alternating sandy-clay and clay layers. The former represent the suspended load of the stream with high transport power; the latter stem from a stream of lower carrying capacity.

The fate of the suspension load brought into the sea cannot be deduced simply from the sediment distribution observed today on the margins of the continental shelf. A a result of major changes in sea level in the most recent geologic past, these sediments are to a considerable extent relicts, deposited under conditions different from those prevailing today. According to estimates by EMERY (1968) about 70% of the present shelf is covered by such relict sediments. The simple scheme of a regular decrease in the mean sediment grain size with increasing distance from the coast is seldom realized. At numerous places on the remote parts of the shelf, especially near its edge, coarse grain sediments occur, which relate back to the lower sea level at the time of the last glaciation. (A good example is described by NOTA [1958] from the shelf off the mouth of the Orinoco. For other examples see KUENEN [1950] and GUILCHER [1963]). Only a relatively narrow strip of ocean bottom off the continental coasts contains fine-grained recent sediments.

Different observations have indicated that most of the suspension load brought into the sea, not already deposited in the delta region, is sedimented near to the coast. This follows, for example, from the observation that the visible water turbidity off the mouths of large rivers does not reach far out to sea. Off the mouths of the Niger and Orinoco the zone of visible turbidity perpendicular to the coast reaches an extent of 30 to 50 km, while it reaches many hundreds of km in a narrow strip parallel to the coast (PORRENGA 1967). The clay turbidity can remain for a long time in the moving water of the current, but disappears rapidly in deep water. We have mentioned earlier the small amount of suspended matter in sea water and the very low sedimentation rates of deep sea sediments. In spite of their high velocities and appreciable turbulence, open ocean currents carry only very small amounts of suspension load into the distant ocean basins.

Near-coastal sedimentation of suspension load in the ocean originates, as in lakes, from the decrease in stream velocity, turbulence, and exchange. In the ocean flocculation of fine clay particles by ions in sea water if effective. Suspended clay particles carry a predominantly negative surface charge, which is ompensated for by a cloud of positive ions, in the so-called diffuse double layer. The thickness of the double layer is the greater, the more dilute is the solution in ions. Since the electric double layers repel each other, clay particles can approach each other closer, the lower the thickness of the ion cloud, thus the approach is closer in concentrated electrolyes than in dilute ones. Sufficient approach leads to flocculation, that is, to the formation of larger aggregates, because the particles exert mutual attractive forces at shorter interparticle separations. In dilute electrolyte solutions suspensions of slowly settling individual particles can be maintained, while an addition of salts causes flocculation, with the formation of larger and rapidly settling aggregates. This phenomenon which is a well-known characteristic of all clay suspensions, should play a role when river water enters the sea with its suspension load. The mixing of river and salt waters must cause flocculation, forming larger aggregates, leading to rapid sedimentation.

The salt concentration necessary for complete flocculation is at constant clay concentration greater the smaller the particle size and the greater their specific surface charge. Cation exchange capacity gives a qualitative measure of the latter. Since individual clay minerals differ appreciably in their particle size and exchange capacities, as well as origin and occurrence, it is likely that the flocculating effects of certain salt solutions depend on the kind of clay mineral particles present. WHITEHOUSE, JEFFREY & DEBBRECHT (1960) investigated the effect of sea water of different concentrations on the settling velocities of suspended particles of illite, kaolinite, and montmorillonite. In all suspensions monodisperse systems formed by flocculation, that is, particles of a single given settling velocity or size. Illite and kaolinite were flocculated completely in solutions of 0.5 ‰ chloride, so that the settling velocity of kaolinite particles rose only from 80 to 81 cm/min; for illite particles it rose from 89 to 110 cm/min when the concentration was increased up to that of sea water (18 ‰ chloride). On the other hand montmorillonite suspensions were much more stable. For chloride contents from 0.5 to 18 ‰ the settling velocity rose continuously from 0.2 to 8.8 cm/min. These observations correspond to theoretical expectations, since montmorillonite particles are generally especially small and carry a higher surface charge than other clay minerals, as reflected by its high exchange capacity. This is related also to experience gained in the technical use of clay suspensions. For example, if one needs as a drilling mud, a clay suspension that is flocculated least by addition of salts, one uses montmorillonitic clays.

Off the mouths of some large rivers, clays near the mouth consist predominantly of illite and kaolinite, while montmorillonite is prevalent further out on the sea bottom (PORRENGA 1967). Off the mouth of the Niger the montmorillonite content increases quite regularly with water depth or distance from the delta. Similarly sediments directly off the mouth of the Orinoco contain less montmorillonite and more kaolinite and illite than the deeper water sediments. These observations probably can be explained, as were similar ones from the Gulf of Paria (VAN ANDEL & POSTMA 1954), by differential flocculation of the different clay minerals. That these findings may not be generalized, is shown, for example, by Gulf of Mexico sediments. Here the clay mineral assemblage supplied by streams and consisting chiefly of montmorillonite with some kaolinite, is distributed relatively uniformly in the bottom sediments (JOHNS & GRIM 1958; GRIFFIN 1962). Also it has not been possible yet to observe directly flocculation by the ions in sea water of the mud carried by streams into the sea. Investigations of salt content and amounts of suspended matter at different river mouths on the North American Atlantic coast have resulted in no indication of an enhanced flocculation at the points where fresh and salt water meet (MEADE 1968 a).

Sediments deposited on inclined slopes can, as noted in section 3.164, give rise to landsliding and suspension currents, which lead to the formation of special deposits, the turbidites. The suspension current is a turbulent flowing mass of water and particles of different sizes. Particle distribution in the current is determined by turbulence and settling velocity in the same way as deduced for more dilute suspensions. Each given grain size is distributed approximately exponentially over the current cross section, wherein the concentration of the smallest particles decreases little with distance from the bottom, while the coarser particles are concentrated next to the bottom. The suspension current leads to the deposition of a turbidite, when velocity and corresponding turbulence and exchange decrease. These processes are the same as were depicted for the formation of flood sediments from a slackening stream current. The difference consists of the higher suspension concentration and the greater spread of constituent grain sizes. If one focuses on a given point of the path of a suspension current, with decreasing velocity, chiefly the coarsest material, mixed with all the fine components, is deposited. Later sediments follow whose coarsest grain size becomes increasingly finer, until finally, when the current comes to a standstill, the last finest clay is deposited. The result is a layer showing so-called graded bedding. Each bed consists of a poorly sorted sediment characterized by a certain upper grain size limit, but always containing some of all of the finer fractions. The upper grain size limit decreases continuously from the bottom upward, so that the whole graded bed ranges from a poorly sorted sand at the base to a fine clay at the upper border. Since the suspension current, continuously slowing down during its course of forward motion, becomes impoverished in coarse constituents, a corresponding graded sequence is to be expected also in its direction of flow (in addition to the literature citations in section 3.164, see also KUENEN & MENARD 1952, and DOTT 1963 for mechanisms of deposition).

Proceeding from a simple model for the decay of turbulence in a gradually decelerating suspension current, SCHEIDEGGER (1965) attempts to calculate the vertical grain size distribution of the resulting sediment. This results in a continuous distribution depicted qualitatively. The mean grain size decreases gradually upward from the base. The trend of this function, that is, whether the mean (or modal) grain size decreases linearly with height above the base (or is depicted by a concave or convex curve), is determined by the primary grain size distribution of material suspended in the current. From many fossil turbidites a positive correlation between thickness of a graded bed and grain size has been noted. The layers on the average are thicker, the coarser the maximum grain size of their constituents. According to POTTER & SCHEIDEGGER (1966) this correlation can be explained, in that a suspension current can transport more and coarser material, the higher its turbulence. The greater its turbulence, the more time is necessary for its complete decay. If from the point of origin of the current sufficient material with the same size distribution is supplied subsequently, more and coarser material will be deposited from a current of high turbulence than from one of less turbulence.

In those deep sea basins which are sufficiently remote from the continents or shielded by trenches and rises, so that they cannot be reached by suspension currents, pelagic sediments are deposited. These are different in some respects from the previously discussed deposits (see ARRHENIUS 1963). In comparison to sedimentary processes on the continents and margins of shelves, deposition in the deep sea takes place extraordinarily slowly. According to available data the accretion of terrigenous material in the deep sea amounts only to $5 \cdot 10^{-5}$ to $5 \cdot 10^{-4}$ cm/year (ARRHENIUS 1963).

The very low rates of deposition determine all of the special characteristics of pelagic sediments. Clay carried by ocean currents as dilute suspensions of very fine particles often certainly form a small part of the detrital portion of these sediments. Significant amounts are transported as fine dust by winds (aeolian dust); another portion stems from the activity of volcanoes (pyroclastic constituents). Since these clastic constituents are supplied very slowly, components of other origins fall into pelagic sediments in greater percentages than occur in the rapidly forming deposits of the shallow seas. In this category belong especially the neoformations which occur on the sea floor, such as the concentrations of manganese oxides, phosphorites, sulfates, zeolites, and glauconites. In addition silicates and metallic spheroids of meteoritic origin are found in higher concentrations in pelagic than in normal sediments.

Frequently the structures of pelagic sediments are influenced considerably by the burrowing activity of benthonic organisms, which produce a thorough mixing of the deposits.

Where conditions permit the preservation of calcium carbonate, a high concentration of carbonate shells from planktonic foraminifera and coccolithophorids can occur on the deep sea bottom (pelagic carbonate sediments).

3.2 Transport and deposition in air

3.21 General

The same principles which apply to the transport of solid particles in flowing water apply also to transport in air currents. The fundamental principles of water transport can be used here also. Nevertheless significant differences do exist between wind and water transport, which are related firstly to the low density and viscosity of air, and secondly to the fact that air currents have neither sharp lateral nor sharp vertical boundaries.

Table 3-20 gives data on the specific gravity and viscosity of air at different temperatures. The ratio of specific gravity to viscosity (ϱ/μ) is about 7 for air, but about 100 for water. The turbulence which results when a certain critical Reynold's number ($R = u \varrho m/\mu$) is exceeded, by otherwise similar dimensions in an air current, is initiated at higher velocities than for a water current. For example, at a critical Reynold's number $R = 1500$, for a current of 10 m depth, the critical velocity in air is about 2 mmsec^{-1}, and in water about 0.15 mmsec^{-1}. As this example shows, all winds of any consequence can be considered as turbulent currents.

Table 3-20. Specific gravity (ϱ) and viscosity (μ) of dry air at 1 at and different temperatures.

Temperature in °C	ϱ	μ in centipoises
0	0.001293	0.0171
10	0.001247	0.0176
20	0.001205	0.0181
30	0.001165	0.0186

In air, as in water, suspension and bottom transport are distinguished. The suspension load consists of the finest particles, held in suspension by the swirling action of the turbulent current. The bed load consists of grains, which barely can be removed or not removed at all from the surface as well as those grains saltating over the surface and falling back again.

The two kinds of natural wind transport are easily observed. Suspension transport results in dust storms, which rise as dense clouds and sometimes obscure the sun. Bottom transport takes the form of sand storms, as they occur primarily in desert regions as dense layers of saltating sand grains moving over the land surface. They have a sharp upper boundary, generally not over 2 m above the ground. Since the air above this layer is clear, in contrast to a dust storm, one often sees the head and shoulders of people protruding above the sand storm cloud (BAGNOLD 1965).

3.22 Transport in suspension (suspended load)

Equation (3.20), derived in section 3.13 for all turbulent currents, relates also the dependence of wind velocity $\bar{u}_y$ on distance y above the ground surface, as long as the wind is transporting no bed load:

$$\bar{u}_y = 5.75 \, u_r \, \lg \frac{y}{y_0} \qquad (3.135)$$

Accordingly the wind velocity depends not only on distance above the ground surface, but in addition very significantly on y_0, whose magnitude is determined by the surface roughness. If the surface consists of a sand of diameter D, then $y_0 = D/30$. However, if other irregularities such as small rocks or sand ripples are present, a correspondingly higher value of y_0 applies. The transport power of a given wind is determined fully, only if the shear velocity u_r and roughness numbers are known. To designate the wind velocity means little, if the distance above the ground surface where measured is not indicated also.

Fig. 3-32 shows some measurements by BAGNOLD (1965) for velocity distribution of air currents as measured in a wind tunnel. According to equation (3.20) straight line plots are formed when velocities ($\bar{u}$) are plotted against the logarithm of height

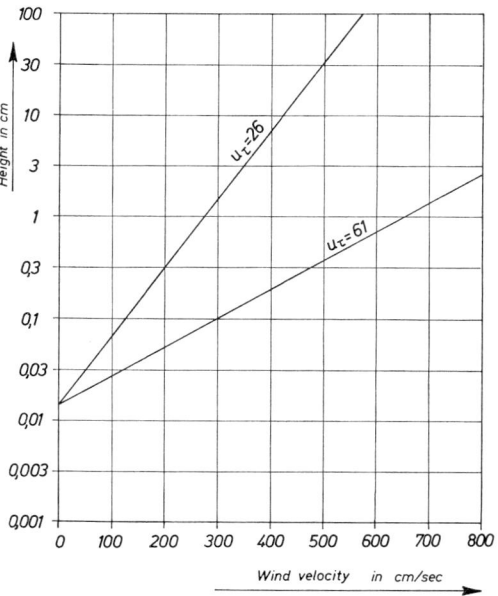

Fig. 3-32. Velocity distribution in air currents as measured in a wind tunnel (after BAGNOLD 1965).

above the ground surface (y). The shear velocity u_τ is given by the slope of those curves:

$$u_\tau = 0.174 \frac{d\bar{u}}{d \lg y} \qquad (3.136)$$

Particle settling velocities in air can be calculated from the formulae developed in section 3.13. The settling velocity of a particle in air is about 100 times that in water. Therefore, the boundaries, defining the range of applicability of the STOKES, OSEEN & NEWTON formulae in air, correspond to smaller diameters than in water. Table 3-21 gives these values for spheres of quartz. The settling velocities of spherical quartz grains in air are plotted in fig. 3-4.

Table 3-21. Applicability ranges for the different formulae for settling velocities of spherical quartz particles in air (20 °C).

Reynolds number	Diameter in cm	
< 1	< $6 \cdot 10^{-3}$	STOKES
1	$6 \cdot 10^{-3}$	OSEEN
5	$1 \cdot 10^{-2}$	Transitional
1000	$1.4 \cdot 10^{-1}$	NEWTON
> 1000	> $1.4 \cdot 10^{-1}$	

Fine particles can be transported as suspended load by the wind, when the turbulent exchange is capable of holding the particles in suspension in opposition to their settling velocity. For the height distribution of particles suspended in wind currents, equation (3.50) from section 3.13 applies:

$$\frac{n_y}{n_a} = \left(\frac{a(h-y)}{y(h-a)}\right)^p \tag{3.137}$$

$$p = 2.50\, v/u_\tau$$

n_y and n_a are the number of particles per cm³ at the heights y and a above the ground, h is the thickness of the streaming layer, which, in the case of a dust storm, is the height above which no particles are in suspension. v is the settling velocity. If h is very large in relation to y and a, the equation can be simplified to the following:

$$\frac{n_y}{n_a} = \left(\frac{a}{y}\right)^p \tag{3.138}$$

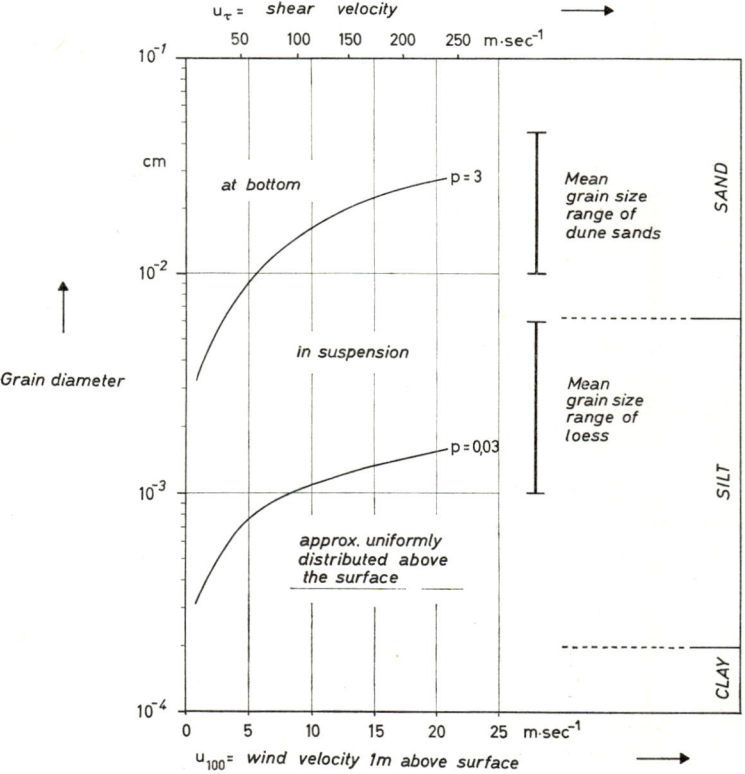

Fig. 3-33. Modes of transport of quartz spheres at different wind strengths.

As was shown in fig. 3-8, the nature of the particle distribution depends on the value of the exponent p. Whether particles of a given size still can be transported as suspended dust by a wind current, is determined by the ratio of the particle settling velocity to shear velocity of the wind. Total transport in suspension, that is, essentially uniform distribution over the entire height of the air current, is to be expected when $p \leq 0.03$ (when $v \leq 0.01\ u_r$). Essentially no suspension transport takes place, if $p \geq 3$ (when $v \geq u_r$). In fig. 3-33 these boundaries for different modes of transport are plotted for spherical quartz grains at different wind velocities. The winds are characterized by their shear velocities and velocities at 1 meter distance above the ground. The latter parameter was calculated from equation (3.135), letting $y_0 = 1$ cm. If the ground surface roughness is less, higher velocities prevail at 1 m.

3.23 Bottom transport (bed load)

The force acting on a planar sandy surface by a flowing medium is given by the shear stress τ_0 at the bottom. For the flow of water in streams one can derive τ_0 from the hydraulic radius and the stream gradient. In the case of a wind τ_0 is derived from the velocity gradient measured over the ground surface, or from the shear velocity u_r, according to the relations which generally apply for turbulent flow:

$$\tau_0 = \varrho_1 u_r^2 = 0.0303\, \varrho_1 \left(\frac{d\bar{u}}{d\lg y} \right)^2 \tag{3.139}$$

In the case of air (20 °C) the following expression applies:

$$\tau_0 = 3.63 \cdot 10^{-5} \left(\frac{d\bar{u}}{d\lg y} \right)^2 \tag{3.140}$$

If an air current of increasing velocity is allowed to act upon a planar sand surface of uniform grain size, the movement of sand grains begins when a certain critical shear velocity is exceeded. As was the case for water transport, it is useful to relate this critical shear stress, expressed in units of force, to the weight of the uppermost sand layer, which is proportional to the expression $(\varrho_2 - \varrho_1) gD$ (see section 3.14). Sand grains of diameter D just begin to move when the shear stress constitutes a certain fraction of the weight of the uppermost layer of grains. One obtains a dimensionless quantity for the critical shear stress:

$$\Theta = \frac{\tau_0}{(\varrho_2 - \varrho_1) gD} \tag{3.141}$$

For completely developed turbulence wind tunnel experiments have given the value

$$\Theta \approx 0.01 \tag{3.142}$$

a value lower than that measured for water ($\Theta \cong 0.046$) (BAGNOLD 1965). From this we obtain for the critical shear velocity:

$$u_r^* = 0.1 \sqrt{\frac{\varrho_2 - \varrho_1}{\varrho_1} gD} \tag{3.143}$$

For quartz spheres in air:

$$u_r^* = 147\sqrt{D} \tag{3.144}$$

According to equation (3.135), one obtains for the velocity, measured at height y, just necessary to mobilize grains of diameter D, assuming that $y_0 = D/30$:

$$\bar{u}_y^* = 0.575 \sqrt{\frac{\varrho_2 - \varrho_1}{\varrho_1} gD} \lg \frac{30\,y}{D} \tag{3.145}$$

For quartz spheres:

$$\bar{u}_y^* = 846 \sqrt{D} \lg \frac{30\,y}{D} \tag{3.146}$$

If the sand surface in some way has irregularities greater than the grain size of the sand (containing larger immobilized grains, rocks, or sand ripples) a correspondingly higher value must be substituted for D in the log term. The critical wind velocities, calculated from equation (3.146) for heights of 1 m and 10 cm are plotted in fig. 3-34.

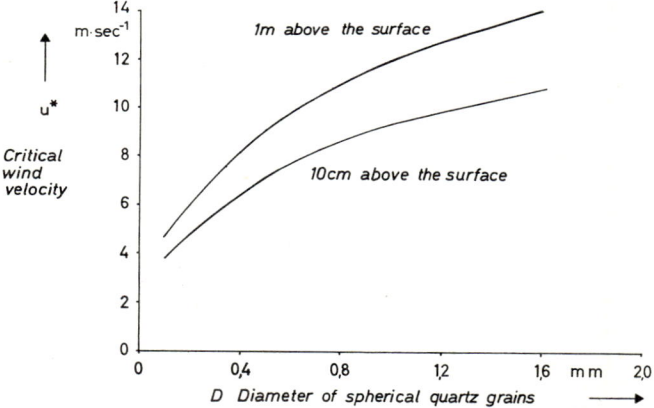

Fig. 3-34. Critical wind velocities (measured at 10 cm and 1 m heights above the surface) for bottom transport of quartz spheres.

As was the case for water flow also, the equations (3.142—3.146) apply only to sand grains above a certain grain size. In air this is about 0.01 cm, a somewhat smaller diameter than in water (0.02 cm). For smaller grains the critical shear stress

3.23 Bottom transport (bed load)

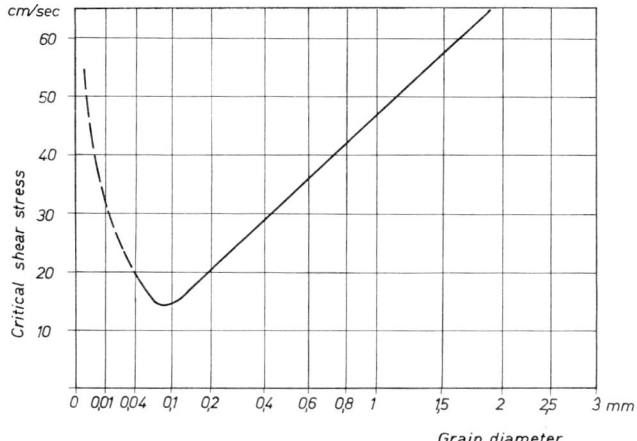

Fig. 3-35. Critical shear stress at the surface for bottom transport of quartz grains in air (BAGNOLD 1965).

rises greatly with decreasing grain diameter, as shown in fig. 3-35 (BAGNOLD 1965). It is interesting to compare this diagram with the corresponding one for bottom transport in water (fig. 3-12).

The movement of grains initiated at the bottom consists on the one hand of grain saltation, and on the other of a creeping of the uppermost surface layer. According to the studies by BAGNOLD (1965) the saltation take place in a manner quite different from the saltation of sand grains at the bottom of a water current. In a wind tunnel it can be observed that grains falling back to the bottom impact with such a high velocity that they either rebound elestically or produce small craters in the loose sand, from which one or two new grains are ejected into the air. In this way trajectory heights are attained which correspond in order of magnitude to thousands of grain diameters. In contrast grains in water must be lifted from the bottom through the water current itself. This leads to trajectory heights of only a few grain diameters. Grains returning to the bottom with low velocities can exert only slight mechanical effects. The reasons for these differences are the lower viscosity and specific gravity of air. A sand grain flying through the air has much greater energy than one moving in water. Once it acquires its higher velocity its energy is expended much more slowly due to friction than it is in water. Therefore, sand grains saltating in a wind current can attain a height where the flow velocity is much greater than at the surface. Upon falling they are able to maintain this high horizontal velocity, so that upon impact at the surface they impart a significant impulse through which new grains can be thrown upwards. In air the saltating grains are involved in transport from the zone of high velocity to the surface, while in water the flowing fluid immediately above the bottom must mobilize the bed load. In water bottom transport begins at the first threshold value Θ_1, increases with increasing current velocity, and becomes fully developed only when the shear stress at the bottom rises above the second threshold value Θ_2. This happens when the shear stress is sufficient to overcome the frictional resistance of the dense suspension of saltating grains at the bottom. In wind by contrast, fully developed bottom transport is achieved immediately as soon as the threshold value Θ is exceeded.

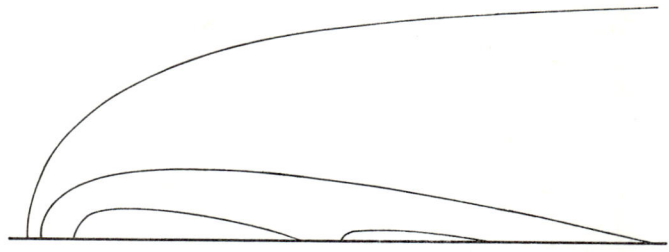

Fig. 3-36. Trajectories of saltating sand grains in wind currents (after BAGNOLD 1965).

In fig. 3-36 are shown typical trajectories of saltating sand grains, as recorded photographically by BAGNOLD (1965) in wind tunnel experiments. Individual trajectories differ in direction and in the force with which a grain impacts another. The mobilized grain rises upwards with a vertical initial velocity v_0 and would attain in vacuum the height $h = v_0^2/2\,g$. Actually the grain is slowed down by air resistance and attains a lesser height. In addition its trajectory is directed more and more in the wind direction, since it gets into wind layers of increasing velocity. At the highest point of the trajectory the vertical velocity is zero, the grain is moved at the prevailing wind velocity and then begins to fall, retaining, however, the horizontal component attained at the highest point of the trajectory. At the instant when the grain attains its maximum height, its vertical component of velocity equals the settling velocity v, and the horizontal component is equal essentially to the wind velocity u at the upper limit of the trajectory. The impact angle γ is given by the ratio of these two velocities:

$$\operatorname{tg} \gamma = v/u \qquad (3.147)$$

According to the observation of BAGNOLD, the descending path of the grain is always quite flat, creating the impression in a sand storm that grains are flying almost horizontally. The angle γ lies in the range between 10° and 16°. This indicates that the vertical and horizontal components of grain velocity have ratios between 0.18 and 0.29. That this ratio varies between such narrow limits is based on the fact that as the vertical path is diminished, and along with it the horizontal velocity, the vertical component during descent is also diminished. This is because the descent time available to attain a steady terminal velocity is short.

The trajectory heights of sand grains also vary under constant conditions and uniform grain sizes over wide limits. In a sand storm the maximum trajectory height is defined as the upper boundary of the moving sand cloud. The mean grain trajectory height is much lower. It is only about 1/10 that of the maximum. As a result the greater amount of surface transport results from saltating grains in a layer near the surface.

Trajectory heights depend, under given wind conditions, on the nature of the transport surface (BAGNOLD 1965). The coarser the grain size of the surface material, the less energy is lost by mutual movement of grains on the surface, and the higher into the air are saltating grains ejected. Over a surface of coarse gravel or solid rock grains saltate higher and farther than over a sandy surface.

Since the saltating grains extract energy from the wind, the wind velocity over the surface is diminished and the velocity profile modified as soon as surface transport sets in. This is elaborated schematically in fig. 3-37, which is based on the studies of BAGNOLD (1965).

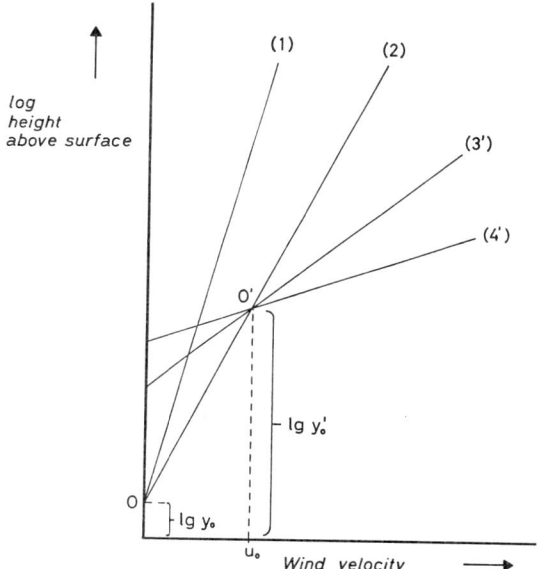

Fig. 3-37. Surface transport in wind (explanation in text).

Curve (1) corresponds to a shear velocity which lies below the critical value for bottom transport. Therefore, the simple equation (3.135) applies here. Curve (2) corresponds exactly to the threshold values for bottom transport. Once bottom transport is initiated, the velocity profile is defined by curves such as (3') or (4'), which may be expressed by an equation of the form

$$\bar{u}_y = 5.75\, u_\tau \lg \frac{y}{y_0'} + u_0 \tag{3.148}$$

These curves intersect at point O' rather than the origin (O); that is, in a wind laden with sand, at a certain height y_0' above the surface, the same wind velocity

(u_0) always prevails, independent of wind strength. According to observations made by BAGNOLD (1965) both in wind tunnels and in desert studies, y_0' is approximately equal to the height of ripples which form after initiation of bottom transport.

In fig. 3-38 the results of BAGNOLD's (1965) measurements of sand storms in the Libyan desert are reproduced. For a sand of mean grain size $D = 3.2 \cdot 10^{-2}$ cm, a critical shear velocity of 23 cmsec^{-1} was obtained (26 cmsec^{-1} according to equation 3.144); the values $y_0' = 1$ cm and $u_0 = 4$ msec^{-1} were likewise obtained.

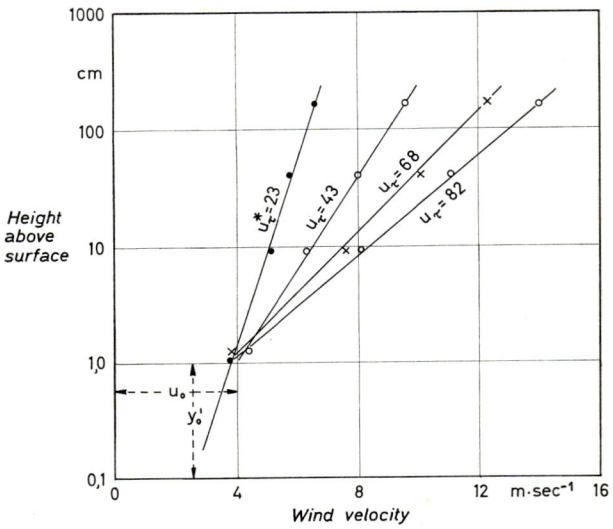

Fig. 3-38. Wind velocities in sand storms in the Libyan desert (after BAGNOLD 1965).

Saltating grains constitute the greater part but not all of the bed load of an air current. Saltating grains falling back on the surface utilize a portion of their energy, in that grains at the surface are put into motion by them. This induced movement leads to an overall slow creeping of a surface layer of grains in the wind direction. The energy for this transport is not supplied directly by the wind but rather from the impacts of saltating grains.

According to the observations of BAGNOLD (1965) this creeping sand makes up about one quarter of the total bottom transport. Particles too large to be moved directly by the wind are transported as a result of bombardment by descending high impact sand grains. For a sand consisting of different grain sizes the creeping grains will be coarser on the average than those transported by saltation. Since creep takes place very much slower (according to BAGNOLD 1965: creep is measured in fractions of cmsec^{-1}, saltation in one to several cmsec^{-1}), a sorting of material can result.

As soon as creeping and saltating bottom transport has set in, ripples begin to form on sandy surfaces. Their crests are oriented perpendicular to the wind direction.

They migrate slowly in the wind direction. Upon exceeding a certain wind velocity they disappear, and a flat surface is formed again.

Wind ripples are flatter than water-formed current ripples. As BAGNOLD (1965) showed, they form differently from water ripples. As soon as transport by saltation takes place, a planar sand surface is no longer stable. A sand surface may contain fortuitously a shallow depression. The leeward slope of the depression will be less impacted by grains descending at low angle than the windward slope. Therefore on the windward slope more grains are set into motion in the wind direction. As a result the depression is deepened and a sand accumulation is formed from the creeping sand. The depression has a new leeward slope, at the foot of which deepening is renewed. The majority of the saltating grains starting from the windward slope travel a distance equal to the mean trajectory distance. This results in a down wind accumulation of sand with a windward slope which in due course again becomes a source of saltating grains. From unevenly distributed surface irregularities there occurs gradually in this manner a system of ripples oriented perpendicular to the wind direction. Their distance of separation, as BAGNOLD (1965) showed, is equal to the mean trajectory distance of the grains.

The forward migration of ripples results from the mobilization of sand grains from the windward slope, partly by saltation and partly as a creeping surface layer. At the same time at the crest of the slope creeping sand rolls down the leeward slope and is deposited there. The upward growth of the crest of the ripple is limited by the effect of the higher wind velocity there. If the sand consists of different size grains, the coarser grains are concentrated at the crests, offering the greatest resistance to forward movement. Correspondingly more fine material is found in the troughs.

Since mean trajectory height and distance increase with wind velocity, distance between ripples grows with wind intensity. Correspondingly ripple spacing should, at equal wind intensity, be smaller, the greater the mean grain size.

3.24 Transport capacity of winds

In order to derive a relationship for the transport capacity of a wind current we can refer back to similar considerations for streams (section 3.15). Therefore equation (3.62) applies to the total energy of a wind per sec and unit surface area:

$$\omega = \tau_0 = U = \varrho_1 u_r^2 U \tag{3.149}$$

where τ_0 is the shear stress at the bottom, u_r the shear velocity, and U a mean wind velocity, which would be calculated from the total air mass moved (cm³ cm⁻² sec¹ = cmsec⁻¹). ϱ_1 corresponds to the specific gravity of air. The velocity U is proportional always to the shear velocity, so that

$$\omega = \text{const.} \, u_r^3 \tag{3.150}$$

For bottom transport a fraction of this energy is available, so that for the weight m_B of saltating and creeping grains, one obtains, per cm² and per sec:

$$m_B = K_B \cdot u_r^3 \tag{3.151}$$

The constant K_B depends on grain size and specific gravity of the mobilized grains, as well as the nature of the surface. For sands between 0.1 and 1 mm diameter, according to BAGNOLD (1965), K_B increases with the square root of the diameter. For quartz sands BAGNOLD experimentally found the following values:

	K_B
Approx. uniform sand	$1.2 \cdot 10^{-6} \sqrt{D}$
Dune sands	$1.4 \cdot 10^{-6} \sqrt{D}$
Sand with very broad grain size distribution	$2.2 \cdot 10^{-6} \sqrt{D}$

K_B is greater, the farther a grain saltates. Therefore it is greater for a gravel or solid rock surface than for a sandy one. The above values relate to a sandy surface.

For suspension transport m_s in g/cm² and sec is as follows:

$$m_S \sim \omega U/v = K_S u_r^4 / v \tag{3.152}$$

v is the settling velocity of the grain and K_s a constant. According to equation (3.148)

$$u_r = \frac{(\bar{u}_y - u_0)}{5.75 \lg y/y_0'} \tag{3.153}$$

For most actual cases y_0' can be assumed to be approximately equal to 1 cm. If the wind velocity is measured at 1 m above the surface, then

$$u_r = 0.087 (u_{100} - u_0) \tag{3.154}$$

Then instead of equations (3.151) and (3.152), we get the following for the wind velocity 1 m above the surface (u_{100})

$$m_B = 6.58 \cdot 10^{-4} K_B (u_{100} - u_0)^3 \tag{3.155}$$
$$m_S = 5.72 \cdot 10^{-5} K_S (u_{100} - u_0)^4 / v \tag{3.156}$$

Transport power of winds, therefore, changes markedly with wind velocity. The amount of material moved at the bottom (sand storms) or in suspension (dust storms) increases for a given grain size with the third and fourth power of wind velocity (reduced by u_0). It is presumed of course that the surface character does not change and that adequate material is supplied. The value of u_0 can be estimated from equation (3.135), if the critical shear velocity is substituted for u_r, y_0' for y and D/30 for y_0. The calculation of absolute values for m_s and m_B is of course not yet possible, since the constants K_B and K_S are not accurately known.

3.25 Mechanical effects of aeolian transport

In air as in water mechanical effects can be exerted only on particles moved as bed load. Wind-mobilized bed load consists of sand grains creeping and saltating at the surface. While saltating grains are elevated only a few grain diameters in water and return to the bottom with low velocity, grains in air saltate very much higher, so that they impact obstacles or the surface with a much higher kinetic energy. Thus it is to be expected that the mechanical effect of saltating grains is greater in air than in water.

Rock, stones, and other obstacles are attacked much in the manner of sand blasting, by the saltating grains of sand storms, which fly at a low angle to the surface. This results in characteristic erosion shapes and wind-carved pebbles (windkanter) known from desert and arid regions. We cannot elaborate on these here. The sand grains themselves are abraded when they strike such obstacles. Similarly descending grains impart high energy impact to grains they encounter lying on the surface. To a portion of these, this energy may be exchanged in elastic collisions, so that no mechanical effect is noted. For another portion the collision energy will alter the form and surface of the sand grains during surface transport.

It has been known for some time that dune sands, especially from desert regions look dull and opaque, while the grains of many river and beach sands are lustrous and appear to be smoothly polished. Immersion in a fluid with an index of refraction similar to quartz renders an apparently cloudy grain clearly transparent. This indicates that the phenomenon is related to a roughened surface.

CAILLEUX and co-workers (CAILLEUX 1942; CAILLEUX & TRICART 1959) attributed this sort of frosting of the surfaces of sand grains to abrasion experienced by grains during aeolian transport. The fine shattering of the surface should result from the collision of grains, such as occurs during wind transport. Abrasion during fluvial transport results from more of a grinding action, leading instead to smooth surfaces. Therefore, frosted quartz grains should serve as a criterion for aeolian sand transport.

This criterion is certainly not unequivocal, since frosted surfaces can result from other causes (KUENEN & PERDOK 1962). For example, it results from etching by solutions, or by the reverse process, the overgrowth of tiny, oriented quartz crystallites. Both processes occur in deserts where SiO_2 is mobile in solutions which occassionally occur there. This is indicated in some desert soils where quartz and opal precipitation occurs near the surface.

Abrasion of quartz, limestone, and feldspar has been studied experimentally in wind tunnels by KUENEN (1960). Abrasion increases with particle size, grain surface roughness, and with wind velocity. The lower limit for abradability of quartz grains in a wind tunnel according to these experiments lies at 0.05 mm. Smaller grains were not abraded. Under like conditions the abradability of quartz grains in air is 100 to 1000 times greater than over an equal distance in water. The mechanical attack of unrounded grains first affects the corners and edges and then proceeds to the faces, until the grain assumes a rounded shape. Therefore, the attrition per unit distance travelled decreases as roundness increases.

3.26 Deposition in air

Particles transported by the wind are deposited when the wind velocity and corresponding transport power decrease. From the bed load, transported as in sand storms, sandy deposits result. From the suspended load, carried as in dust storms, silty sediments occur.

Two kinds of sand accumulation derived from the bed load can be distinguished, according to conditions of formation: sand drifts and dunes.

Local sand drifts occur wherever obstacles (shrubs, rocks, or projections) lower the wind velocity. Geologically they play only a minor role. One can refer to BAGNOLD (1965) for further details.

Dunes are formed essentially unrelated to surface relief and more perfectly on quite flat surfaces. They are larger accumulations of sand which is piled up into individual ridges or mounds. They occur either individually or in groups and chains. Also dunes act as obstacles to the wind, and as such are molded by the wind. Therefore, they can migrate, grow, be broken up or completely destroyed under the influence of winds. They are on the one hand more or less transient transport bodies of sand moving by the wind on the surface; on the other hand they are large sand accumulations which are preserved occasionally as sandstones from the geologic past.

As shown in section 3.22, the velocity distribution of winds near the surface is influenced strongly by the uneveness of the surface. A change in the surface configuration can change greatly the transport power of near-surface winds. A sudden decrease in transport power can lead to local deposition of bed load which under certain conditions is formed into large dunes. The principles underlying the formation of such sand accumulation may be explained by two simple models (BAGNOLD 1965). In both cases a gravel strewn surface is assumed, on which a strip of sand is found whose surface is much smoother than the pebbly surface. This is true as long as the sand contains no ripples. In the first model the sandy strip runs perpendicular and in the second model parallel to the prevailing wind direction. It will be shown, that under certain conditions dunes develop from both models. The ridges of these dunes lie in the first case (transverse dunes) perpendicular and in the second case (longitudinal dunes) parallel to the wind direction.

In the first model the wind passing over the ground surface encounters the sand strip which is oriented perpendicular to the wind direction. The wind's velocity over gravel can be represented by the straight lines PA, PB & PC in fig. 3-39. The distance OP corresponds to the uneveness of the gravel surface. A very weak wind, designated by the velocity profile PA assumes the profile OA' above the sandy strip. OA' is parallel to PA because the shear velocity must remain constant; OA' passes approximately through the origin O, since the roughness of the sandy surface is very low. The wind velocity at all heights above the immobilized sand is greater that over the gravel surface. Upon leaving the sandy strip the velocity again decreases, corresponding to a translation of profile OA' to PA.

A somewhat stronger, but still gentle wind has the velocity profile PB over the gravel surface. Above the sand surface this wind would assume first a velocity profile

3.26 Deposition in air

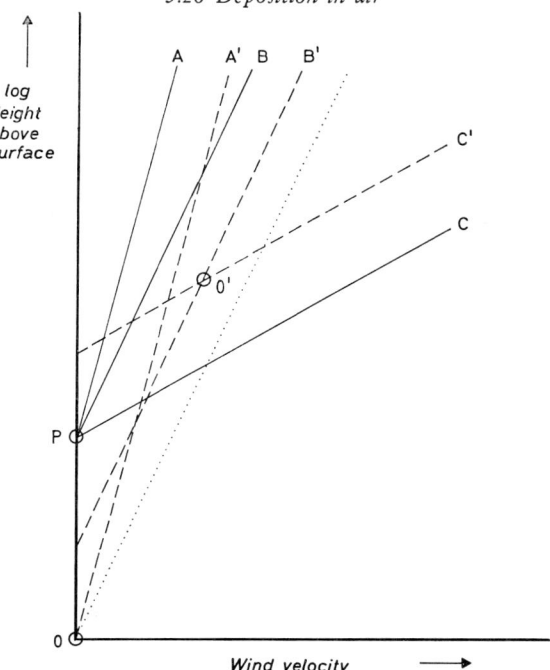

Fig. 3-39. Explanation of dune formation (see text).

parallel to PB, but passing through O. Such a wind will mobilize the grains at the surface of the sand, forming ripples. Then the velocity profile above the sand must pass through the point O', resulting in the profile O'B'. At all heights above the sand greater wind velocities prevail than above the gravel. Upon leaving the sandy strip the velocity distribution is displaced again from O'B' to PB, the velocities diminishing. Correspondingly deposition of the previously mobilized sand takes place on the sand-free surface beyond the sandy strip. Such a wind results in sand migration down wind.

A stronger wind is represented by the velocity distribution PC above the gravel. Above the sand surface the distribution O'C' is assumed, resulting in a velocity decrease at all heights above the ground. If this wind is carrying sand along with it, sand accumulates at the edge of the sandy strip forming a dune at this point. Upon leaving the sandy strip the wind velocity again increases, corresponding to a displacement of the profile from O'C' to PC. Grains mobilized on the sandy strip, if they are still a part of the bed load, cannot be deposited on the gravel surface beyond the sandy strip, but are carried away.

In the second model the sandy strip runs parallel to the wind direction. A gentle wind which does not move sand has a higher velocity above the sand surface than over the adjacent surface. At the border between the sandy strip and its surroundings, eddies occur, because of the change in wind velocity. These are directed so that they have tendency to distribute the sand laterally on the surrounding ground surface. A stronger wind, interacting with grains in the sandy strip, has the opposite effect, since its velocity above the sandy strip is less than above the surrounding area. The eddies at the borders of the two currents are then directed so that sand passing over the surrounding area is directed toward the sandy strip, where it is sedimented in part in the slower wind current prevailing there. In this way dunes form which run parallel to the wind direction.

The cross section through a completely formed transverse dune exhibits a flatter (5—12°) windward slope and a steeper (30—34°) leeward slope. The wind velocity and its corresponding shear stress acting on the dune surface increases on the windward side to a point beneath the dune ridge, and then decreases again over the leeward side. Relative to their effects one must distinguish again between "gentle" and "strong" winds. The velocity of a gentle wind is higher over the dune surface than over the surrounding uneven surface. On the windward slope of the dune sand is mobilized, and is deposited again on the leeward slope. A strong wind produces such extensive sand movement on the dune surface that its velocity is less above the dune than above its surroundings. If such a wind carries no sand along with it, sand is removed from the windward side and is deposited in part again on the leeward side. Such a dune migrates down wind, but becomes constantly smaller, since in passing on to the area beyond the dune, the wind picks up velocity and carries part of the sand away. However, if the wind is carrying sand with it, deposition takes place at the foot of the windward slope, where its velocity is diminished. Such a dune can migrate, growing in volume, if the deposition on the windward side exceeds the amount removed on the leeward side.

Sand deposition on the leeward slope produces crossbedding there as a depositional structure and leads to a gradual increase in the leeward slope angle. If the angle of repose (approx. 34°) of sand is reached, sudden slumping of the sand takes place along a less steeply inclined shear surface. Then renewed deposition of sand falling over the dune crest sets in, until the critical angle of repose is exceeded again and slumping happens again. Within a transverse dune are found both crossbedding, which originates from deposition on the leeward slope, and chaotic structures, which arise from gravity sliding. Crossbedding is observed as a rule predominantly in the upper parts of dunes, since slumping, which destroys crossbedding, usually does not take place at the dune crest, but only somewhat lower down.

Transverse dunes generally do not extend any great distance perpendicular to the wind direction. They occur mostly as isolated bodies. Especially characteristic are the barchan or crescent dunes, whose crescent-like shapes result from the more rapid migration of the lower flanks of the dune relative to the main mass. Barchans always occur in clusters or groups.

On longitudinal dunes sedimentation and erosion are controlled not so much by the prevailing winds as they are by the occasional cross winds. Each cross wind, depending on its direction, changes the leeward slope from one side to the other. In this case also deposition leads to a steepening of slope, resulting in slumping. The interiors of longitudinal dunes likewise exhibit crossbedding as well as relatively structureless parts, which are related to slumping.

The height of longitudinal dunes can be considerable. BAGNOLD (1965) reported on 210 m high dunes of this kind in south Iran and on up to 100 m high dunes in Egypt. Along the crests were observed repetitions of individual high points whose separations varied between 20 and 500 m. Chains of longitudinal dunes extend in straight lines over considerable distances. The longest and highest longitudinal dunes occur on quite flat and structureless ground surfaces. They can be followed without interruption for 60 to 100 km.

The dimensions of barchans vary over wide limits. While the maximum height is about 30 m, maximum length and breadth of such dunes lie at about 400 m.

Material deposited from the suspension load of dust storms is called loess. Periglacial and desert loess can be distinguished. Periglacial loess develops through the action of wind on vegetation-free soils of the Arctic and on glacial moraines, formed especially during the Pleistocene. Desert loess consists of fine material blown about from desert regions.

The mean grain size of loess lies between 0.01 and 0.06 mm, falling in the silt range. As was indicated in fig. 3-35, silt grains are smaller than the grain size (0.08 mm) which can be mobilized by the minimum wind velocity. In the silt size range, in contrast to sand, the wind velocity necessary to lift up a grain lying on the surface increases with decreasing diameter. A once-deposited loess particle is mobilized again with more difficulty, the smaller it is.

Silt particles are produced by the grinding action of glaciers and are deposited along with the coarser detritus in moraines, especially ground moraines. If these moraines are without vegetation and exposed to the action of severe storms, sand and silt particles are mobilized. As is shown in fig. 3-33 the grain size range for loess corresponds to the grain sizes transported at all wind velocities as supended matter; in contrast dune sands are moved along predominantly at the surface as saltating grains in sand storms. Where the wind becomes weaker, it deposits bed and suspended load together. Accumulation of silt cannot occur as long as saltating sand repeatedly stirs up the bottom deposits. Therefore, sand dunes are quite free of finer components. Deposition of silt is only possible when the wind has diminished to the extent that sand transport ceases. Then there often forms downwind from the dune zone a very widely distributed loess cover, which is very stable against renewed wind erosion. The thickness of the loess cover, that is, the amount deposited in a given time, decreases, as does the grain size in the wind direction. In central Illinois the loess thickness decreases regularly over a distance of about 100 km from over 20 m to about 5 m. Simultaneously the mean grain size decreases regularly from above 0.03 mm to about 0.015 mm (SMITH, after KRUMBEIN & SLOSS 1963).

3.3 Distinguishing aeolian and aquatic clastic sediments

Transport mechanisms in wind and water currents are, on account of their distinctive densities and viscosities, so different that it should be possible from a clastic sediment to ascertain whether it acquired its sorting and the derivation of its constituents by aquatic or aeolian transport.

Sediments derived from suspended load of water currents consist of particles which were distributed somewhat uniformly over the cross section of the stream ($<$ 0.05 mm, clays and fine silt) in addition to grains which were transported preferentially near the bottom (0.05—0.2 mm, coarse silt, fine sand) — see fig. 3-13. Clays ($<$ 0.002 mm) are especially widely distributed as deposits of aquatic suspension load. In contrast only coarser grain loess is known as distinctive sediments derived from aeolian suspension load. It consists mainly of grains between 0.06 and 0.01 mm. According to fig. 3-32 these grains are transported near the surface. Most loess contains also grains below 0.01 mm; the more so the farther they are deposited from their source. Purely aeolian sediments are lacking, which represent deposition of grains uniformly distributed over the air stream; that is, there are no aeolian clays.

Clay dunes, such as occur in Australia, Texas, Algeria, and at the mouth of the Volga, are not derived from the suspended load of dust storms. They form as follows: Thin clay crusts form during the drying up of salt lakes and saline lagoons. These are broken up into small sand size particles, which are seized by the wind and gathered into dunes. In the dunes these particles meld together into a more or less homogeneous and porous clay mass which can no longer be moved by the wind (PRICE 1963).

Neglecting the very fine-grained deposits of volcanic tuffs, aeolian clay deposits do not form because the finest material in dust storms is elevated to very great heights and is then distributed over very wide distances. In the "sea of air" no narrowly defined "basins" exist in which turbulence suddenly ceases. Also there is no flocculation which plays such an important role in clay deposition in the marine shelf regions. Therefore, the finest fraction of aeolian suspension load forms no individual deposits. It becomes a quantitatively minor component of different aquatic sediments.

Bottom transport is initiated in water at grain sizes above 0.2 to 0.5 mm, and in air above 0.05 to 0.3 mm. Correspondingly dune sands are on the average somewhat finer grained (0.1—0.5 mm) than aquatic sands ($>$ 0.2 mm). The grain size ranges for aeolian and aquatic sand overlap, so that wind or water transport cannot be concluded from grain size alone.

On the other hand under certain circumstances the transport mechanism can be deduced from the shape of the size distribution curve. FRIEDMAN (1961) investigated the grain size distributions of 267 dune, river, and beach sands. It turned out that the skewness of grain size distribution in Φ-units is always positive for dune and river sands (excess of fines), and that it is negative for beach sands (excess of coarser particles). This difference results from the fact that in air and water currents sediment derived from the bed load is admixed usually with somewhat finer material from the suspended load. This fine material is deposited in the pore space between coarser grains and cannot be mobilized again. In contrast wave action at the coasts winnows out the finer material, so that a symmetrical distribution or an excess of coarser components results.

In this way it is possible in certain instances to distinguish dune and beach sands.

3.3 Distinguishing aeolian and aquatic clastic sediments

Certain differentiation of all aquatic and aeolian sediments is not possible from the grain size distribution.

Other possibilities arise from comparing the grain size distributions of the light and heavy components of sands. In section 3.181 the relationship was derived which relates specific gravity and size of different heavy mineral grains which are sedimented together. For deposition from the bed load the following expression applies:

$$D_{ar} : D_{br} = (\varrho_b - \varrho_0) : (\varrho_a - \varrho_0) \qquad (3.157)$$

D_{ar} and D_{br} are the diameters of grains with specific gravities ϱ_a and ϱ_b. ϱ_0 is the density of the medium (air or water). For grains sedimented together from the suspension load, the STOKES formula applies:

$$D_{as} : D_{bs} = \sqrt{(\varrho_b - \varrho_0) : (\varrho_a - \varrho_0)} \qquad (3.158)$$

With increasing grain diameter this ratio changes until finally for quartz grains above 2.6 mm the same relation applies as for bottom transport.

Since in the above formulae the specific weight of air can be disregarded, while for water $\varrho_0 = 1$, the respective ratios have different values for aquatic and aeolian transport. From the grain size distributions of the light and heavy minerals in a sand it should be possible to deduce whether it was formed in air or water, assuming that in the material supplied heavy minerals of the corresponding grain size classes were present in sufficient amount.

It has been shown that the grain size distributions of light and heavy constituents of sands do in many cases actually correspond to the derived relations (v. ENGELHARDT 1940 a; RITTENHOUSE 1943; WALGER 1966).

The size relations of different minerals according to formulae (3.157) and (3.158 are given in table 3-22.

The size difference between co-sedimented quartz and heavy mineral grains, therefore, should ordinarily be greater in aquatic than in aeolian sediments.

Table 3-22. Size ratios for minerals of different specific gravity in aquatic and aeolian sands.

Heavy mineral	Specific gravity	$D_{quartz} : D_{heavy\ mineral}$			
		aeolian sediment		aquatic sediment	
		Bottom transport	Suspension transport in STOKES range	Bottom transport	Suspension transport in STOKES range
Magnetite	5.2	1.96	1.40	2.55	1.60
Pyrite	5.0	1.89	1.38	2.43	1.56
Garnet	4.2	1.59	1.26	1.94	1.39
Zircon	4.6	1.74	1.32	2.18	1.48
Rutile	4.25	1.61	1.27	1.97	1.40
Tourmaline	3.15	1.19	1.09	1.30	1.14

Conclusions regarding the transport medium, based on shape and surface character of the constituents of a sediment are generally not unequivocal. As shown above, the mechanical effects of grains flying in a sand storm are much greater than of those moving at the bottom of a stream of water. Wind carved pebbles are always a reliable index for the aeolian environment. Aeolian sands often have frosted grains. Still the frosting, as indicated above (section 3.25), is not always a clear sign of an aeolian sediment. Dune sands as a rule are well and better rounded than aquatic sands. However, well-rounded sand grains form also on beaches. Therefore rounding alone does not distinguish between aeolian and aquatic sands.

3.4 Transport and deposition by ice

The flow of a glacier is a complicated process which depends on the translation capability or plasticity of ice crystals as well as the slippage of large ice masses along glide planes. For the latter process partial melting of ice under pressure plays a role. If glacial flow were restricted to the plasticity of ice alone, the relation between velocity gradient and effective shear stress would be the same as depicted in fig. 3-21. As for all plastic bodies, a certain minimum stress, the so-called flow limit, must be exceeded before flow takes place.

Ignoring a flow limit, that is, assuming that ice flows like a normal fluid, different authors have deduced viscosities between 10^{10} and 10^{15} poises from the flow velocity of ice at temperatures between 0° and −30 °C at certain shear stresses (DRYGALSKI & MACHATSCHEK 1942, p. 76). The basis for the large range in values is not understood fully. In any case the apparent viscosity of ice is many orders of magnitude greater than for water or air. Therefore, rocks and solid particles of all kinds settle very slowly in ice. Solid constituents of less than a certain mass should not be able to settle in ice as long as the shear stress produced is less than the flow limit for ice.

On account of the viscosity and flow limit the transport mechanism of ice is quite different from that of air and water. All material transported by a glacier is found essentially either at the base of the glacier (ground moraine) or on its surface (superficial moraine).

Superficial moraine consists of debris, which falls upon the glacier from rock slides, land slides, mud flows, rock falls or avalanches, or in some cases from torrential streams from higher terranes. Since this material cannot sink into the ice the amount of debris transported on the glacial surface increases downstream, when rubble is supplied continuously from the slide slopes. The constituents of the superficial moraine experience no mechanical reworking or sorting during transport, if they are not exposed occasionally to melt water. Very large blocks are mixed up with fine material and all fragments or grains maintain essentially the same shape they had when first dumped on the glacier: No polishing or scratching or surfaces and no rounding takes place.

Ground moraine forms a mass of debris more or less permeated by ice in the boundary region between the rock surface and pure glacial ice. It consists of all grain size classes ranging from finest clay to very large blocks. Giant blocks, such as occur in the superficial moraine, seem usually to be lacking in ground moraine. Therefore, ground moraine is richer in the finest materials, which are formed anew by the slow movement of the glacier. The constituents of ground moraine are in certain instances subject to quite high pressures at the base of the glacier (20 to 40 kg/cm^2 in Alpine glaciers; 70 to 80 kg/cm^2 in Greenland glaciers). In addition shear forces are active as a result of the glacial movement. Least resistant constituents are crushed to fine sand or clay. Harder rocks are rounded, polished, or scratched, with the finer material acting as a polishing agent. The striated or scratched boulders which form in this way in the ground moraine are characteristic. The abrading action of the moving ground moraine makes an impression also on the bed rock. A polished and smoothed bedrock with numerous striations and coarse grooves occurs, as does glacial fluting especially typical of glacial erosion.

The amount of fine grain material in the ground moraine increases with the size and distance of advance of the glacier. Simultaneously the less resistant rocks are eliminated gradually, and a decrease in average boulder size takes place. This can be recognized only in the moraine deposits of the far reaching Pleistocene glaciers.

As the flow velocity slackens, the glacier can no longer dispose of the debris of the ground moraine in the same amount as it is supplied. Where the gradient is decreased the debris piles up at the base of the glacier and the glacier bed is elevated correspondingly; the glacier "swims" on its own debris. In this way thick deposits of ground moraine can form.

As soon as a glacier leaves a region that it has covered for a long time, the debris carried along by it remains behind. In this discarded material the angular constituents of the superficial moraine should lie above and on the mechanically severely stressed constituents of the ground moraine: Frequently both components are chaotically mixed together. Insofar as no later action of flowing water takes place, moraine deposits exhibit neither bedding nor sorting: Large blocks lie immediately adjacent to fine mud.

For the most important ice transport at the glacier's base, it is characteristic that much fine clay and silt is produced by polishing and scratching stresses. This forms the fine turbidity of glacial streams (glacial milk), which may be deposited as lacustrine banded clays or as marine varve clays. Also periglacial loess stems from the fine material produced by the action of ice.

4. Formation of Chemical Sediments

4.1 General

Those chemical constituents released during chemical weathering, and which are not fixed by neoformation or crystallization in soils and in recently deposited clastic sediments, are transported in solution by streams. They collect to a small extent in continental lakes and then primarily in the ocean. In addition to the relatively soluble substances supplied by weathering, lake and sea waters contain also volatile and soluble components which stem from the outgasing of the earth and from the atmosphere. In addition sea water contains some constituents residual from the primordal atmosphere of the earth.

Increases in concentration of these solutions by evaporation or some other process leading to supersaturation, in terms of the solubility product, result in the chemical precipitation of minerals. Collectively the result is the formation of chemical sediments, regardless of whether the precipitation has taken place inorganically or by the action of organisms.

The nature of chemical sediments, and their place of formation and occurrence have been described in the second volume of this series. The following presentation will be limited to equilibria and processes of formation for those minerals which make up the main constituents of chemical sediments. Mineral equilibria of some additional phases in aqueous solution have been considered already in section 2.33 (Al- and Fe-oxides and hydroxides, SiO_2, phyllosilicates).

The quantitatively most important chemical sediments must consist of those constituents which are supplied continuously from weathering solutions. Carbonate and sulfate are the predominant anions in river water; calcium is the main cation (table 2-4). Therefore, the relatively insoluble carbonates and sulfates of calcium, that is, calcite, aragonite, dolomite, gypsum and anhydrite, are the main constituents of chemical sediments. The formation of these minerals from aqueous solution will be considered in some detail in the following section. Since crystallization and dissolution of these phases play a role also in soils and especially during diagenesis, this information will have wide applicability.

Because the vast majority of chemical sediments originate in the great oceanic collecting basins, factors affecting precipitation from sea water are especially important. As will be shown, sea water is saturated or nearly so with respect to calcium carbonate and dolomite. Precipitation of calcium carbonate does not require marked increase in concentration by evaporation. Minor changes in temperature and CO_2-pressure are all that are required. Dolomite formation is inhibited for reasons based on reaction kinetics. It does not form by direct precipitation, but can form only by the interaction of more or less concentrated sea water with freshly formed calcium carbonate sediments.

All other precipitation involving the major components of sea water can result only when concentration is raised considerably by evaporation. The composition of sea water at the point when gypsum, halite, and K-Mg-salts begin to crystallize is indicated in table 4-1 (after Braitsch 1962, p. 66).

Table 4-1. Compositional changes in sea water during isothermal evaporation (25 °C) and when precipitation of gypsum, halite, and K-Mg-salts is initiated (after Braitsch 1962). (Moles / x Moles H_2O).

	Sea water	Initial precipitation of		
		gypsum	halite	K-Mg-salts
H_2O (x)	70700	21100	1450	1170
K_2	6.4	6.4	6.4	6.4
Mg	69.0	69.0	69.0	69.0
SO_4	24.6	24.6	24.6	24.6
Na_2Cl_2	303	303	303	26
$CaCO_3$	1.5	—	—	—
$CaSO_4$	11.7	11.7	1.8	—
Weight of solution	1000 g	322.4 g	121.3 g	24.46 g

Calcium sulfate precipitates first as gypsum (25 °C) when the amount of water is reduced by evaporation by a factor of 3.4. Precipitation of halite is initiated when the water content of sea water is reduced by a factor of 11. Only when the water content is reduced below about 1/60 its original amount does the precipitation of the relatively soluble K-Mg-salts begin. Therefore, rather extreme conditions are necessary for marine evaporites to occur. Such conditions are realized only seldom, and then only within narrowly bounded parts of basins. As a result evaporite sediments form only a minor portion of all marine sediments.

The crystallization sequence of the very soluble salts of sea water, because of the many possible phases which can form, is dependent in complex ways on the prevailing conditions of formation. The reader is referred to the book by Braitsch (1962) for an explanation of the equilibria which apply and their use in predicting the evaporation of sea water under natural conditions. A prolific previous literature is cited also. Here only some general remarks will be added.

The salt series established by natural observation deviates for a number of reasons from the sequence observed when in the laboratory one isothermaly evaporates sea water. The following factors can play a role.

First of all, most natural salt series are not the result of evaporation of normal sea water. As the absence of all but small amounts of anticipated magnesium salts shows in many cases, precipitation of mineral salts takes place from a magnesium depleted sea water. The cause of this magnesium deficiency is the early diagenetic formation of dolomite by reaction of sea water with early formed calcium carbonate sediment.

Secondly, the natural salt sequence results by evaporation from a solution whose composition and concentration was changed continuously by numerous influxes of fresh sea water.

Thirdly, the evaporation need not have taken place at a constant temperature.

Fourthly, the precipitated salts have not had the opportunity always upon continuing evaporation to react with the concentrated solutions, and in addition did not always precipitate the stable phase, but rather metastable crystalline species.

Fifthly, subsequently the early crystallized salts have been considerably altered by diagenetic or metamorphic transformations by the interaction of supervening solutions.

4.2 Limestone

The ions Ca^{+2}, CO_3^{-2}, HCO_3^-, H^+, OH^- and the partial pressure of CO_2 determine the equilibrium between solid calcium carbonate and aqueous solution in the presence of a gas phase. CO_2 is dissolved in water as H_2CO_3. The following equilibria define the ion and carbonic acid concentrations according to the law of mass action:

$$[Ca^{++}] \cdot [CO_3^-] = K_a \qquad (4.1)$$

$$\frac{[H^+][HCO_3^-]}{[H_2CO_3]} = K_b \qquad (4.2)$$

$$\frac{[H^+][CO_3^{--}]}{[HCO_3^-]} = K_c \qquad (4.3)$$

$$[H^+][OH^-] = K_d = 10^{-14} \qquad (4.4)$$

$$\frac{[H_2CO_3]}{P_{CO_2}} = K_e \qquad (4.5)$$

P_{CO_2} is the partial pressure of CO_2 in the gas phase with which the solution is in equilibrium.

The above equations are applicable strictly only for activities a_{Ca}, a_{CO_3}, etc. of the components. The activity of the hydrogen ion is determined by pH measurement. The analytically determined concentrations of the other ions m_{Ca}, m_{CO_3}, etc. are equivalent essentially to activities only in very dilute solutions. In concentrated solutions the analytical concentrations are higher than the activities. In general the following applies:

$$a_i = \gamma_i \, m_i \qquad (4.6)$$

The activity coefficient γ is equal to 1 in infinitely diluted solutions and less than 1 when ions are dissolved. As a measure of electrolyte content of a solution the so-called ionic strength is used, which is defined as follows:

$$I = \tfrac{1}{2} \sum m_i \, z_i^2$$

m_i is the concentration of ion species (i) expressed in mol/l; z_i is the ionic charge. The expression is summed over all positive and negative ions present in the solution. In fig. 4-1 the activity coefficients for the ionic species most important in calcium carbonate equilibrium are plotted versus ionic strengths.

The constants K_a, K_b, K_c, and K_e are temperature dependent. Values for temperatures between 0° and 80° are given in table 4-2.

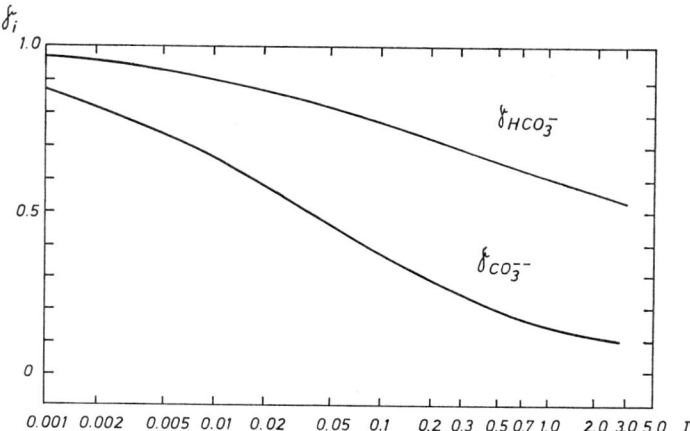

Fig. 4-1a. HCO_3^- and CO_3^{--} activity coefficients (γ) and its dependence on ionic strength (I) of a solution (GARRELS & CHRIST 1965).

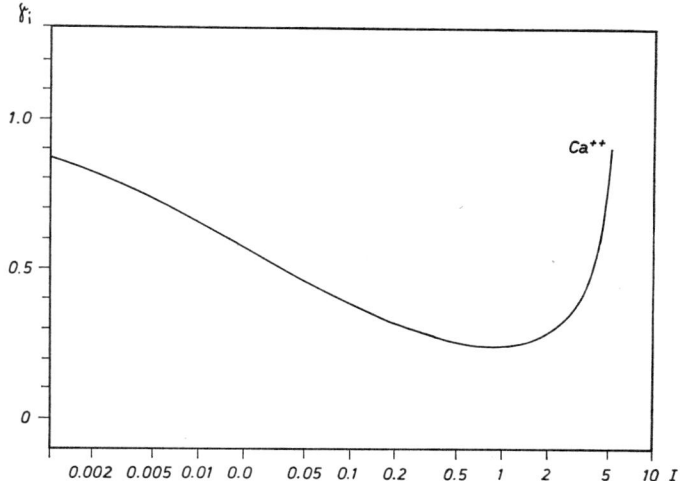

Fig. 4-1b. Ca^{++} activity coefficient (γ) and its dependence on the ionic strength (I) of a solution (GARRELS & CHRIST 1965).

The simplest system is one in which no gas phase is present, and in which calcite is dissolved in pure water. Then equation (4.5) above can be ignored. To calculate the activities of 6 ions, the four equations (4.1—4.4) are pertinent. Two additional equations relate to the stipulation that the sum of all positive charges must be equal to the sum of all negative charges and that all carbonate constituents found in solution stem from the dissolution of calcite. Therefore:

$$2\, m_{Ca^{++}} + m_{H^+} = 2\, m_{CO_3^-} + m_{HCO_2^-} + m_{OH^-} \quad (4.7)$$

$$m_{Ca^{++}} = m_{CO_3^-} + m_{HCO_2^-} + m_{H_2CO_3} \quad (4.8)$$

If we equate concentrations with activities, which is valid for very dilute solutions, one obtains the following concentrations, using the numerical values for equilibrium constants at 20 °C from table 4-2, by solving equations (4.1—4.4) and (4.7—4.8):

Table 4-2. Equilibrium constants for the system $CaCO_3$—CO_2—H_2O (from GARRELS & CHRIST 1965, after measurements of different authors).

Temperature °C	$-\lg K_a$	$-\lg K_b$	$-\lg K_c$	$-\lg K_e$
0	8.02	6.58	10.62	1.12
5	8.09	6.52	10.56	
10	8.15	6.47	10.49	
15	8.22	6.42	10.42	
20	8.28	6.38	10.38	
25	8.34	6.35	10.33	1.47
30	8.40	6.33	10.29	
40	8.52	6.30	10.22	1.64
50	8.63	6.29	10.17	
80	8.98	(6.32)	(10.12)	

Calcite in equilibrium with pure water, $P_{CO_2} = 0$, (25 °C):

$$m_{Ca^{++}} = 10^{-3.9}$$
$$m_{CO_3^-} = 10^{-4.4}$$
$$m_{HCO_2^-} = 10^{-4.05}$$
$$m_{H_2CO_3} = 10^{-7.6}$$
$$m_{H^+} = 10^{-9.95}; \quad pH = 9.95$$

At equilibrium with pure water without CO_2 and at 25 °C, only $1.3 \cdot 10^{-4}$ moles = 13 mg/l of $CaCO_3$ is dissolved.

If a gas phase is present, in which a given CO_2-partial-pressure prevails, the amount of $CaCO_3$ dissolved at equilibrium is greater. The activities of ions in solution are determined by solving equations (4.1—4.5) and (4.7). By substitution one obtains the following relation between CO_2-partial-pressure and hydrogen ion concentration:

$$P^2_{CO_2} + \frac{10^{-14}[H^+] - [H^+]^3 P_{CO_2}}{2 K_b K_c K_e + K_b K_e [H^+]} - \frac{2 K_a [H^+]^4}{K_b K_c K_e (2 K_b K_c K_e + K_b K_e [H^+])} = 0 \quad (4.9)$$

4.2 Limestone

or
$$P_{CO_2} = -\frac{A}{2} + \sqrt{C + \frac{A^2}{4}} \qquad (4.10)$$

noting that
$$A = \frac{L}{2M + N[H^+]}; \qquad C = \frac{O[H^+]^4}{M(2M + N[H^+])}$$
$$L = 10^{-14}[H^+] - [H^+]^3; \quad M = K_b K_c K_e; \quad N = K_b K_e; \quad O = 2K_a.$$

By these equations the hydrogen ion activity which prevails in a $CaCO_3$-saturated solution in equilibrium with CO_2 of a given partial pressure is determined. This relation between CO_2-pressure of the gas phase and hydrogen ion concentration in solution is plotted in fig. 4-2.

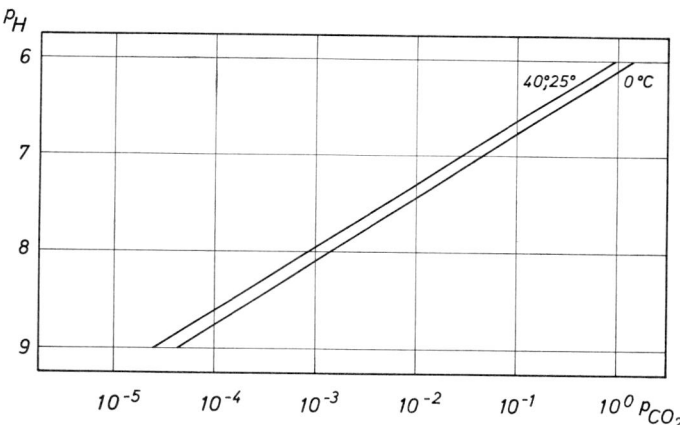

Fig. 4-2. pH in $CaCO_3$-solutions as a function of CO_2-partial-pressure of the gas phase.

By substituting the hydrogen ion activity for a given partial pressure and the corresponding dissolved H_2CO_3-concentration, the activities of Ca^{+2}, CO_3^{-2}, and HCO_2^- in saturated solution can be calculated in relation to CO_2-partial-pressure. The activity of dissolved Ca (or $CaCO_3$) is of special interest. In fig. 4-3 the Ca-activity calculated in this way is plotted in relation to pH at CO_2-partial-pressures in the range from below 10^{-4} to 1, at 0°, 25°, and 40 °C.

For Ca-ion activity it follows from the above equations:

$$[Ca^{++}] = \frac{K_a[H^+]^2}{M P_{CO_2}} \qquad (4.11)$$

The lengthy calculations can be simplified significantly for CO_2-partial-pressures above 10^{-3}, because individual terms become so small that they can be ignored without significant error. It follows:

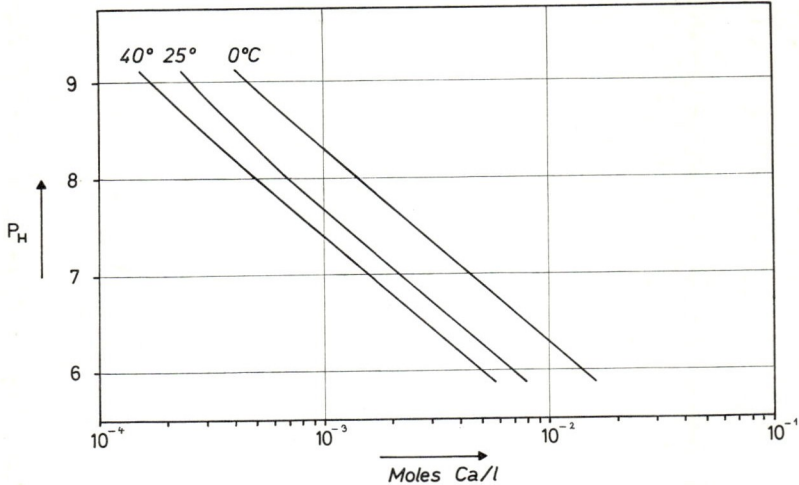

Fig. 4-3. Ca^{++}-ion activity (mole/l) in $CaCO_3$-solutions as a function of pH.

For $P_{CO_2} > 10^{-3}$

$$C = \frac{O[H^+]^3}{MN}; \qquad P_{CO_2} = \sqrt{C}.$$

A simple equation results for Ca-ion activity:

$$[Ca^{++}] = \left(\frac{K_a K_b K_e P_{CO_2}}{4 K_c}\right)^{1/3} \tag{4.12}$$

and for hydrogen ion activity:

$$[H^+] = \left(\frac{K_b^2 K_c K_e^2 P_{CO_2}^2}{2 K_a}\right)^{1/3} \tag{4.13}$$

If Ca^{+2} and H^+ activities are determined from these equations, activities for the carbonate species can be calculated by substitution in equations (4.2—4.5). An example may illuminate this: In question is the composition of a saturated solution in equilibrium with a gas phase in which $P_{CO_2} = 10^{-2.00}$. The temperature is 25 °C. Then the equilibrium constants are:

$$K_a = 10^{-8.34}, \; K_b = 10^{-6.35}, \; K_c = 10^{-10.33}, \; K_e = 10^{-1.47}$$

According to equation (4.12) and (4.13) we obtain

$$[Ca^{++}] = 10^{-2.41}$$
$$[H^+] = 10^{-7.11}$$

Substituting in equations (4.2, 4.3, 4.5) gives:

$$[CO_3^{--}] = 10^{-5.93}; \; [HCO_3^-] = 10^{-2.71}; \; [H_2CO_3] = 10^{-3.47}.$$

4.2 Limestone

The activities of the carbonate constituents can be calculated of course directly from the constants if the CO_2-partial-pressure is given. Then the following equations apply (when $P_{CO_2} > 10^{-3}$):

$$[CO_3^{--}] = \left(\frac{4 K_a^2 K_c}{K_b K_e P_{CO_2}}\right)^{1/3} \tag{4.14}$$

$$[HCO_3^{-}] = \left(\frac{2 K_a K_b K_e P_{CO_2}}{K_c}\right)^{1/3} \tag{4.15}$$

$$[H_2CO_3] = K_e P_{CO_2} \tag{4.16}$$

At CO_2-partial-pressures below 10^{-2}, that is, for solutions containing less than about $1.5 \cdot 10^{-3}$ mol/l = 150 mg $CaCO_3$/l, activities and concentrations can be considered equal. In this concentration range fig. 4-3 gives directly the solubility of $CaCO_3$ at different pH.

For sediment formation the solubility at a CO_2-pressure corresponding to a normal atmosphere is especially important ($P_{CO_2} = 10^{-3.5} = 0.0003$). The following values can be taken from figs. 4.2 and 4.3:

Calcite in equilibrium with pure water and normal atmosphere ($P_{CO_2} = 10^{-3.5}$):

		0°	25°	40°
m_{CaCO_3}	Mol/l	$8.8 \cdot 10^{-4}$	$5.0 \cdot 10^{-4}$	$3.6 \cdot 10^{-4}$
	mg/l	88	50	36

At CO_2-partial-pressures above 10^{-2} (pH < 7.3) solution concentrations are so high that activity coefficients are noticeably less than 1. The amount of $CaCO_3$ dissolved is actually greater than indicated in fig. 4.3 and is derived from activities calculated by the simplified equations above. If one inserts the activity coefficients for individual species (equation 4.6) in the above equations, the following formulae are obtained for calculating the analytically ascertained solution concentrations:

$$m_{Ca} = \left(\frac{K_a K_b K_e P_{CO_2}}{4 K_c} \cdot \frac{1}{\gamma_{Ca}\gamma_{HCO_3}^2}\right)^{1/3} \tag{4.17}$$

$$m_{CO_3} = \left(\frac{4 K_a^2 K_c}{K_b K_e P_{CO_2}} \cdot \frac{\gamma_{HCO_3}}{\gamma_{Ca}^2 \gamma_{CO_3}^3}\right)^{1/3} \tag{4.18}$$

$$m_{H_2CO_3} = K_e P_{CO_2} \cdot \frac{1}{\gamma_{H_2CO_3}} \tag{4.19}$$

If concentrations are to be calculated with these equations, activity coefficients must be calculated or estimated. For this the ionic strength of the solution must be known; but this depends in turn on the concentrations to be determined. The calculation must be carried out, therefore, by a series of successive approximations.

The empirically determined solubility of calcite, carried out at 25 °C and different CO_2-partial-pressures by different investigators, is plotted in fig. 4-4 (after GMELIN 1961). With increasing CO_2-pressure the solubility becomes increasingly larger than the activity. In equilibrium with pure CO_2 ($P_{CO_2} = 1$) analytically determined $CaCO_3$ in solution is greater by a factor of 1.4 than the calculated equilibrium activity.

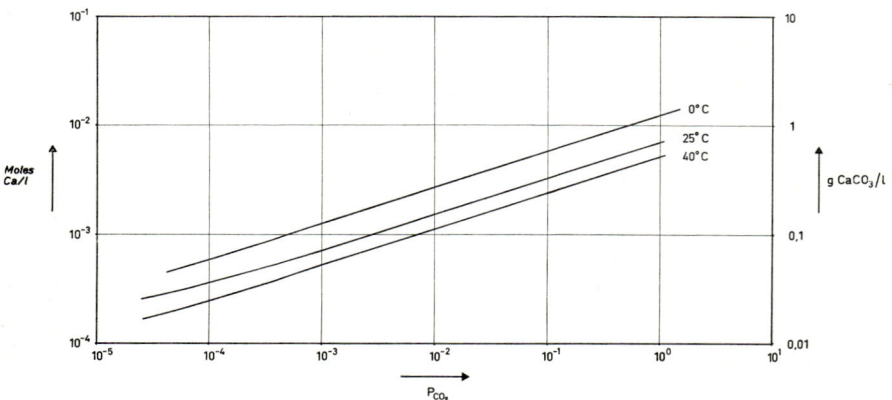

Fig. 4-4. Analytically established solubility of calcite in pure water as a function of CO_2-partial-pressure (GMELIN 1961).

For diagenetic processes the solubility of calcium carbonate at higher temperatures and CO_2-pressures can be quite important. Fig. 4-5 shows the solubility of calcite at temperatures between 100° and 300 °C and at CO_2-pressures between 1 and 62 atmospheres, as determined by ELLIS (1959).

For precipitation and dissolution at the earth's surface and under diagenetic conditions the dependence of calcite solubility on CO_2-partial-pressure and temperature is important. These relations can be deduced from fig. 4-3 or the above equations.

At constant temperature solubility depends upon CO_2-partial-pressure. According to equation (4-12) it increases with the cube root of the partial pressure, when the latter is greater than 10^{-3}. Therefore in this range the log plot of fig. 4-3 is a straight line. An increase in CO_2-partial-pressure from 0.0003 the value for normal air, to 0.003, increases the solubility by a factor of 2.15, that is, at 25 °C from 50 mg/l to 107 mg/l.

The influence of temperature follows from the temperature dependence of the constants, K_a, K_b, K_c, and K_e. As table 4-2 shows, each of these constants changes differently with increasing temperature. K_a becomes smaller with increasing temperature; K_b and K_c increase, that is, the dissociation of carbonic acid becomes greater. K_c becomes less, that is, the amount of carbonic acid dissolved in water at a given CO_2-pressure decreases with rising temperature. The solubility of $CaCO_3$, according to equation (4.12) depends on the expression $K_a K_b K_e / K_c$, which becomes smaller

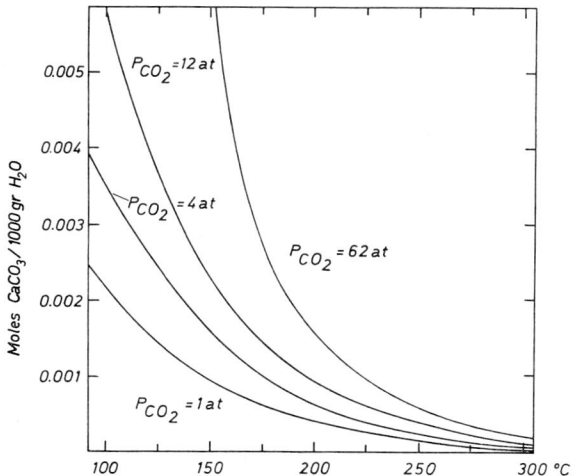

Fig. 4-5. Calcite solubility at elevated temperatures and different CO_2-pressures (ELLIS 1959).

with increasing temperature, so that the solubility of $CaCO_3$ decreases with temperature. For this effect the decrease in water-dissolved carbonic acid with rising temperature (K_e) is not primarily responsible. The temperature dependence of the other constants has a greater influence. As can be seen from the figures, at a given CO_2-pressure 2.4 times as much $CaCO_3$ is dissolved at 0° C as at 40 °C.

From a $CaCO_3$-saturated solution limestone will precipitate, when the CO_2-partial-pressure drops or when the temperature increases, or when both influences operate. Conversely, limestone is dissolved, when the temperature decreases and/or CO_2-pressure rises. Thus assimilation of CO_2 by plants causes precipitation of calcium carbonate. Effusion from springs leads to formation of calcarous sinter as a result of heating of spring water and the escape of dissolved carbonic acid.

For many processes the following case is of interest. A solution is in equilibrium with a gas phase and with solid $CaCO_3$ and contains a substance which produces a given hydrogen ion concentration. Then P_{CO_2} and $[H^+]$ are fixed. The concentrations of the ionic species in solution at CO_2-partial-pressures above 10^{-3} are given as follows:

$$[Ca^{++}] = \frac{K_a}{K_b K_c K_e} \frac{[H^+]^2}{P_{CO_2}} \tag{4.20}$$

$$[CO_3^{--}] = K_b K_c K_e \frac{P_{CO_2}}{[H^+]^2} \tag{4.21}$$

$$[HCO_3^-] = K_b K_e \frac{P_{CO_2}}{[H^+]} \tag{4.22}$$

It is seen that at a fixed CO_2-pressure, $CaCO_3$-solubility increases with increasing hydrogen ion concentration. On the other hand an increase in CO_2-pressure decreases solubility, if simultaneously the hydrogen ion concentration is held essentially constant by some sort of buffering action. Such a buffering action could arise chemically, for example, from decomposing silicates (feldspar weathering).

Deposition of carbonate rocks takes place primarily in shallow seas. The supply of dissolved calcium carbonate in ocean water is continuously supplemented by additions by streams. Geochemical estimates lead to the following inventory:

Volume of the ocean	$1.35 \cdot 10^{21}$ l
Total dissolved Ca in the ocean (0.41 g/l)	$5.50 \cdot 10^{20}$ g
Stream flow into the ocean	$3.23 \cdot 10^{16}$ l/yr
Mean Ca-content of river water	$1.5 \cdot 10^{-2}$ g/l
Supply of Ca into the ocean	$4.85 \cdot 10^{14}$ g/yr
(all data from TUREKIAN 1969)	

It follows from these data that the total calcium dissolved in the ocean today was introduced by streams in $5.50 \cdot 10^{20} : 4.85 \cdot 10^{14} = 1.16 \cdot 10^{6}$ years, geologically a relatively short time. More recent precipitation of carbonates from sea water must, therefore, counter-balance the addition by streams.

Table 4-3 gives the amounts of dissolved ions in sea water. On account of the high concentration and large number of dissolved components, equations (4.1—4.5) cannot be used directly for calcium carbonate-carbonic acid equilibria in the sea. Calcium and carbonate ions enter into varying interactions with the other ions in sea water, so that they are not freely disposed for equilibrium considerations. The result is that the product K_a', calculated from analytically determined contents of calcium ($m_{Ca^{++}}$) and carbonate ($m_{CO_3^{--}}$), is higher than the equilibrium constant K_a.

Table 4-3. Mean composition (dissolved ions) in surface sea water. $Cl^- = 19\ ‰$, $25°$, $pH = 8.15$ (after GARRELS & CHRIST).

	Mol/l
Na^+	0.48
Mg^{++}	0.054
Ca^{++}	0.010
K^+	0.010
Cl^-	0.56
SO_4^{--}	0.028
HCO_3^-	0.0024
CO_3^{--}	0.00027

In very dilute solutions equation (4.1) applies:

$$K_a = m_{Ca^{++}} \cdot m_{CO_3^-} = 4.6 \cdot 10^{-9}$$

With the Ca^{++}- and CO_3^{--}-contents of surface sea water at 25 °C, one obtains as a product

$$K'_a = m_{Ca^{++}} \cdot m_{CO_3^-} = 2.7 \cdot 10^{-6} \tag{4.23}$$

4.2 Limestone

This value is by a factor of about 6000 higher than the solubility product in infinitely dilute solution. In order to obtain the true solubility product in sea water, the analytical concentrations $m_{Ca^{++}}$ and $m_{CO_3^{--}}$ must be multiplied by their activity coefficients. These take into account the fact that because of electrostatic interactions, ion pairing, and complex formation, only a small portion of the Ca^{++} and carbonate ions are freely distributed at equilibrium. The difficulty is to determine the magnitudes of these activity coefficients.

For moderately concentrated solutions ion activities can be calculated from the DEBYE-HUCKEL theory, inasmuch as they result from eletrostatic interactions. The activities calculated in this manner for the ionic strength of sea water (0.7) are always too high, because the solubility product, in spite of this correction comes out too high. GARRELS, THOMPSON & SIEVER (1961) could show that the $CO_3^=$-ion activity in NaCl- and $MgCl_2$-solutions is very much less than in other salt solutions of equal ionic strength. The reason is the formation of little-dissociated $MgCO_3$-complexes, which lower specifically CO_3^{---}-ion activity. GARRELS & THOMPSON (1962) developed a model of dissolved ion species and complexes for sea water, from which the following activity coefficients result (25 °C):

Activity coefficients in sea water (after GARRELS & THOMPSON):

Ca^{++} 0.25
HCO_3^- 0.47
$CO_3^=$ 0.018

The $MgCO_3$-complex is much more tightly bound than the $NaCO_3$-complex (according to GARRELS & Thompson the dissociation constant for the former is $10^{-3.4}$; for the latter $10^{-1.87}$). Therefore, the $CO_3^=$-ion activity in solutions of variable composition is lower, the higher the Mg:Na ratio. In sea water this ratio is constant. $CaCO_3$ must precipitate from $CaCO_3$-saturated sea water, when Mg-ions are removed from it.

BERNER (1965) developed a method to determine $CO_3^=$-ion activity by measuring the pH of sea water samples at given values of CO_2-partial-pressure. He determined also the Ca^{++}-ion activity through determination of titration alkalinity $A = m_{HCO_3^-} + m_{CO_3^{--}}$ of sea water equilibrated at given CO_2-partial-pressures with solid $CaCO_3$. The values so-determined agree quite well with those derived by GARRELS & THOMPSON (25 °C):

Activity coefficients in sea water (BERNER):

Ca^{++} 0.22
HCO_3^- 0.56
$CO_3^=$ 0.024

From these values one obtains for the solubility product in seawater (25°):

$$K'_a = m_{Ca^{++}}\gamma_{Ca^{++}} \cdot m_{CO_3^-}\gamma_{CO_3^-} = 12.5 \cdot 10^{-9} \quad (4.24)$$

Since this value is greater by a factor of 2.7 than the equilibrium constant K_a, it follows that sea water at the ocean's surface clearly is supersaturated with respect to $CaCO_3$. This had been assumed even earlier by WATTENBERG (WATTENBERG & TIMMERMANN 1936; WATTENBERG 1936, 1937).

That in sea water no thermodynamic equilibrium prevails with respect to solid $CaCO_3$, but rather supersaturation in $CaCO_3$, follows also from the fact that the carbonate sediments of recent shallow seas consist predominantly of aragonite and Mg-rich calcite. Only minor amounts of low Mg-calcite, stable only at the earth's surface, occur (see volume II of this series, FRIEDMAN 1964). In the laboratory also aragonite precipitates from sea water as a result of loss of CO_2 (GEE 1932) or by precipitation with Na_2CO_3 (MONAGHAN & LYTLE 1956). The metastable phases aragonite and Mg-rich calcite must have a greater solubility than pure calcite. For their formation a certain supersaturation is necessary.

At 25 °C aragonite is stable only at pressures above 4000 atm (JAMIESON 1953). Therefore, it is found at the earth's surface and in the diagenetic milieu always outside its stability field.

According to the data from GARRELS & CHRIST (1965) the free energy difference $\triangle F = 250$ cal/Mol. For the solubility product of calcite (K_{aCa}) and aragonite (K_{aAr}) we obtain:

$$\begin{aligned} \triangle F &= RT \ln K_{aAr}/K_{aCa} \\ 250 &= 1364 \ln K_{aAr}/K_{aCa} \\ K_{aAr} &= 1.2 \cdot K_{aCa} = 1.2 \cdot 4.6 \cdot 10^{-9} = 5.5 \cdot 10^{-9} \end{aligned} \quad (4.25\ a)$$

From different values for the difference in free energy, BERNER (1965) calculated:

$$K_{aAr} = 7.1 \cdot 10^{-9} \quad (4.25\ b)$$

Therefore, in sea water aragonite must have a solubility 20 to 50% greater than calcite. Solubility differences of this order of magnitude have been established in fact by different observers. Since the activity product in sea water is actually $12 \cdot 10^{-9}$, the surface water of the ocean is supersaturated with respect to aragonite as well as calcite.

The factor which makes difficult or inhibits precipitation of the stable calcite phase in spite of supersaturation, must affect the kinetics of nucleation and crystal growth. Nuclei of higher energy crystal species can develop and grow only when nucleation and growth of the stable species is prevented.

As early as 1910 LEITMEIER had emphasized that the presence of Mg-salts induced the formation of aragonite. MONAGHAN & LYTLE (1956) investigated the precipitates generated by Na_2CO_3 from $CaCO_3$-solutions in the presence of different constituents of sea water. Calcite always precipitates from pure solutions as well as from solutions with additions of NaCl, KCl, and sulfate. An addition of $MgCl_2$ in concentrations corresponding to that of sea water, induced the formation of aragonite in addition to $CaCO_3 \cdot H_2O$. LIPPMANN (1960 b) precipitated $CaCO_3$ from $CaSO_4$-solutions by adding KOCN, which decomposes slowly in aqueous solution, releasing $CO_3^=$-ions. At room temperature in solutions containing only Ca-ions, the precipitate was calcite. Addition of $MgSO_4$ induces the formation of aragonite along with calcite, when the molar ratio Mg:Ca falls between 1 and 3. At molar ratios above 3 a pure aragonite precipitate forms. Experiments with additions of Na_2SO_4 showed that $SO_4^=$-ions are not responsible for this effect. BISCHOFF & FYFE (1968) also have studied calcite and aragonite crystallization in the presence of different dissolved ingredients. The growth of calcite was observed in solutions with additions of NaCl, KCl, $CaCl_2$, (OH), and CO_2. By comparison even small amounts of $MgCl_2$ exert a strong inhibiting action on calcite nucleation and growth.

Inhibition of calcite crystal growth by Mg^{++}-ions is regarded as the cause of aragonite formation from sea water. According to LIPPMANN (1960 b) this results because Mg^{++}-ions are preferentially adsorbed on calcite surfaces, by virtue of calcite-magnesite isomorphism. The small Mg^{++}-ions are strongly hydrated, carrying a tight hull of water dipoles. They must be removed in order for the calcite structure to continue to grow. Since dehydration necessitates a high activation energy, the calcite crystals "poisoned" with Mg^{++}-ions remain as tiny nuclei, while aragonite nuclei, not influenced by magnesium, can continue growing.

Mg-rich calcite, which is metastable at low temperatures with respect to pure calcite, occurs as a constituent part of the skeletons of many recent shallow sea organisms (CHAVE 1954; GOLDSMITH, GRAF & JOENSUU 1955). Here we are probably dealing with a biochemical influence on the kinetics of phase formation. Nevertheless, occurrences of apparently inorganically precipitated fine-grain Mg-calcite do occur in different recent and sub-recent shallow sea carbonate sediments (see FÜCHTBRAUER in the second volume of this series, p. 355 and 357). These show that under certain unknown conditions not involving organisms, aragonite formation does not occur and the adsorption of Mg-ion can lead to its inclusion in the growing calcite structure.

Certainly caused by biochemical influences is the precipitation in sea water of calcite exoskeletons and shells. While formation of aragonitic hard parts corresponds to the inorganic precipitation from sea water, organic influences must be active for precipitation of calcite tests. Organic membranes, for example, which are not permeable to hydrated Mg-ions, may be responsible in screening out these ions which inhibit calcite formation.

If one wants to express the supersaturation of sea water in $mgCaCO_3/l$, closed and open systems must be distinguished. In a closed system, no adjustment with a CO_2-bearing atmosphere can occur; that is, changes in concentration of carbonate constituents (H_2CO_3, HCO_3^-, $CO_3^=$) result only from dissolution or precipitation of $CaCO_3$. In an open system an equilibrium is established with the CO_2-content of the atmosphere, when $CaCO_3$ is dissolved or precipitated. BERNER (1965 b) calculated the following formulae, in which the excess dissolved $CaCO_3$ appears as a function of titration alkalinity $A = m_{HCO_3^-} + 2\, m_{CO_3^{--}}$ and hydrogen ion activity a_H:

4.2 Limestone

$$\text{closed system:} \quad \Delta_{closed} = \frac{10^5 A\,(K_a/K_a' - 1)}{2 + 1/B} \quad [mg/l]$$

$$\text{open system:} \quad \Delta_{open} = \Delta_{closed} + \frac{10^5 A\,(\sqrt{K_a/K_a'} - 1)}{2 + 4B} \quad [mg/l] \qquad (4.26)$$

$$B = \frac{K_b \gamma_{HCO_3^-}}{a_H + \gamma_{CO_3^-}}$$

K_a' is the actual product $\gamma_{Ca^{++}} m_{Ca^{++}} \cdot \gamma_{CO_3^{--}} m_{CO_3^{--}}$ for super-saturated sea water; K_a is the equilibrium constant at saturation. With these formulae one can ascertain the maximum amount of $CaCO_3$ that can precipitate from sea water under given conditions. If exchange with the atmosphere is assmued (open system), more $CaCO_3$ can precipitate than when exchange is prevented (closed system). For a small body of shallow sea water in equilibrium with the atmosphere at 25 °C a supersaturation of $\Delta = 52$ mg/l results.

Reduction in hydrogen ion concentration and CO_2-partial-pressure (or amount of CO_2 dissolved in water), and increase in temperature, increases, as explained above, the supersaturation and promotes precipitation of $CaCO_3$. Carbonate precipitation takes place, therefore, predominantly in warm shallow seas. Local increases in pH by biological processes, or by the decomposition of organic matter can bring about carbonate deposition.

Pressure and hydrogen ion concentration increase with depth in the ocean, while temperature decreases. Ion activity coefficients do not change significantly with pressure and temperature (BERNER 1965 b). In contrast the solubility product K_a depends on pressure and temperature. The temperature dependence is known and is shown in table 4-2. K_a increases with decreasing temperature. K_a must increase likewise with increasing pressure, since the molar volumes of Ca^{++} and $CO_3^{=}$ are greater in solution than in the solid state. Since pressure increases and temperature decreases with depth in the ocean, $CaCO_3$-solubility must increase with depth. Exact determination of solubility product K_a at different ocean depths is still not possible, because experimental data on the pressure dependence of K_a are lacking. From thermodynamic data BERNER (1965 b) has calculated K_a at different depths for an ocean model. The results are reproduced in table 4-4 and in fig. 4-6. The solubility product of $4.9 \cdot 10^{-9}$ at the surface (20°) increases to $21.2 \cdot 10^{-9}$ at a depth of 5000 m. In order to estimate the saturation state for the deeper parts of the ocean these solubility products must be compared to the concentration product $K_a' = \gamma_{Ca^{++}} m_{Ca^{++}} \cdot \gamma_{CO_3^{--}} m_{CO_3^{--}}$ actually prevailing at different depths. Assuming that the total salinity and titration alkalinity A are constant over the entire depth range, the dependence of the prevailing product K_a' on pH can be represented by the following formula:

$$K_a' = \gamma_{Ca^{++}} m_{Ca^{++}} \cdot \gamma_{CO_3^-} m_{CO_3^-} = \frac{A \gamma_{Ca^{++}} m_{Ca^{++}}}{pH/K_c \gamma_{HCO_3^-} + 2/\gamma_{CO_3^-}} \qquad (4.27)$$

The values calculated in this manner by BERNER for normal profiles in the north Pacific (MOORE et al. 1962) and equatorial Atlantic are given in table 4-4 and

Table 4-4. Saturation of sea water in $CaCO_3$. $Cl^- = 19‰$. $A = 2.3 \cdot 10^{-3}$ mol/l, $m_{Ca^{++}} = 0.103$ (after BERNER 1965).

Depth m	Pressure atm.	Temperature °C	$K_a \cdot 10^9$	Pacific (MOORE et al. 1962)		Atlantic (HARVEY 1955)	
				pH	$K'_a \cdot 10^9$	pH	$K'_a \cdot 10^9$
0	0	20	4.9	8.2	12.8	8.2	12.8
500	50	13	6.2	7.8	5.5	7.6	3.6
1000	100	6	7.4	7.6	3.2	7.7	4.0
2000	200	5	9.7	7.6	3.7	7.9	6.8
3000	300	4	12.8	7.6	4.2	7.8	6.3
5000	500	3	21.2	7.6	5.3	7.8	7.9

$K_a = \gamma_{Ca} m_{Ca} \cdot \gamma_{CO_3} m_{CO_3}$ for saturation in $CaCO_3$ (calcite) (solubility product)
$K'_a = \gamma_{Ca} m_{Ca} \cdot \gamma_{CO_3} m_{CO_3}$ in sea water

fig. 4-6. The thermodynamic calculation of solubility product is sufficiently accurate to justify the conclusion that only the uppermost few hundred meters is supersaturated. The bulk of the underlying ocean water is undersaturated with respect to $CaCO_3$. In this way the early developed picture of WATTENBERG (1937) of $CaCO_3$-saturation of sea water is confirmed.

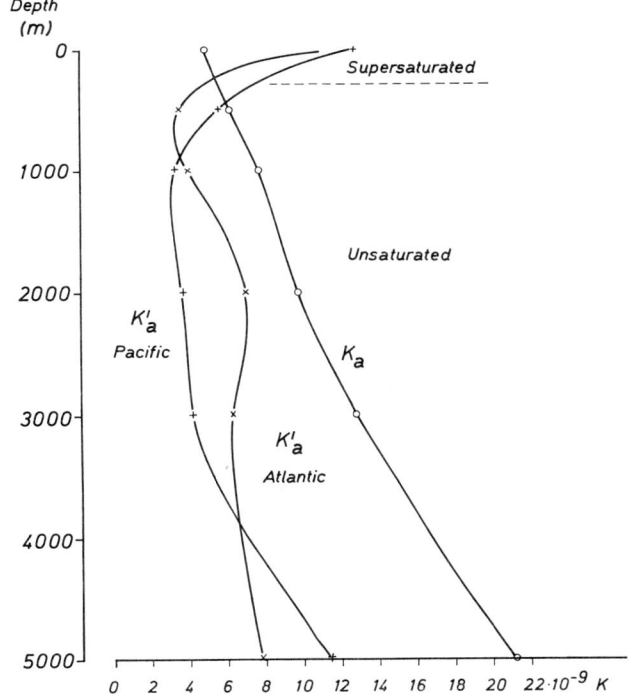

Fig. 4-6. Saturation of sea water with respect to $CaCO_3$ (after BERNER 1965).
K_a', solubility product calculated from the analytically determined concentrations
K_a, thermodynamically calculated solubility product

The $CaCO_3$-saturation of sea water determines the formation and distribution of marine carbonate sediments. Marine limestones are formed primarily in relatively limited shallow seas, where the supersaturation is high. Carbonate sediments in deeper basins stem from carbonate remains of planktonic organisms, mainly foraminifera (calcite), coccolithophorids (calcite), and pteropods (aragonite). The calcareous shells of these organisms undergo the dissolution effect of undersaturated deep waters. The present distribution of calcareous sediments on the deep sea bottom is influenced by different modifying factors. For example, in the Atlantic dissolution is especially intensive in the Brazilian basin, where a strong current of cold water is brought up from the Antarctic (WATTENBERG 1933, 1937; CORRENS 1939). Complete dissolution of planktonic carbonate leads to a carbonate-free sea bottom. Where the supply exceeds the dissolution rate calcareous deposits can accumulate in the deep sea. Since the dissolution rate of metastable aragonite and Mg-calcite is greater than that of calcite, selective dissolution results. While the shallow sea carbonates consist predominantly of aragonite and Mg-rich calcite, a minor amount of such metastable phases occur in recent deep sea carbonate sediments. That aragonite is preferentially dissolved is demonstrated by impressions of pteropods dissolved out of calcareous deep sea sediments (FRIEDMAN 1964, 1965).

4.3 Dolomite

An exact value for the solubility product in water is very difficult to determine at low temperature, since all reactions between dolomite and aqueous solutions take place extremely slowly. According to the present state of knowledge, the following applies at 25 °C:

$$[Ca^{++}] \cdot [Mg^{++}] \cdot [CO_3^{--}]^2 \approx 10^{-17} \tag{4.28}$$

This value is based on older data by HALLA & RITTER (1935), the calculations by KRAMER (1959), more recent measurements in the laboratories of LANGMUIR (1964) and BERNER (1967), as well as the composition of cave waters (HOLLAND, KIRSIPU, HUEBNER & OXBURGH 1964) and ground waters (HSU 1963) in dolomite-limestone rocks.

The solubility product of dolomite, like calcite, decreases with increasing temperature. For equilibria in solutions containing CO_2 equations (4.2—4.5) must be considered. Like calcite the solubility of dolomite increases with rising CO_2-pressure and decreasing temperature. Temperature elevation and diminishing CO_2-pressure should lead to dolomite precipitation.

Since in nature dolomite usually occurs in close association with calcite, solutions which are saturated with respect both to calcite and dolomite, are of special interest. In such solutions equilibrium occurs with respect to the reaction:

$$2\,CaCO_3 + Mg^{++} \leftrightharpoons CaMg(CO_3)_2 + Ca^{++} \tag{4.29}$$

For this reaction the equilibrium constant calculated from the solubility products of calcite and dolomite apply (at 25 °C):

$$K_{Cal \cdot Dol} = \frac{[Ca^{++}]}{[Mg^{++}]} = \frac{K_{Cal}^2}{K_{Dol}} = 10^{0.32} \tag{4.30}$$

Although this equation applies strictly only to ion activities, even for concentrated solutions can concentrations be substituted directly, as long as it can be assumed that the Ca- and Mg-activity coefficients are not significantly different from one another.

In solutions which are saturated with respect to both calcite and dolomite the molar concentration at 25 °C of Ca^{++} is about twice that of Mg^{++}. From solutions with $[Ca^{++}]:[Mg^{++}] > 2$, calcite should precipitate; from those with $[Ca^{++}]:[Mg^{++}] < 2$, dolomite. For temperatures between 50° and 180° determinations of molar Ca:Mg ratios for simultaneous dolomite and calcite saturation are available from USDOWSKI (1967); for temperatures between 275° and 420° we have the data of ROSENBERG & HOLLAND (1964). The results of these investigators are summarized in fig. 4-7. According to these data equilibrium solutions become richer in calcium

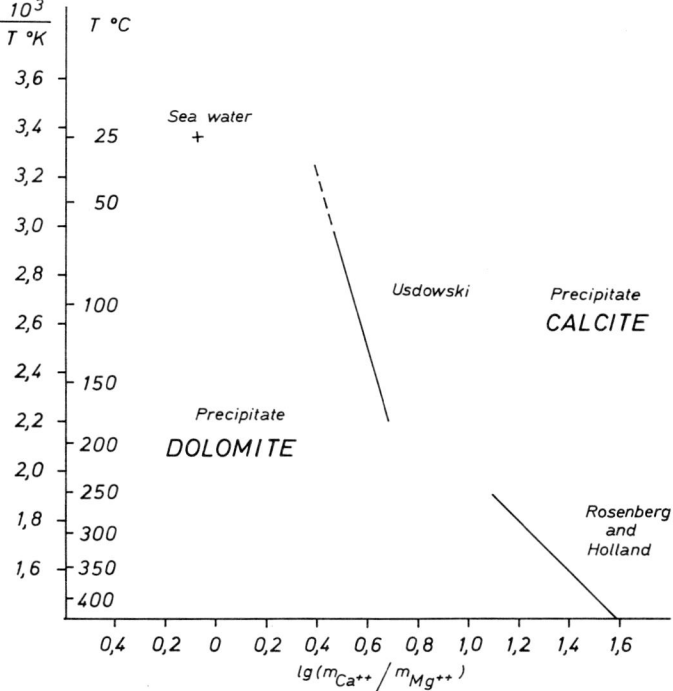

Fig. 4-7. Ca:Mg-ratio in aqueous solution for saturation with respect to dolomite and calcite as a function of temperature (USDOWSKI 1967; ROSENBERG & HOLLAND 1964).

with rising temperature. Apparently the dolomite solubility product (K_{dol}) decreases more with rising temperature than that for calcite (K_{cal}). A given solution which is in equilibrium with calcite at a lower temperature, could be in equilibrium with dolomite at higher temperatures. This fact can play an important role during late-diagenetic dolomitization of primary limestones.

A portion of the Mg released by weathering from primary Mg-minerals is bound by newly formed phyllosilicates during the course of subaerial weathering. The Mg-content of river waters shows, however, that Mg is transported to the ocean also in solution. Geochemical estimates lead to the following inventory:

Volume of the ocean	$1.35 \cdot 10^{21}$ l
Total Mg dissolved in ocean (1.29 g/l)	$1.74 \cdot 10^{21}$ g
Stream flow into ocean	$3.23 \cdot 10^{16}$ l/yr
Mean Mg-content of river water	$4.1 \cdot 10^{-3}$ g/l
Supply of Mg into ocean	$1.32 \cdot 10^{14}$ g/yr

(all data from TUREKIAN 1969)

It follows from these data that the total magnesium dissolved in the ocean today was introduced by streams in $1.74 \cdot 10^{21} : 1.32 \cdot 10^{14} = 1.3 \cdot 10^7$ years. During the course of geologic time Mg must have been abstracted from sea water. The removal of Mg by chlorite formation on the sea bottom plays a role which is difficult to estimate. A large amount of Mg temporarily is included in Mg-calcite. The major part of Mg removed from sea water ends up in dolomite rocks, as they occur in marine carbonate series of all geologic formations.

For the ions Ca^{++}, Mg^{++}, and $CO_3^=$, the following activities apply in sea water (GARRELS & CHRIST 1965):

	a
Ca^{++}	$2.5 \cdot 10^{-3}$
Mg^{++}	$1.94 \cdot 10^{-2}$
$CO_3^=$	$4.9 \cdot 10^{-6}$

These give the product:

$$a_{Ca} \cdot a_{Mg} \cdot a_{CO_3} = 1.2 \cdot 10^{-15} \tag{4.31}$$

This value is greater by a factor of 100 than the solubility product of dolomite ($K_{dol} = 10^{-17}$). Sea water, therefore, is strongly supersaturated with respect to dolomite also. Also the molar ratio $m_{Ca} : m_{Mg} = 0.19$ in sea water is so low that the precipitation of dolomite rather than calcite should be expected (see fig. 4-7).

In spite of the high supersaturation, in nature the direct precipitation of significant amounts of dolomite from sea water has never been observed. Neither has it been possible in the laboratory to crystallize dolomite from sea water at low temperatures. The occurrence of individual dolomite crystals in deep sea clays shows that rare and quantitatively minor amounts of dolomite can be formed from sea water, as it corresponds to thermodynamic equilibrium conditions. This is, however, probably possible only on the deep sea bottom, where sedimentation of clastic material is so slight that a nucleus, once formed, has sufficient time to grow very slowly into a macroscopic crystal.

Dolomite formation, like $CaCO_3$-precipitation, cannot be understood fully from thermodynamic equilibria. From the solubility products one can deduce probably which phases should be able to form under given conditions, when thermodynamic equilibrium is attained. As to whether this equilibrium is indeed attained, and in what manner, is a question of reaction kinetics.

It is logical to presume the same cause for the kinetic inhibition of dolomite crystallization from sea water as is presumed to influence precipitation of $CaCO_3$-phases from sea water. Also in the case of dolomite the strong hydration of Mg-ions at low temperature may have a restrictive effect on the growth rate of nuclei. This is because the dehydration energy of Mg-ions is large, in comparison to thermal energy available at low temperature and to the amount of energy released by the deposition of a Mg-ion in the growing dolomite crystal. For the same reasons it is also not possible to crystallize magnesite from aqueous solutions (see LIPPMANN 1973).

The discrepancy between equilibrium theory and observation constitutes the so-called dolomite problem, which has engaged geologists and petrographers for many years. On the one hand equilibrium theory precludes the primary formation of dolomite from sea water. On the other hand we observe the wide distribution of dolomite rocks in marine carbonate series of geologic formations.

From the structure and geologic association of older dolomite rocks, and from more recent observations of recent dolomitization in different coastal and shallow sea areas, it has become apparent that dolomite rocks always form and occur by the action of Mg-bearing solutions on primary carbonates. Dolomite rocks, therefore, are formations of early or late diagenesis. This will be considered in a later section in connection with the diagenesis of limestones.

4.4 Gypsum and anhydrite

Calcium sulfate occurs in sedimentary rocks as gypsum ($CaSO_4 \cdot 2 H_2O$) and anhydrite ($CaSO_4$). The occurrence of the hemi-hydrate (bassanite, $CaSO_4 \cdot 1/2 H_2O$), which is unstable under all conditions in relation to gypsum or anhydrite, is uncertain in evaporites.

For the equilibrium between gypsum and anhydrite in the presence of water, the following reaction applies:

$$CaSO_4 \cdot 2 H_2O \text{ (solid)} \rightleftarrows CaSO_4 \text{ (solid)} + 2 H_2O \text{ (fluid)}$$

For the equilibrium constant of this reaction one can write:

$$(K_{G-A})_{P,T} = \frac{a_{anh.} \cdot a_{H_2O}^2}{a_{gypsum}} = a_{H_2O}^2 \qquad (4.32)$$

Since the activity of a solid phase can be set equal to unity, the equilibrium depends only on the activity of water in this system. Expressed differently, at a given temperature and pressure gypsum and anhydrite co-exist stably at a given activity of water. At water activities above these critical values gypsum is the stable phase, at lower activities anhydrite.

Over the pressure-temperature range under consideration, water vapor can be considered as an ideal gas. The activity of water in a given solution can be measured by the water vapor pressure above the solution:

$$a_{H_2O} = P_{H_2O} : P_{H_2O}^\circ \qquad (4.33)$$

P_{H_2O} is the partial pressure of water over the solution at given values of total pressure and temperature; $P_{H_2O}^\circ$ is the vapor pressure over pure water at the same pressure and temperature. The activity of pure water is equal to unity. The activities of solutions are always less than 1, since dissolved substances lower the vapor pressure.

Whether from a given solution gypsum or anhydrite precipitates as the stable phase, when solubility is exceeded, depends only on the partial water vapor pressure, not on the nature of the dissolved components. From knowledge of the critical water vapor pressure for the coexistence of both phases, it can be deduced for all aqueous solutions, whether gypsum or anhydrite is stable, when only the water vapor pressure is known.

HARDIE (1967) established for H_2SO_4- and Na_2SO_4-solutions of known water vapor pressure at 1 atm pressure, the temperature dependence of the activity of water (a_{H_2O}), at which gypsum and anhydrite coexist stably. He found the values:

°C	a_{H_2O}
55	0.960
39	0.845
23	0.770

These data are plotted in fig. 4-8.

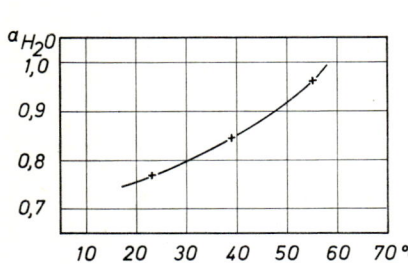
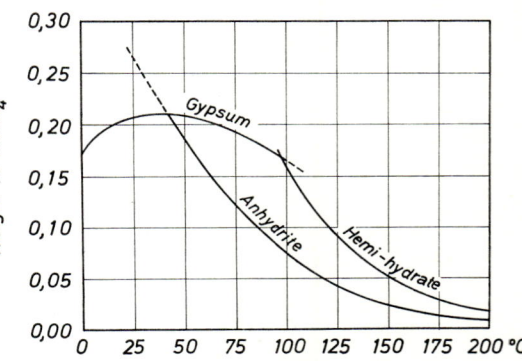

Fig. 4-8. Activity of water in solutions simultaneously in equilibrium with gypsum and anhydrite, as a function of temperature (HARDIE 1967).

Fig. 4-9. Solubility of gypsum, anhydrite, and hemi-hydrate (POSNJAK 1940).

4.4 Gypsum and anhydrite

The equilibrium temperature in pure water is 58°. Gypsum should precipitate from pure water below 58°, anhydrite above 58°. (On account of the low solubility of $CaSO_4$, a $CaSO_4$-saturated solution can be treated as pure water.) This temperature is clearly higher than deduced from solubility curves and the acknowledged equilibrium temperature of about 42° (see below). HARDIE's experimental runs involved time spans up to a year and attained equilibrium both from the gypsum and anhydrite sides of the equation. On the other hand solubility determinations always are carried out from unsaturated solutions. It is possible, therefore, that in the latter case, the true saturation concentration was not attained due to the sluggishness of the solution process. If the solubilities of gypsum and anhydrite were somewhat higher than indicated in fig. 4-9, so would the equilibrium temperature derived from the solubilities be displaced to higher values.

Fig. 4-9 shows how the solubility of anhydrite, gypsum and the hemi-hydrate in pure water at 1 atm pressure depends on temperature. The more recent measurements by HARDIE indicate, as noted, that the true solubilities are somewhat greater. The solubility of gypsum changes only a little with temperature and passes through a maximum at about 40 °C. Anhydrite solubility decreases with increasing temperature as shown in fig. 4-10. This has important implications in diagenetic processes. Pressure rise increases its solubility, although the effect is slight (DICKSON, BLOUT & TUNEL 1963).

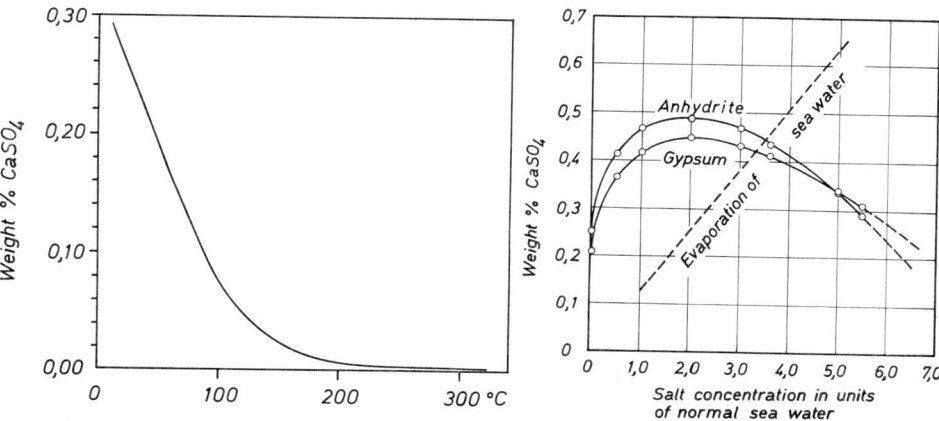

Fig. 4-10. Solubility of anhydrite at elevated temperatures (DICKSON et al. 1963).

Fig. 4-11. Solubility of gypsum and anhydrite in sea water at 30 °C (POSNJAK 1940).

The solubility of calcium sulfate is influenced by additional ions. In general foreign ions increase solubility, while Ca^{++} and $SO_4^=$ reduce it. In fig. 4-11 solubility curves for anhydrite and gypsum in sea water of different concentrations are plotted for 30 °C. At first the foreign ions in sea water increase the solubility. At higher concentrations the inhibiting influence of $SO_4^=$ and Ca^{++} takes over. Therefore, the solubility curve exhibits a maximum.

The dashed line in fig. 4-11 shows the change in concentration as sea water evaporates. The precipitation of calcium sulfate begins as soon as this line intersects the corresponding solubility isotherm. From the solubility curve of Posnjak (1940) this happens at 30 °C at 3.4 times the initial concentration of sea water. Then gypsum should precipitate, while the solution changes composition along the solubility curve. At a concentration about 5 times the initial concentration the solubility curve of gypsum intersects that of anhydrite. Further isothermal evaporation should precipitate anhydrite.

In contradiction to these conclusions based on the solubility curves, numerous crystallization experiments have produced only gypsum (hemi-hydrate at higher temperatures). Primary gypsum is formed from sea water also in the stability field of anhydrite, since nucleation and growth rates are greater for the former. In the case of calcium sulfate also, reaction kinetics is more determinant on the nature of the primary precipitation than is equilibrium.

Although in some regions of recent calcium sulfate precipitation from sea water, conditions prevail under which anhydrite should form, one finds, in agreement with laboratory studies, gypsum almost exclusively as the primary formation. An exception is an occurrence in the Persian Gulf (Trucial Coast, Curtis et al. 1963). Here also it is not clear whether anhydrite forms primarily or from gypsum.

In general it is likely that at the higher natural temperatures gypsum may be the primary precipitate formed by the evaporation of sea water. Anhydrite forms only secondarily, either at an early stage of diagenesis (if temperature and salinity are high enough), or as a result of subsidence into regions of higher temperature.

5. Diagenesis
5.1 General

The term diagenesis has been defined in various ways at different times by different authors. Universally all processes are considered to be diagenetic which alter a sediment after its deposition, but before metamorphism begins. Defining and separating diagenesis from pre- and post-diagenetic processes is difficult. This is especially so for subaquatic weathering and metamorphism, because in both cases a continuous transition is involved. Only arbitrary boundaries can be drawn, but this should be done on the basis of practical considerations and according to a clear definition.

All important diagenetic processes take place in porous sediments and sedimentary rocks. The solid mineral particles, as well as the fluids and gases which fill the pore space, take part in these processes. The presence of pore space is a prime characteristic for diagenesis. Metamorphism begins when the pore space becomes closed. Metamorphic processes are essentially solid state reactions, resulting from solid diffusion or diffusion along grain boundaries.

A second characteristic of diagenesis is the very slow movement of fluids and gases in the pore space. Therefore, for diagenetic reactions only limited amounts of solution are available, and their compositions and concentrations are changed by the diagenetic process. In this way diagenesis is distinguished from subaerial and subaquatic weathering, where mineral particles react in an open system with practically unlimited amounts of solution agent. Related to these differences, initiation of diagenesis is determined by that point in time at which, as a result of burial by later deposits, free exchange between the depositional medium and sediment pore fluid is cut off completely or restricted very significantly.

Therefore, diagenesis is defined as the total of all processes affecting a sediment from the time it is sealed off through burial by new deposits until its pore space disappears.

In the following section the properties of the pore space of sediments and sedimentary rocks will be considered first. This will be followed by consideration of the principles of flow of solutions and gases in pores. The compositional and diagenetic changes of formation waters which fill pore space will be described and explained in a following section. Further sections will then deal with the diagenesis of the major sedimentary rock types (sands, clays, carbonates).

5.2 Pore space in sediments and sedimentary rocks

5.21 General

Except for those chemical sediments which, like many evaporites, precipitate as compact masses, the formation of sediments comes about in such a manner that individual grains accumulate as more or less loose masses. These become solid rocks during the course of time only as a result of different diagenetic processes. An unconsolidated rock consists of two components of volume, the solid particles and pore space, which is filled at first with the medium in which deposition took place — air or aqueous solution. Diagenesis produces a reduction in pore space by processes involving both the solid constituents of the sediment and fluids and gases filling the pores. On the one hand chemical reactions play a role, on the other flow of mobile phases within the volume of the sediment. A theory of diagenesis has to consider, therefore, not only the solid constituents, but also the pore space and the fluids and gases contained therein.

In the following sections, pore space and the relations of fluids and gases in it will be considered.

5.22 Porosity

The total volume of a rock (V_r) is made up of the volume of the solid constituents (V_s) and the volume of the pore space (V_p):

$$V_r = V_s + V_p \tag{5.1}$$

Porosity (ε) is defined as that fraction of the total volume of a rock which is pore space:

$$\varepsilon = \frac{V_p}{V_r} = \frac{V_r - V_s}{V_r} \tag{5.2}$$

Instead of porosity the relative pore space (E), the ratio of pore volume to solid volume, is used also.

$$E = \frac{V_p}{V_s} \tag{5.3}$$

Both quantities are related as follows:

$$\varepsilon = \frac{E}{1+E}; \qquad E = \frac{\varepsilon}{1-\varepsilon} \tag{5.4}$$

If ϱ_s is the specific gravity of the solid constituents and ϱ_r that of the total rock, whose pores are filled with air, then:

$$\varepsilon = 1 - \frac{\varrho_r}{\varrho_s} \qquad E = \frac{\varrho_s}{\varrho_r} - 1 \tag{5.5}$$

The so-defined porosity ε (or relative pore space E) includes the total void space of the rock. One speaks, therefore, of the total porosity. From this one can distinguish the effective or useful porosity. It measures only that portion of the total void space which is interconnected. Indeed it is conceivable that during consolidation completely isolated voids occur, although experience has shown the pore space in sediments as a rule to be interconnected. In the following, discussion of porosity will be always in the connotation of effective porosity.

The different methods for measuring porosity depend on determining both of the volumes expressed in equation (5.1). These methods were described in the first volume of this series and will not be dealt with here.

Among the clastic sediments two main types can be distinguished, which differ basically with respect to the geometry of the pore space and the manner of its reduction during diagenesis. The first type includes the coarse grain sediments and is represented by sands. The second type is illustrated by clays. Both types grade into one another, the differentiation lying at a grain size of about 0.02 mm.

A sand of equal size grains to a first approximation can be compared to homogeneous packed spheres, that is, packing in which all spheres assume geometrically equivalent positions and are contiguous with an equal number of neighboring spheres (coordination number). Densest homogeneous packing of spheres has a coordination number 12 and a porosity of 0.26. If deposition of sand were to result only according to the principle of densest packing, homodisperse sands of somewhat spherical grains should have a porosity of about 0.26.

Actually the porosities of natural unconsolidated and homodisperse sands lie between 0.40 and 0.45. Also by shaking a container of sand grains or spherical particles one can not achieve the porosity of densest packing of spheres. Such artificially packed particles have porosities no less than 0.35 to 0.45 (for details see v. ENGELHARDT 1960). Densest packing, which represents the state of lowest potential energy in the gravity field, is attained neither in nature nor through experiment. This is because the friction at the points of contact of individual grains is so large that the movement of grains against one another necessary to produce densest packing is not possible.

Natural sands are never homodisperse. Their grain size distribution curves usually exhibit a more or less broad maximum. Well sorted sands with a narrow maximum have the highest porosity. With increasingly poorer sorting the porosity diminishes, since smaller grains lodge completely or in part in the space between larger grains. An especially low porosity is to be expected for sands with asymmetrical distribution curves.

In addition to grain size distribution the shape of sand grains has an influence on porosity. Sands composed of irregular particles have higher porosities than those of spherical grains. For example, the addition of mica platelets to spherical sand grains increases the porosity.

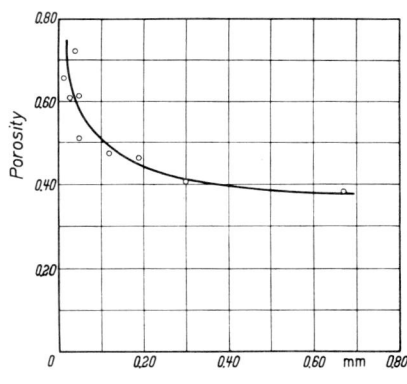

Fig. 5-1. Porosities of recent sediments off the coast of California (HAMILTON & MENARD 1956).

Porosity of a homogeneous packing of spheres does not depend on the absolute size of the spheres. Accordingly the primary porosity of freshly sedimented sands extending into the fine sand size range is little dependent on grain size. For 200 sand samples from the breaker zone, from sand flats, and from the shallow bottom of the southern North Sea, with median diameters between 0.1 and 0.3 mm, FÜCHTBAUER & REINECK (1963) found porosities between 0.37 and 0.42, independent of median grain size. Fig. 5-1 shows the results of HAMILTON & MENARD (1956) for recent sediments off the California coast. The samples were collected by divers and come from 30 m deep water. Down to median diameters of about 0.2 mm the porosity of these sands is independent of grain size and gives values at about 0.40. In contrast a strong rise in porosity results when the mean grain size is smaller than 0.1 mm.

The increase in porosity in passing into the fine sand range shows that for finer grain sediments a simple model of packed spheres does not apply. For smaller grains increasing frictional forces come into play. These are proportional to the grain surface area, and, therefore, the square of the grain diameter. At the same time the effect of gravity in producing dense packing is proportional to grain volume. The ratio frictional force: force of gravity, therefore, increases with decreasing grain size (with spheres it is inversely proportional to sphere radius). As the grain size dependence of porosity shows, the influence of friction becomes noticeable for deposition of sands of 0.2 mm diameter or less.

The increasing porosity with decreasing grain diameter for fine sands extends over the silt range and into clays, whose porosity in the freshly deposited state is very much higher than for sands. According to available data one can assume that the porosities of freshly deposited clays fall in the range between 0.7 and 0.9 (v. ENGELHARDT 1960, p. 33). The high porosity of clays is explained by the unique shape of clay particles. Clay particles, consisting essentially of layer silicates, have the shapes of very thin platelets. Pure geometry would suggest that these platelets could pack together in parallel configurations forming a compact sediment with essentially no porosity. Porosity of clays could be significantly less than that of sands, whose spherical particles always develop porosity, even when densely packed. The relations of clay particles in a sediment cannot be interpreted in terms of purely geometric considerations. Surface forces are noticeably active. For platy clay particles the ratio of surface to volume is so great that all of the effective surface forces by far outweigh the volume related forces. Therefore, as the particles arrange themselves, particle weight plays only a minor role, in contrast to the interaction of surface forces. As a result they form very voluminous clay muds, whose porosities essentially can withstand also the burial of several meters of sediment. EMERY & RITTENBERG (1952) found that the porosities of marine clays off the California coast (Santa Barbara Basin, 530 m water depth) decreased from 0.82 at the surface to only 0.73 at 5 m depth below the surface.

The porosity of freshly deposited clays should depend on electrolyte content of the water from which the clay has deposited. We would expect theoretically that more porous clay muds would be deposited from saline than from fresh water. This stems from the theory of the influence of ions on the mutual interactions of clay particles. The mutual interaction of two particles is based on the one hand on attractive forces, and on the other hand on repulsive forces. The attractive forces are the van der Waals forces between positively charged edges and negatively charged basal surfaces of phyllosilicates. Electrostatic repulsive forces result from the collection of ions around each clay particle, which, as an electric double layer, compensates the prevailing negative surface charge. The thickness of the double layer is greater, the lower the electrolyte concentration. The attractive forces decline with distance more rapidly than the repulsive forces. Therefore, at not too high electrolyte concentrations maximum repulsion develops at a finite distance, preventing the particles from approaching close enough to come together and coagulate. With increasing electrolyte concentration, or decreasing double layer thickness, the repulsion maximum diminishes and is displaced simultaneously to smaller distances. At sufficiently high electrolyte concentrations the repulsive barrier disappears or is reduced sufficiently that the particles approach close enough to attract one another. Then flocculation can occur. This results in the formation of complexes of numerous particles which are relatively voluminous, if constituent particles adhere to ane another with positively charged edges attached to negatively charged basal surfaces (see fig. 5—47, p. 270). In highly concentrated salt solutions like sea water clays should form such coagulates. Thus it would be expected that a clay deposited in sea water would form a more porous sediment than a clay deposited in fresh water, where many more individual particles sediment.

Different experiments as well as natural observations confirm these expectations. ROSENQUIST (1962) found, when sedimenting illite clays from NaCl-solutions (0.002—0.5 n), that sediment volume increased with increasing salt concentration. HSI & CLIFTON (1962) obtained the same result in sedimenting kaolinite and metahalloysite clays from solutions of different salts of Na, Ca, Fe, and Al. Also MEADE (1964) observed the same phenomenon with mixed illite-kaolinite clays. From the porosity of different deeply buried illite-kaolinite-chlorite Tertiary clays from the Rhine Valley graben, HELING (1969) could deduce that their initial porosities, for like mineral content and grain size, were higher the higher the prevailing salinity at time of deposition.

All these observations related to illite, kaolinite, or chlorite clays. From experiments by HOFMANN & HAUSDORF (1945) is appears that montmorillonites behave differently. With montmorillonite suspensions they observed a decrease in sedimentation volume with increasing NaCl- and KCl-concentration, while $CaCl_2$ and $MgCl_2$ had no effect. Apparently flocculation of montmorillonite clays does not lead always to an especially porous sediment.

5.23 Homogeneous flow processes

Gases or fluids which fill the interconnected pores of a sediment can flow through the pore space if a pressure gradient is operative. Velocity and the flow direction depend on the properties of the flowing medium (density, viscosity, compressibility) as well as on the permeability of the rock.

By way of defining permeability (see v. ENGELHARDT 1960; SCHEIDEGGER 1960), consider the flow of an incompressible fluid through a rock cylinder which is inclined at an angle γ to the vertical (fig. 5-2), and whose cross-sectional area is F cm². For the fluid contained in the pore space to flow through the cylinder, a pressure gradient must prevail in the direction of the cylinder axis x (there is a pressure drop, that is, a negative value for dp/dx in the flowdi rection). The amount of fluid flowing past a given point in the cylinder per second per unit cross section depends on the viscosity μ, the pressure gradient dp/dx at this point, the density ϱ of the fluid and the inclination angle γ of the cylinder. The inclination angle and density enter the picture, because only that portion of the pressure gradient produces flow

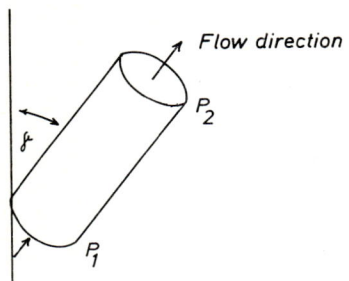

Fig. 5-2. For the definition of permeability (explanation in text).

which does not stem from the weight of the fluid. This model leads to the so-called Darcy equation:

$$q = \frac{Q}{F} = -\frac{k}{\mu}\left(\frac{dp}{dx} - g\varrho \cos \gamma\right) \tag{5.6}$$

The proportionally factor k is called permeability.

In actually making permeability measurements in the laboratory the pressure gradient is large with respect to the pressure difference due to the weight of the fluid. Besides, if the pressure gradient is constant over the distance involved, one can use the following simplified equation:

$$q = \frac{Q}{F} = \frac{k}{\mu} \frac{\Delta p}{l} \tag{5.7}$$

($\Delta p = p_2 - p_1$). p_1 and p_2 are the outlet and inlet flow pressures, and l is the cylinder length between these two points. In connection with natural rocks where long flow distances and great differences are known to exist between inlet and outlet points, equation (5.7) applies only for horizontal flow, that is, when $\gamma = 90°$.

From equation (5.6) it follows that permeability has the dimension cm², when all quantities are measured in cm-g-sec units. If the pressure gradient terms in equations (5.6) or (5.7) are expressed in atm/cm, the viscosity in centipoises, and the flow velocity in cm/sec, the permeability is expressed in a unit, which is called a Darcy, as suggested by WYCKOFF, BOTSET, MUSKAT & REED (1934). The following conversion factors apply:

$1 \text{ cm}^2 = 9.81 \cdot 10^7$ Darcy (p in tech. atm $= 9.81 \cdot 10^5$ dyn cm^{-2})
$1 \text{ cm}^2 = 1.013 \cdot 10^8$ Darcy (p in phys. atm $= 760$ mm Hg)

For all practical purposes it is sufficient to assume:
 1 Darcy $= 10^{-8}$ cm²

It is customary to subdivide the Darcy:
 1 Darcy $= 1000$ millidarcy (md)

In the case of gases flowing through the pore space it is necessary to take into consideration that the gas density depends on the pressure. As a result the volume flowing becomes greater as the pressure in the direction of flow decreases. The following expression applies then instead of equation (5.7):

$$q_1 = \frac{Q_1}{F} = \frac{k}{\mu} \frac{1}{p_1} \frac{p_2^2 - p_1^2}{2\,l} \tag{5.8}$$

Q_1 is the volume of gas following per sec at the outlet of the cylinder at pressure p_1. p_1 and p_2 are the outlet and inlet pressures.

With $\bar{p} = 1/2\,(p_1 + p_2)$ and $\triangle p = p_2 - p_1$, one can write:

$$\frac{Q_1}{F} = \frac{k}{\mu} \frac{\bar{p}}{p_1} \frac{\triangle p}{l} \tag{5.9}$$

The methods for measuring rock permeabilities in the laboratory are described in volume I of this series.

The validity of the Darcy equation for the flow of gases and fluids through porous media is subject to certain limitations. These must be considered if this relation is to be used for flow through rocks. High flow resistance is to be expected, if flow velocity rises above a certain critical value. Observations to date, however, indicate that such high velocities are not encountered in natural flow in sediments (see v. ENGELHARDT 1960, p. 64).

Other deviations from the Darcy relationship occur for the streaming of gases through fine grain sediments and for the flow of aqueous solutions through clays and clay-bearing sands and sandstones.

While fluids flowing through a capillary adhere strongly at contact with the capillary wall, gases glide at the wall. Therefore, in the case of gases the viscous flow, described by the Darcy equation, is augmented by gliding, whose contribution to total flow increases with decreasing capillary radius (see CARMAN 1956; SCHEIDEGGER 1960; v. ENGELHARDT 1960). A significant increase in flow velocity of gases as a result of gliding is to be expected when the pore radius is about 100 times the mean free path of the gas molecule. At a pressure of 1 atm at 0 °C this is of the order of magnitude of $5 \cdot 10^{-6}$ cm.

Deviations from the Darcy equation related to greater transport may be expected by gas flow through rocks whose pores are smaller than about 10^{-3} cm. At higher pressures this limit is displaced to smaller pore diameters. If rock pores are smaller than the mean free path, the idea of viscosity looses its meaning and the gas flows in a diffusion stream along the pressure gradient. This so-called Knudsen flow is initiated in rocks with 10^{-5} cm diameter pores. Gliding and Knudsen flow produce a more intense gas flow than the viscous flow of the Darcy relation. The result is that gases flow more rapidly through fine grain sands and sandstones, and especially through clays than expected from the Darcy equation. Gliding and Knudsen flow depend on the molecular weight of the gas. Heavy gas molecules flow slower than light ones. As a result of such gas transport a separation of mixed gases into their components could result over long distances of migration.

Laboratory studies and experience with the injection of water into petroleum reservoirs have shown that the flow of water in clay-bearing sands and sandstones does not follow the Darcy equation (see v. ENGELHARDT 1960, p. 69; SWARTZENDRUBER 1962). Aqueous solutions deviate in two respects from gases and "normal" fluids like non-polar hydrocarbons or CCl_4. First of all aqueous solutions in argillaceous rocks have a lower permeability than is calculated from the Darcy equation from the viscosity of water and for gases or non-polar fluids. The rock has for aqueous solutions apparently a lower permeability, or else the viscosity of water appears to be increased in it. Secondly aqueous solutions in argillaceous rocks do not exhibit the linear proportionality between pressure gradient and flow velocity required by the Darcy equation. The flow velocity increases exponentially rather than linearly with the pressure gradient in such rocks. Both effects depend on the electrolyte content of the solution. They are strongest for pure water and decrease with rising salt content. Concentrated salt solutions behave quite similarly to non-polar fluids and gases.

The origin of this phenomenon is still not clear. It is related undoubtedly to a series of interactions between the surfaces of clay mineral particles in the rock pores and the water molecules, with the result that the pore water has properties quite different from free water. The electrolyte concentration plays a role in that the dissolved salts reduce the activity of the water, and therefore, the number of free water molecules available for interaction with the clay particle surfaces.

In defining permeability flow through a cylindrically bound porous body was considered. This simple model for the Darcy relation can be used in real natural situations only when the flow lines run parallel to each other. It does not apply when the flow lines converge or diverge. Also the permeability must be the same throughout the rock. In such cases a family of equations must be developed from the Darcy equation from which three-dimensional flow relations can be derived. This problem has been dealt with in the literature (MUSKAT 1946; SCHEIDEGGER 1960; v. ENGELHARDT 1960). Here will be considered the case of three-dimensional variability in permeability, which is especially important in diagenesis.

Void space in a rock is isopermeable if the permeability is constant throughout. In heteropermeable pore space permeability changes from place to place. In addition the permeability at a particular place can be independent of flow direction or change with flow direction. With respect to permeability rocks are either isotropic or anisotropic.

A rock complex consisting of parallel layers of fine and coarse materials is as a whole heteropermeable. If a flowing fluid encounters such a complex so that the flow direction is exactly perpendicular or parallel to the layer boundaries, no change in flow direction takes place. The mean flow permeability of the layer packet, however, is different in each case. If $k_1, k_2, \ldots k_n$ are the permeabilities of individual layers, and $l_1, l_2, \ldots l_n$ the relative proportions in the profile under consideration, the mean permeability for perpendicular incidence is:

$$\bar{k}_\perp = \frac{1}{l_1/k_1 + l_2/k_2 \ldots + l_n/k_n} \tag{5.10}$$

5.23 Homogeneous flow processes

In contrast, if the flow is parallel to the layers, the mean permeability is:

$$\bar{k}_{\shortparallel} = l_1 k_1 + l_2 k_2 \ldots + l_n k_n \tag{5.11}$$

With parallel incidence the high permeability layers exert a greater influence and the mean permeability is higher than with perpendicular incidence. This can be illustrated by an example. A layer packet consists of four layers of equal thickness ($l_1 = l_2 = l_3 = l_4 = 0.25$) with permeabilities $k_1 = 0.2$, $k_2 = 0.1$, $k_3 = 0.05$, and $k_4 = 0.025$ Darcy. Flow parallel to the layers gives $\bar{k}_{\shortparallel} = 0.095$ Darcy; for perpendicular flow $\bar{k}_{\perp} = 0.053$ Darcy.

If flow encounters the boundary between two layers of differing permeability at an oblique angle, the flow direction is changed, analogous to refraction of light at the interface between two optically different media. If k_1 and k_2 are the permeabilities of contiguous layers, α_1 and α_2 the angles between the flow direction and the normals to layers 1 and 2, the following law of refraction applies to the flow directions (fig. 5-3):

$$k_1 : k_2 = \operatorname{tg} \alpha_1 : \operatorname{tg} \alpha_2 \tag{5.12}$$

As can be seen in fig. 5-3 the flow direction is more strongly bent from the layer normal in the more permeable layer (for the derivation of this formula see v. ENGELHARDT 1960).

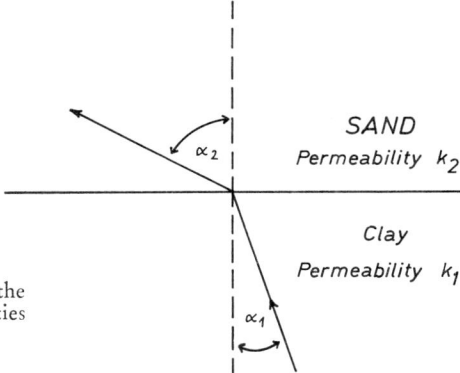

Fig. 5-3. Refraction of flow lines at the boundary of two media with the permeabilities $k_1 < k_2$.

The refraction of flow directions is of considerable importance for flow in packets of layers of different permeability. If the permeabilities are very different, flow takes place, even when the flow directions form a very small angle with the normal to the interface, predominantly in the high permeability layer. This can be clarified by an example:

A sand layer is interbedded within a clay. The permeability of the clay $k_1 = 10^{-6}$ Darcy; that of the sand $k_2 = 10^{-1}$ Darcy. The flow moves from the clay to the clay-sand interface at an angle which is only $\alpha_1 = 10'$ with respect to the normal. Then, according to equation (5.12), the flow direction in the sand $\alpha_2 = 89°$

48'; that is, the flow in passing into the sand is refracted almost 90° and proceeds almost parallel to the layers. If the thickness of the sand layer is 10 m, an individual flow line will run for 3000 m in the sand, before it intersects the upper surface. Upon entering the overlying clay it would be refracted again into the original direction, almost perpendicular to the bedding planes. Therefore, flow through a heterogeneous sequence of layers will greatly lengthen the transport distance and the highly permeable beds are inundated more than the less permeable ones.

Most sediments are anisotropic with respect to permeability; that is. at a given point the permeability depends on the direction of flow or pressure gradient. Generally the lowest permeability is for flow directions perpendicular to the bedding. The highest permeabilities prevail in the bedding planes. Differences in permeability within the bedding surfaces were established also (see, for example, POTTER & MAST 1963). The permeability of such an anisotropic sedimentary rock can be characterized by three main permeabilities. k_1 is the lowest permeability for flow perpendicular to the bedding. The intermediate permeability k_2, and the maximum k_3 apply to two mutually perpendicular directions parallel to the bedding planes. In most cases it can be assumed that in all directions within this plane the same permeability applies, that is, $k_2 = k_3$.

Only for flow in the direction of the three main permeabilities, that is, only parallel to the bedding, or perpendicular to it, does flow take place in the same direction as the effective driving force. In all other directions the flow takes place at some angle with respect to the driving force. Permeability in general or arbitrary directions of flow and the relation between flow direction and direction of force is shown by means of a simple construction arising from the application of tensor analysis. The Darcy equation described in vectorial form is as follows:

$$\vec{q} = \frac{k}{\mu} \vec{\Phi} \qquad (5.13)$$

$\vec{q}$ is the magnitude and direction of flow per unit cross-sectional area (flow vector). The force vector Φ is the difference between the vector for pressure gradient and the downwardly acting gravity vector g:

$$\vec{\Phi} = -\left(\frac{d\vec{p}}{dx} - \varrho \vec{g}\right) \qquad (5.14)$$

The permeability, characterized by k_1, k_2, and k_3, is, as the coupling of two vectors, a second degree tensor. This permeability can be represented in two ways as triaxial ellipsoids. This is shown in fig. 5-4 in a two dimensional section through the k_1—k_2 plane. One ellipsoid has, in the direction of the major permeabilities, the semi-major axes k_1, k_2, and k_3. The second ellipsoid has the semi-axes $1/\sqrt{k_1}$, $1/\sqrt{k_2}$ and $1/\sqrt{k_3}$ in the same directions. OP becomes the direction of the force Φ producing the flow. The corresponding flow direction is obtained by constructing the tangent at point P, which is the intersection of the force vector and the ellipsoid defined by $1/\sqrt{k_1}$, $1/\sqrt{k_2}$, and $1/\sqrt{k_3}$. The flow vector is drawn perpendicular to this

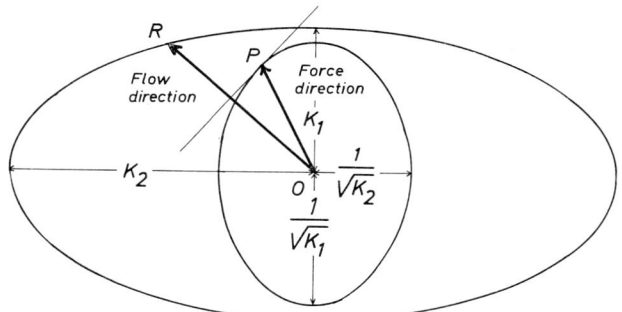

Fig. 5-4. Permeability tensors (explanation in text).

tangent as shown in fig. 5-4. The flow vector intersects the ellipsoid defined by k_1, k_2, and k_3 at point R. The magnitude of OR is equal to the permeability in this direction. It can be noted from this construction that the flow and force directions coincide only in the major permeability directions; also the angle between them is greater, the greater the differences in major permeabilities.

Permeability anisotrophy of direction of flow through sediments has in addition a quite similar effect as the refraction of flow lines in alternating beds of different permeability. If the direction of the pressure gradient is somewhat inclined to the bedding, the flow direction is deflected toward the bedding plane from the pressure direction.

There is no simple connection between the porosity of a sediment and its permeability. Sediments with similar porosities can have very different permeabilities. Porosity measures total pore space; permeability, however, depends on the shapes of individual flow channels. Very fine grain rocks have a lower permeability than coarse grain rocks of equal porosity.

For unconsolidated sands a perceptual model leads to a formula useful for interrelating porosity, grain size distribution, and permeability.

For flow through a cylindrical capillary the HAGEN-POISEUILLE relation applies:

$$q' = \frac{r^2}{8\mu} \cdot \frac{\Delta p}{l_e} \qquad (5.15)$$

q' is the volume flow per unit cross-sectional area of the capillary, r is the capillary radius, Δp the pressure differential, l_e the capillary length. In order to deal with other than circular cross sections, the hydraulic radius m is introduced:

$$m = \frac{\text{volume of capillary}}{\text{surface of capillary}}$$

One obtains then a generalized form for the HAGEN-POISEUILLE equation for generalized cross sections:

$$q' = \frac{m^2}{c_o} \cdot \frac{1}{\mu} \cdot \frac{\Delta p}{l_e} \qquad (5.16)$$

The term c_o has different values for different cross-sectional shapes. From experiment c_o for circular, elliptical, and rectangular capillaries falls between 2.00 and 2.65. Flow through individual capillaries makes up the total flow through a sand. In order to obtain the volume flow q through the total cross section, q' must be multiplied by the porosity ε since the porosity denotes the sum of all capillary cross sections. If l is the thickness of the total packed layer, it is to be taken into consideration, that the capillary length l_e will be greater than the thickness l. Therefore, $\varepsilon q'$ must be multiplied by the factor l/l_e. We obtain then for the volume flow through a unit cross section of a packed sand:

$$q = q' \cdot \varepsilon \cdot \frac{l}{l_e} = \frac{m^2 \varepsilon}{c_o} \left(\frac{l}{l_e}\right)^2 \cdot \frac{1}{\mu} \cdot \frac{\Delta p}{l} \tag{5.17}$$

Now the hydraulic radius of the capillary can be expressed as the quotient of the total volume of the capillary, which equals the porosity, and the total capillary volume:

$$m = \frac{\varepsilon}{S} = \frac{\varepsilon}{(1-\varepsilon) S_o} \tag{5.18}$$

S is the internal surface in a unit volume of packed sand, which can be expressed in terms of S_o, the specific surface of the solid grains.

$$S = S_o (1 - \varepsilon)$$

One obtains the following equation derived by KOZENY (1927) and CARMAN (1937, 1948, 1956).

$$q = \frac{\varepsilon^3}{(1-\varepsilon)^2} \cdot \frac{1}{S_o^2} \cdot \frac{l^2}{c_o l_e^2} \cdot \frac{1}{\mu} \cdot \frac{\Delta p}{l} \tag{5.19}$$

The KOZENY-CARMAN equation has been verified by different investigators using packed particles of different shapes. For packed spheres the quantity $c_o l_e^2 / l^2 \approx 5$. For unconsolidated sands consisting of spherical grains this gives the following relation:

$$q = \frac{\varepsilon^3}{5(1-\varepsilon)^2 S_o^2} \cdot \frac{1}{\mu} \cdot \frac{\Delta p}{l} \tag{5.20}$$

As comparison with equation (5.7) shows, for permeability the following should apply:

$$k = \frac{\varepsilon^3}{5(1-\varepsilon)^2 S_o^2} \tag{5.21}$$

A value for the internal surface S_o can be calculated for unconsolidated sands from the grain size distribution in the following manner. If n_e is the number of spherical grains of radius r_e in 1 cm³ of solid material, the volumetric or gravimetric fraction G_e of the corresponding grain size class in the total sand is given by:

$$G_e = \frac{4}{3} \pi r_e^3 n_e \tag{5.22}$$

For the total surface S_0 of all grains in 1 cm³ of solid material with radii r_1, $r_2 \ldots r_e$:

$$S_0 = 4\pi(n_1 r_1^2 + n_2 r_2^2 \ldots + n_e r_e^2) \tag{5.23}$$

Then one can write:

$$S_0 = w \left(\frac{G_1}{r_1} + \frac{G_2}{r_2} \ldots + \frac{G_e}{r_e} \right) \tag{5.24}$$

with $w \gtrless 3.0$. If the grains are spheres, $w = 3.0$. If their shape deviates from spherical form, w is greater than 3.0. For natural quartz sands a value of about 3.5 may be correct. Table 5-1 shows permeabilities of mixtures of different sieve fractions of a natural sand, calculated from equations (5.21) and (5.24) from the grain size distribution and porosity. In addition the measured permeabilities are given (v. ENGELHARDT & PITTER 1951). From these data one can see that useful estimates of permeability can be made from known grain size distributions and porosities.

Table 5-1. For calculating permeabilities of loose sands (after v. ENGELHARDT & PITTER 1951).

No.		S_0^2 (cm⁻¹)	Permeability (Darcy) at $\varepsilon = 0.40$	
			calculated[3]	measured
	Monodispersed sands[1] Mean diameter (cm)			
1	0.0200	175	118	72
2	0.0138	255	55	55
3	0.0113	311	37	37
4	0.00875	399	23	26
5	0.00625	560	11	10
6	0.00437	802	5.6	5.4
7	0.00337	1040	3.3	2.5
	Sand mixtures[1]			
8	50 % 1, 50 % 2	215	78	60
9	50 % 5, 50 % 6	679	7.8	10
10	25 % each of 1, 2, 5, 6	448	18	20
11	75 % 1, 25 % 6	331	33	42

[1] Sieve fraction of a diluvial sand and mixtures of the same.
[2] Calculated from the grain size according to eq. (5.24) with $w = 3.5$.
[3] Calculated from S_0 according to equation (5.21).

Table 5-2. Porosity (ε) and permeability (k) of some sedimentary rocks from well cores in the Federal Republic of Germany. Permeabilities are mean values from several samples measured parallel and perpendicular to the bedding (v. ENGELHARDT 1960).

	ε	K (md)	Median (mm)	Carbonate %	<20 μ %	Depth (m)
Coarse grain sandstones, little lithified						
1. Ampfing sandstone, Lower Oligocene, Ampfing	0.199	4900	0.7	14.6	2.1	1820
2. Middle Kimmeridge, Ostenwalde	0.262	9900	0.50	0	1.3	1535
3. Bentheim sandstone, Valendis, Scheerhorn	0.245	5700	0.35	0	1.1	1105
4. Middle Rhät, Abbensen	0.185	1360	0.31	4	2.3	325
5. Lower Pechelbronn sandstone, Oliogcene, Stockstadt	0.246	3200	0.25	0	2.5	1610
6. Upper Valendis, Barenburg	0.251	3100	0.25	4	1.5	810
7. Bentheim sandstone, Valendis, Rühlermoor	0.295	7500	0.25	0	0.8	785
8. Dogger β, Hankensbüttel	0.278	3250	0.21	1	1.6	1535
9. Bausteinschichten, Upper Oligocene, Schwabmünchen	0.285	2380	0.2	60	10	1300
Fine grain and diagenetically more lithified sandstones:						
10. Dogger ε, Kronsberg (carbonate cement)	0.247	105	0.35	49	6.5	650
11. Lias α 2, Eldingen	0.269	1570	0.16	0	2	1485
12. Bentheim sandstone, Valendis, Scheerhorn	0.274	400	0.11	0	9.5	1120
13. Dogger β, Hankensbüttel	0.228	615	0.09	2	2.6	1605
14. Dogger β, Meerdorf	0.190	100	0.09	3	3.1	1733
15. Wealden, Scheerhorn (carbonate cement)	0.272	180	0.10	26	7.5	1160
16. Upper Pechelbronn sandstone, Oligocene, Stockstadt	0.102	7	0.04	30	15	1550
17. Lias α 1–2, Eldingen	0.245	35	0.04	2	10	1520
Carbonate rocks:						
18. Coralline oolite. Oxford, Hohenassel	0.195	2700	—	—	—	520
19. Schalenkalk, Wealden, Lingen	0.236	260	—	—	—	900
20. Schalenkalk-oolite, Portland, Ostenwalde	0.198	65	—	—	—	1495
21. Main dolomite, Zechstein, Itterbeck	0.130	3	—	—	—	1610

For consolidated rocks the KOZENY-CARMAN equation is not directly applicable, since neither the internal surface area nor the tortuosity (l_e^2/l^2) are known. Independent measurements would be necessary to determine these two quantities. However, there are difficulties in doing this (see v. ENGELHARDT 1960, p. 88, concerning the possibility of determining the tortuosity l_e^2/l^2 from electrical resistance measurements of a rock saturated with a salt solution).

Permeabilities of different sediments vary over wide limits. For clastic sediments the permeability decreases markedly with individual particle diameter. In addition it depends on degree of sorting and especially on the extent of diagenetic deposition in the pore space (cementation, lithification). Some examples are summarized in table 5-2.

Exceptionally high permeabilities are found in slightly consolidated and well sorted sandstones (no. 1 to 9 of table 5-2). The influence of grain size stands out in comparing rocks no. 2, 3, 5, 9. Significant cementation lowers permeability as does poor sorting, especially by admixing clay material. Calcareous sandstones have a lower permeability than non-calcareous ones, if the carbonate occurs as pore cement (no. 10 and 15 of table 5.2). Detrital calcareous constituents have no influence on permeability (no. 9).

The permeabilities of carbonate rocks are quite varied. Detrital limestones like fossiliferous limestones have permeabilities which correspond to granular sandstones, as long as they are not strongly lithified (no. 18, 19, 20 of table 5-2). Lithified and dense limestones are among the impermeable sedimentary rocks. Nevertheless dense limestones and dolomites often contain fine fracture systems which bring about noticeable permeability (no. 21 of table 5-2).

Permeabilities of argillaceous sediments are very low. For moderately compressed clays the may be about 10^{-3}md.

5.24 Heterogeneous flow processes and equilibria in pore space[1]

Immiscible fluids and gases frequently occur together in the pore space of sediments and soils. The flow of this sort of immiscible mixture through a porous medium can be described by means of an expansion of the DARCY equation in which a special Darcy equation is set up for each phase. We obtain, for example, for the simultaneous flow of oil and water, the following equations in place of equation (5.6):

$$\left. \begin{array}{l} q_w = -k \dfrac{k_w}{\mu_w} \left(\dfrac{dp_w}{dx} - g \varrho_o \cos \gamma \right) \\[2mm] q_o = -k \dfrac{k_o}{\mu_o} \left(\dfrac{dp_o}{dx} - g \varrho_w \cos \gamma \right) \end{array} \right\} \qquad (5.25)$$

[1] See v. ENGELHARDT (1960), SCHEIDEGGER (1960).

μ_w and μ_o are the viscosities, ϱ_w and ϱ_o the densities of water and oil. q_w and q_o are the flow rates of water and oil. k_w and k_o are the so-called relative permeabilities. Since oil and water flow together, mutually impeding one another, k_w and k_o are less than unity. p_w and p_o are the pressures measured in water and in oil. Relative permeabilities k_w and k_o depend first of all on the gross saturation of pore space with oil and water. It depends secondly and more importantly on the distribution of the two phases in the pore space. Water is found predominantly in the small and oil in the larger pores, so that the ratio $k_o : k_w$ will be larger than 1, if water at equal gross saturation fills the larger and oil the smaller pores.

The manner in which immiscible phases are distributed in a system of different size pores depends upon their wetting properties. At equilibrium the better wetting phase collects in the narrower, the poorer wetting phase in the larger voids. A measure of wetting is the contact angle which forms where the boundary between the two phases intersects the capillary wall (fig. 5-5). The contact angle is smaller in the better wetting phase. In the limiting case the contact angle in the wetting and non-wetting phases is 0° and 180° respectively. One speaks then of complete wetting or non-wetting. For sediments and soils water is normally the wetting phase with respect to oil. If oil and gas occur together without water, oil is the wetting and gas the non-wetting phase.

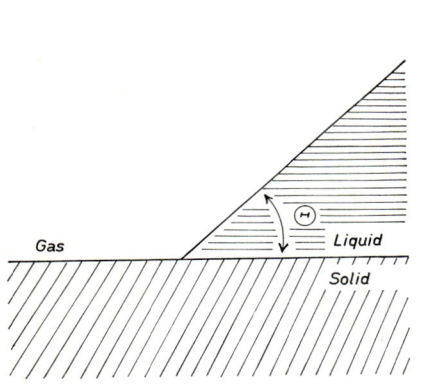

Fig. 5-5. For the definition of contact angle.

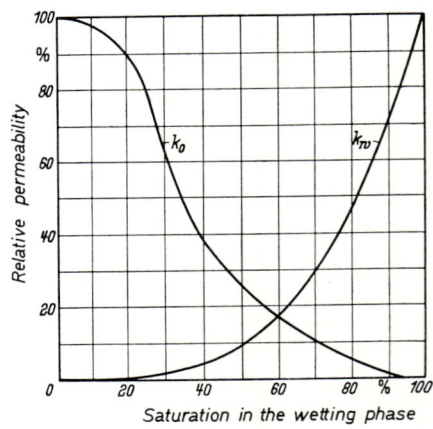

Fig. 5-6. Relative permeabilities of a wetting (k_w) and a non-wetting (k_o) phase as a function of relative proportions of each phase in the pore space.

Fig. 5-6 shows a typical example of the dependence of relative permeability of wetting and non-wetting phases on their relative abundance in pores of a rock. The following can be noted: First, the sum of the relative permeabilities, because of the mutual impedence, is always less than 1; second, since the non-wetting phase chiefly occupies the large channels, its content of the flowing mixture is always relatively higher than corresponds to its contribution in filling the pore space. For example, when the pore space is filled equally with 50 % of both wetting and non-wetting phases, the relative permeability of the non-wetting phase is more than double that of the wetting one.

With increasing concentration of the non-wetting phase the relative permeability of the wetting phase decreases rapidly, since the latter is forced more and more into the smaller voids. At a given saturation with the wetting phase, which lies between 5 % and 35 % of the pore space in sandstones, the permeability of this phase finally becomes immeasurably small. This small amount of fluid cannot move in spite of an external pressure gradient. One speaks then of irreducible saturation or of trapped water. This sort of confined water occurs in almost all fluid and gaseous hydrocarbon deposits. The confined water content is greater, the greater the internal surface area of the rock, that is, the smaller the grain size.

If wetting and non-wetting phases occur together in a cylindrical capillary of radius r, and Θ is the contact angle and γ the boundary surface tension between the movable phases, a curved interface (meniscus) forms between the two phases (fig. 5-7). Between wetting and non-wetting phases a capillary pressure p_c prevails:

$$p_c = \frac{2\gamma \cos \Theta}{r} \qquad (5.26)$$

For non-cylindrical capillaries one can introduce the hydraulic radius $m = r/2$:

$$p_c = \frac{\gamma \cos \Theta}{m} \qquad (5.27)$$

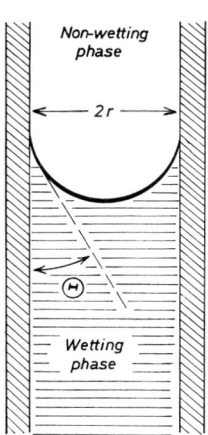
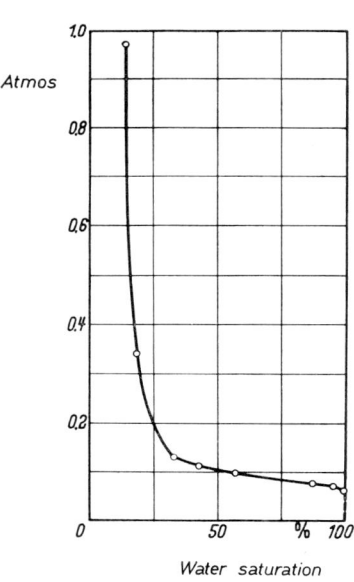

Fig. 5-7. For the definition of capillary pressure (explanation in text).

Fig. 5-8. Capillary pressure curve for a fine grain sandstone.

Since the higher pressure prevails in the poorly wetting phase the capillary pressure is expended in driving the wetting phase out of the capillary. Every rock contains pores of different sizes. The displacement of wetting by non-wetting phase can begin only when a certain critical displacement pressure is exceeded, which in turn depends on the hydraulic radius of the largest pores. Then the displacement continues and the pressure must be elevated long enough for the mobile portion of the wetting phase to be removed from the smallest pores. The course of such displacement can be followed by the so-called capillary pressure curve which is likewise a picture of the pore diameters in the rock under consideration. Capillary pressure curves are shown in fig. 5-8 for a coarse grain and an argillaceous fine grain sandstone. The greater part of the wetting phase is driven out at modest capillary pressure, which corresponds to a medium pore radius. If it is assumed that the mean pore radius of unconsolidated sediments is of the same order of magnitude as grain radii, the capillary pressure for expulsion of water by oil (p_{co}) and of water by gas (p_{cg}) can be calculated. Such data are given in table 5-3. It was assumed that the contact angle in water is 0° (complete wetting) and that the boundary surface tension for water/oil = 25 dyn/cm and for water/gas = 70 dyn/cm.

Table 5-3. Capillary pressures for replacement of water by oil (p_{CO}) and of water by gas (p_{CG}) in sediments of different grain size (eq. 5.26). r = mean pore radius.

Sediment	r	pCO	pCG
	[cm]	[at]	[at]
Clay	10^{-5}	5	14
Silt	10^{-4}	0.5	1.4
Sand	10^{-3}	0.05	0.14

These data are important in understanding the movement of gas bubbles and oil droplets in the water filled pore space of a variable grain size sediment. For example, if an oil droplet is located right at the boundary between clay and sand, it will experience a capillary force, which drives it from the narrow pore of the clay into the wider pores of the sand. The effective pressure results from the difference between capillary pressures in sand and clay. For gas bubbles a pressure of about 14 atm. is in effect; for oil droplets about 5 atm. Once the more poorly wetting phases, oil and gas, are trapped in coarse porous rocks, clays act practically as impermeable barriers, since the high pressure necessary to overcome the capillary pressure generally is not available. Oil and gas are found always to accumulate in coarse grained sedimentary rocks, whereas fine clays contain at most only trace amounts. This is true in spite of the fact that the argillaceous rocks have contained the largest amounts of original organic matter, from which hydrocarbons are derived. The reasons for these facts are the migration processes, whose direction is determined by wetting relations and capillary pressures.

If oil, gas, and water occur together in a homogeneous rock, the immiscible phases arrange themselves according to the magnitude of their densities. Boundaries between individual phases, when at rest, run always horizontal, that is, perpendicular to the force of gravity. An inclined boundary interface between phases of different densities always indicates fluid motion. The angle α, between the horizontal and the interface between two phases of different density, is related to densities, viscosities, and flow velocities of the two phases. The following expression applies (see v. ENGELHARDT 1960, p. 120):

$$\sin \alpha = \frac{1}{gk(\varrho_2 - \varrho_1)} (\mu_1 q_1 - \mu_2 q_2) \qquad (5.28)$$

ϱ_1 and ϱ_2, μ_1 and μ_2 are the densities and viscosities of both media; q_1 and q_2 are components of their flow velocity parallel to the boundary surface.. Subscripts 1 and 2 refer to the light and heavy phases respectively. In the static case $q_1 = q_2 = 0$ and the interface is horizontal.

If the heavy phase (2) is stationary and the upper lighter phase (1) is flowing, then:

$$\sin \alpha = \frac{1}{gk(\varrho_2 - \varrho_1)} \mu_1 q_1 \qquad (5.29)$$

sin α is positive, that is, the phase boundary ascends in the direction of flow and the inclination increases with flow velocity and viscosity. Conversely, sin α is negative, the boundary being inclined downward in the direction of movement, when the heavy phase flows and the light one above is stationary:

$$\sin \alpha = \frac{-1}{gk(\varrho_2 - \varrho_1)} \mu_2 q_2 \qquad (5.30)$$

If both phases are flowing the resulting boundary inclination is defined by equation (5.28). It is to be noted that the boundary plane is horizontal, not only in cases of complete rest, but also when the flow velocities of the two phases are such as just to counter-effect the viscosity differences.

If both phases flow with equal velocity, the phase boundary will be inclined always, if the viscosities are different. In the case, for example, of water and the overlying but higher viscosity oil, the phase boundary is inclined in the direction of flow.

The boundary between the water- and gas-filled areas of the pore space are bounded sharply only in a very coarse-grained sand. In fine grain sediments, because of capillary forces, a more or less broad transition zone exists, in which the different phases occur admixed together. The phase distribution in such a transition zone can be perceived with the help of the capillary pressure curve.

The transition zone between a heavy (water) and a light phase (oil or gas) disappears, when the pores are very large. The capillary pressure is essentially zero. The height of the phase boundary is such a situation is chosen as the reference level ($p_c = 0$; $z = 0$). If the rock contains smaller capillaries, a capillary pressure greater than zero prevails, which indicates how much higher the pressure in the lighter and poorer wetting phase is than in water. The water in this case is elevated in the capillaries as far above the reference level, so that the product of the density difference, gravitational constant, and height z above the reference level is equal to the capillary pressure:

$$g z (\varrho_2 - \varrho_1) = p_c \tag{5.31}$$

By combining with equation (5.26) one obtains for the height z of the phase boundary between water and lighter phases in capillaries of radius r:

$$z = \frac{2 \gamma \cos \Theta}{r (\varrho_2 - \varrho_1) g} \tag{5.32}$$

However, every sediment contains pores of different sizes. In the smaller voids the water is higher, and in the large voids less high above the reference level. The relative saturation with respect to water and light phases within transition zones resulting in this way depends on the pore size distribution in the rock being considered. This is expressed by the capillary pressure curve. The capillary pressure curve reproduces the relations between capillary pressure and saturation with respect to the non-wetting (light) phase. However, since the capillary pressure, in accordance with the density difference (equation 5.31) within the transition zone, increases linearly with height above the reference level, the capillary pressure curve describes the saturation in the lighter phase as a function of height above the reference plane. Capillary pressure curves, such as shown in fig. 5-8 reproduce directly the picture of saturation distribution with increasing height above the reference plane, if one divides the values for p_c (in dyn/cm²) by $(\varrho_2 - \varrho_1)$ g. The scale of the abscissa then gives the height in cm.

In summary, the following applies to the nature of transition zones between water and lighter, poorly wetting phases (oil, gas, also air in soils): sediments with a narrow range of pore diameters (well-sorted sands) give rise to thin transition zones. In argillaceous rocks containing voids of quite varied size, the transition zone can attain very great dimensions. In like sediments the thickness of a transition zone depends on the difference in density. The zone is more extensive the lower this difference. Between oil and water the transition zone is always broader than between water and gas.

5.3 Formation waters[2]

5.31 General

At the time of deposition sands contain about 40 volume % and clays up to 90 volume % water. As such new deposits are buried and subside, the pore volume is reduced under the weight of the overburden, and the enclosed fluid must escape. During diagenetic compaction a flow of pore solutions permeates the sediment. These

---

[2] See the following summaries: CHEBOTAREV (1955), SCHOELLER (1956), CHAVE (1960), v. ENGELHARDT (1960, 1961, 1967), WHITE (1957, 1965), DEGENS & CHILINGAR (1967).

solutions react with constituent minerals and chemical processes take place which, together with the mechanical rearrangement and textural compaction, produce a solid sedimentary rock from the loose accumulation of individual particles. Knowledge of the pore solutions enclosed in porous sediments and of their movement in the sediment is just as important in understanding diagenesis as is the study of structural and mineralogical changes.

While sedimentary rocks in various stages of diagenesis are readily available from outcrops and boreholes, data on the composition of pore solutions are much more difficult to obtain. At the surface porous sediments are desiccated, that is, filled with atmospheric gases, or they are saturated with non-saline ground water. Only at greater depths do pore solutions occur which are taking or have taken part in processes of chemical diagenesis. Such solutions can be obtained only from deep bore holes or mines.

In exploration for petroleum and natural gas, deep drilling has penetrated almost all sedimentary basins of the world. As a result we know that porous sedimentary rocks are saturated below the usually non-saline ground water to depths of several thousand meters. They are saturated predominantly with saline solutions, and occasionally with petroleum and natural gas. A very large number of analyses of formation waters obtained from deep wells are available. In addition electric well log analysis provides additional information. From these we can obtain data on the concentrations of pore solutions in permeable beds.

In evaluating chemical analyses of water samples from deep wells, different sources of error must be kept in mind. Since drilling is accomplished using a drilling fluid containing different chemical additives, and because this fluid penetrates permeable beds, contamination of the sampled formation water is possible. Also it is often technically difficult to close off the sampled horizon, so that entrance of water from other beds is prevented completely. Finally formation water can be extracted only from relatively permeable beds. All water analyses, therefore, refer to solutions contained in permeable beds. Very little is known yet concerning the chemistry of pore solutions in clays an low permeability limestones.

The pore solutions contained in sedimentary beds can not be considered as connate water in the strictest sense. That is, it does not represent unaltered fossil sea water. It does contain still constituents of the water entrapped during deposition, but diagenetic reactions have changed their relative abundances and additional matter has been added. Such reactions involve reactions with minerals, dissolution and neoformation, and cation exchange, as well as ion filtration by passage through argillaceous layers. Because of increasing compaction with depth in a subsiding sedimentary sequence, a slow upward flow of pore solution occurs. The permeable beds serve both as reservoirs and conduits for flow. The pore solution found in a permeable bed originates to a large extent from the compaction of other beds, especially clays, and is not the solution that originally filled the pores of the porous sediment. In addition surface waters can penetrate permeable layers at depth and mix with the solutions in the sediments. In certain instances one must contend with the addition of juvenile solutions from the deep subsurface.

Thus formation waters contained in older sedimentary rocks have experienced always a long history, which depends in each individual basin on its special geologic development, on tectonic structure and hydrologic relations (pressure distribution and flow directions). In spite of all the individual differences, some generally valid conclusions concerning the course of pore water diagenesis can be deduced from the many water analyses. These will be developed in the following sections with the aid of some selected examples.

5.32 Composition of formation waters

As an example of the amount and composition of pore solutions stored in a sedimentary basin, one can consider the well-investigated West Canadian Basin (HITCHON 1968). Between the United States border on the south, the 60th parallel on the north, the Rocky Mountains in the west, and the pre-Cambrian shield in the east, the basin covers an area of 1,260,772 km² (fig. 5-9). The basin is filled with sediments ranging in age from middle Cambrian to Tertiary, having a mean thickness of 1778 m and occupying a volume of 2,242,074 km³. Table 5-4 (HITCHON 1968) shows the calculated volumes of sandstones, shales, carbonates, and evaporites contained in the individual formations, as well as their mean porosities. The basin contains collectively 11% evaporite rocks. The carbonates consist of 44% dolomites, and 56% limestones. Table 5-5 gives a summary of the pore space distribution. The porosity averaged over the entire basin is 12%; that is, in this sediment sequence 265,000 km³ of pore space exists today, and it is essentially filled with aqueous solutions.

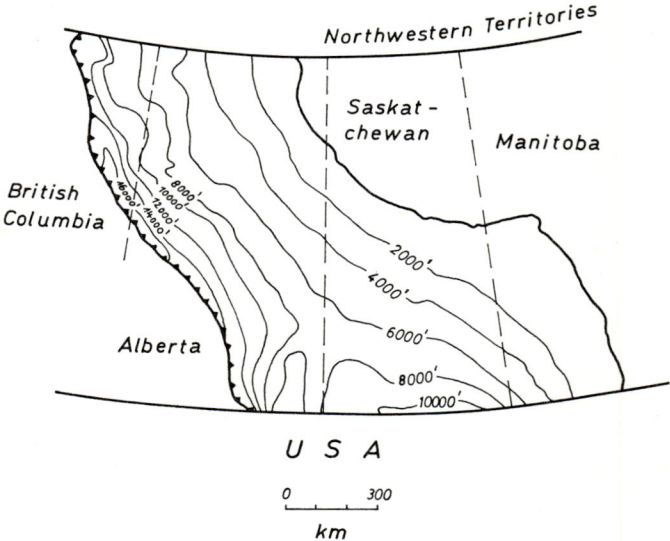

Fig. 5-9. West Canadian sedimentary basin (HITCHON 1968). (Depth contours in feet.)

5.32 Composition of formation waters

The volume of pore solution which was expelled from the basin during diagenetic compaction of the rocks can be estimated, assuming reasonable values for the original sediment porosities. The results are summarized in table 5-6. It was assumed that the original pore space was 40% for sands, 70% for clays and 50% for the limestones. The estimated volume of pore solution expelled from the basin during diagenesis is more than ten times the amount still contained in the sediments today and is somewhat larger than the total volume of the present basin fill.

Since the rocks in this basin are predominantly marine deposits, the original sediments were saturated with sea water. It is informative to compare the concentration and composition of the pore solutions with present sea water, if it is assumed that the chemistry of sea water has not changed essentially since Cambrian time. Table 5-7 shows the mean composition of pore waters contained in the basin rocks. This has been calculated by BILLINGS, HITCHON & SHAW (1969) from numerous analyses of water samples taken in deep wells from individual formations. The volumes and porosities of the individual beds were taken into consideration. The composition of the solutions now filling the pore space no longer corresponds to that of sea water (table 5-8). During diagenesis not only has the amount of pore water been reduced to about 10 percent the original amount, but also changes in the amounts and ratios of the dissolved components have occurred. In comparison to sea water the following differences are noted:

1. The concentration of the pore solutions is higher than that of sea water.
2. The ratios of individual components differ from those of sea water.
   a) Calcium, bicarbonate, and certain trace elements are increased relatively.
   b) Magnesium and sulfate are relatively reduced in amount.
   c) Sodium and chloride ratios are relatively unchanged.

Similar differences in the chemistry of sea water have been noted in many formation waters. This is shown, for example, by the results (table 5-9) of a statistical summary of 1860 formation waters from deep wells in Oklahoma, Kansas, Mississippi, Alabama, and Louisiana (KRAMER 1969). In contrast to the Canadian waters, bicarbonate content in this case is lower than in sea water.

In all formation waters the most important cations are sodium, calcium, and magnesium; the anions are chloride, bicarbonate, and sulfate. Sulfate and magnesium contents in most formation waters tend to be low. The remaining components occur in individual waters in variable ratios to one another. With respect to anions, waters contain predominantly either bicarbonate or chloride. With respect to cations they are either sodium-rich or calcium-rich. Since the main constituents of all formation waters are three cations, Na^+, Ca^{++}, Mg^{++}, and three anions, Cl^-, HCO_3^-, SO_4^{--}, their compositions may be represented by two compositional triangular diagrams. The concentration data (g or mg/l) are recalculated to mole equivalents/l (mole equivalent = moles x valence). Because accessory constituents are minor and the formation waters are neither strongly acid nor alkaline, if the analyses are accurate, the number of cation equivalents/l should equal the number of anion equivalents/l. The total number of equivalents of cations and anions each are set equal

Table 5-4 Volumes and porosities of rocks in the West Canada sedimentary basin (HITCHON 1968).

Formation	Sandstones km³	%	ε	Shales km³	%	ε	Carbonate rocks km³	%	ε	Evaporites km³	%	Total km³	% of whole basin
Tertiary	7 597	25.0	30	22 789	75.0	25	—	—	—	—	—	30 386	1.3
Upper Cretaceous (Post Colorado)	22 558	7.8	30	267 029	92.2	25	—	—	—	—	—	289 587	12.9
Upper Cretaceous (Second Specks — First Specks)	7 824	5.2	12	143 045	94.8	15	—	—	—	—	—	150 870	6.8
Upper Cretaceous (Fish Scales — Second Specks)	14 732	8.7	10	155 476	91.3	15	—	—	—	—	—	170 208	7.6
Lower Cretaceous (Lower Colorado)	8 877	6.9	15	118 888	93.1	15	—	—	—	—	—	127 749	5.7
Lower Cretaceous (Mannville)	74 558	40.8	20	108 012	59.2	15	—	—	—	—	—	182 570	8.1
Jurassic	13 968	26.3	15	30 721	57.7	15	8 157	15.3	10	362	0.7	53 209	2.0
Upper Triassic	2 371	18.7	10	6 215	49.2	15	3 403	26.9	10	654	5.2	12 644	0.6
Middle — Lower Triassic	4 337	13.6	10	27 438	86.3	15	13	0.1	10	—	—	31 787	1.4
Permian	5 891	58.6	10	1 543	15.3	10	2 630	26.1	10	—	—	10 065	0.4
Carboniferous	2 129	1.0	10	71 222	35.5	10	126 076	62.8	10	1 344	0.7	200 770	9.0
Upper Devonian, Wabamun	—	—	—	17 007	19.5	10	62 402	70.6	10	8 637	9.9	87 298	3.9
Upper Devonian, Winterburn	—	—	—	9 012	81.6	10	42 649	81.6	5	633	1.2	52 295	2.3
Upper Devonian, Woodbend	—	—	—	118 668	55.1	5	89 541	41.6	7	7 043	3.3	215 251	9.6
Upper Devonian, Beaver Hill Lake	—	—	—	35 279	25.3	5	91 558	65.6	7	12 640	9.1	139 477	6.2
Middle Devonian	9 350	4.7	10	31 581	15.7	5	68 128	33.8	10	92 083	45.8	201 142	9.0
Silurian	—	—	—	1 487	5.0	5	27 899	93.8	7	358	1.2	29 744	1.3
Ordovician	5 869	8.1	10	10 960	15.0	5	52 991	72.6	7	3 118	4.3	72 938	3.3
Cambrian	67 645	36.8	10	99 628	54.1	5	16 682	9.1	7	—	—	183 956	8.6
Total Basin	247 690	11.0		1 276 002	56.9		591 379	26.4		126 874	5.7	2 241 946	100.0

5.32 Composition of formation waters

Table 5-5. Pore space in rocks of the West Canada basin (HITCHON 1968).

	Sandstones			Shales			Carbonates			Total pore space		
	mean poro-sity	km³	% pore space of basin	mean poro-sity	km³	% pore space of basin	mean poro-sity	km³	% pore space of basin	mean poro-sity	km³	% pore space of basin
Cenozoic	30.0	2 280	0.9	25.0	5 698	2,1	—	—	—	26.3	7 978	3.0
Mesozoic	18.4	27 421	10.4	17.1	147 214	55.5	10.0	1 159	0.4	17.3	175 794	66.3
Paleozoic	10.0	9 086	3.4	6.2	24 762	9.4	8.2	47 494	17.9	6.8	81 343	30.7
Total Basin	15.7	38 787	14.7	13.9	177 674	67.0	8.2	48 653	18 3	11.8	265 114	100.0

Table 5-6. Pore solutions expelled from sediments of the West Canada basin during diagenesis.

	Pore space present today km³	Pore space at time of deposition km³	Pore solution expelled during diagenesis km³
Sandstones	38 787	139 267	100 480
Shales	177 673	2 562 781	2 385 108
Carbonate rocks	48 653	542 726	494 073
Total:	265 114	3 244 775	2 979 661

Table 5-7. Mean composition of pore solutions contained in sediments of the West Canada basin (BILLINGS, HITCHON & SHAW 1969).

	mg/l	Standard variation	Range mg/l	Mol/l	Equiv.%	Conc. in solution: Conc. in sea water
Dissolved solids	46 438	94 654	1 004 — 328 889	...	...	1.4
Cl	26 917	57 813	6 — 190 594	0.759	96.173	1.4
SO₄	350	789	2 — 4 777	0.0036	0.912	0.13
HCO₃	1 496	1 219	18 — 7 749	0.023	2.914	10.7
Na	14 342	26 040	132 — 132 597	0.624	82.438	1.3
K	561	2 711	2 — 8 800	0.0144		1.5
Ca	2 212	9 815	3 — 38 682	0.055	14.205	5.5
Mg	317	1 193	2 — 3 994	0.013	3.357	0.25
Br	114	300	<1 — 1 121	...	...	1.8
I	9	9	<1 — 39	...	...	180
Sr	108	361	<1 — 1 320	...	...	13.5
Rb	0.88	4.55	0.01 — 18.80	...	...	4.4
Li	10.7	27.5	0.1 — 100.0	...	...	56.3
Zn	0.30	3.14	0.03 — 27.50	...	...	30.0
Cu	0.12	0.01	0.02 — 0.49	...	...	40.0
Fe	7.49	16.35	0.01 — 87.00	...	...	749.0
Mn	0.32	5.76	0.02 — 52.00	...	...	160.0

Table 5-8. Mean composition of sea water (CULKIN 1965).

	mg/l	% of dissolved solids	mol/l	Equiv. % anions	Equiv. % cations
Cl	19 353	55.07	0.546	90.35	
SO₄	2 712	7.72	0.028	9.27	
HCO₃	142	0.40	0.0023	0.38	
Na	10 760	30.62	0.4680		78.72
Ca	413	1.18	0.0103		1.78
Mg	1 294	3.68	0.053		17.83
K	387	1.10	0.0099		1.67
Remainder	80	0.23			
Dissolved solids	35 141				

5.32 Composition of formation waters

Table 5-9. Comparison of the mean composition of 1860 formation waters from deep bore holes in Oklahoma, Kansas, Mississippi, Alabama, Arkansas, and Louisiana with present sea water (KRAMER 1969).

(The range of variation corresponds to a standard deviation about the mean value.)

	Mole percent	
	Formation waters	Sea water
Cl	51 — 54	49
SO_4	0.08 — 1.2	3.4
HCO_3	0.07 — 0.06	0.2
Na	36 — 45	42
Ca	3.6 — 7.8	0.9
Mg	0.9 — 2.4	4.8

to 100 and the equivalent percent of each cation and anion calculated. According to these percentage values each formation water is represented by a point in the Na—Ca—Mg- and Cl—HCO_3—SO_4-concentration diagrams.

The cation contents of 250 formation waters from oil fields of the United States are plotted in fig. 5-10 from a compilation by DESITTER (1947). Fig. 5-11 summarizes data from 406 European and North American formation water analyses. The cations of all these formation waters contain no more than 10 equivalent percent magnesium; while the ratios of sodium to calcium vary over wide limits. Usually the waters contain less than 50 equivalent % calcium.

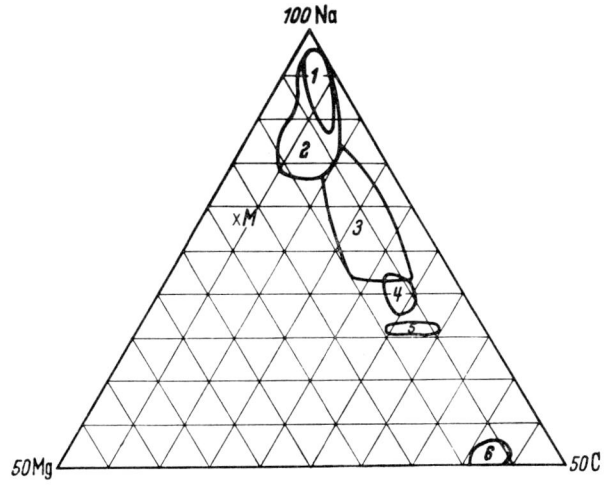

Fig. 5-10. Cation content (equivalent percent) of formation waters from United States oil fields (DE SITTER 1947). 1: Woodbine sand (Cret.), Texas; 2: Tertiary, California; 3: Paleozoic, Kansas and Oklahoma; 4: Mississippian, Appalachians; 5: Upper Devonian, Appalachians; 6: Smackover limestone (Jurassic), Arkansas. M: Sea water.

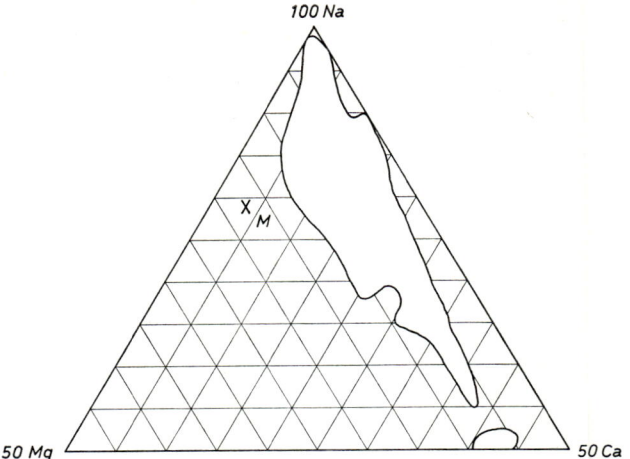

Fig. 5-11. Cation content (equivalent percent) of 406 waters from the following formations: Paleozoic of the United States (250, DE SITTER 1947); Paleozoic of Illinois (30, MEENTS et al. 1956); Permian to Cretaceous, northwest Germany (42, v. ENGELHARDT 1960); Tertiary of the Rhine Valley (23, v. ENGELHARDT 1960); Tertiary of the Vienna Basin (35, KREJCI-GRAF et al. 1957); Tertiary of the Po River Basin (26, v. ENGELHARDT 1960).

Occasionally calcium-rich formation waters occur. As an example the cation contents of waters from Paleozoic beds of the Michigan Basin are plotted in fig. 5-12. These rocks are limestones and sandstones of Silurian, Ordovician, Devonian, and Mississippian age. These are chloride waters containing only minor amounts of bicarbonate and sulfate.

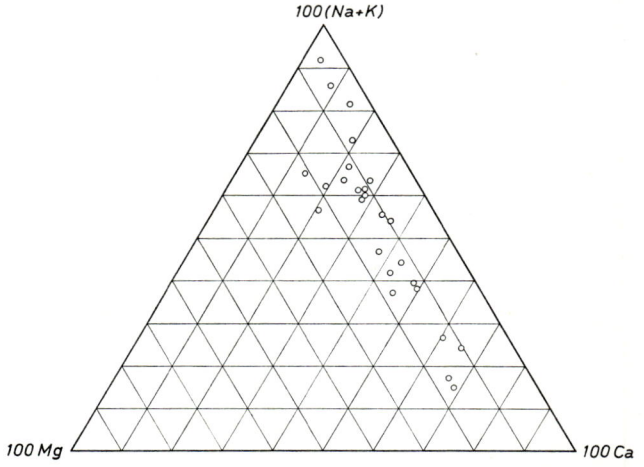

Fig. 5-12. Cation content (equivalent percent) of formation waters from the Michigan Basin (Silurian, Ordovician, Devonian, Mississippian) (after GRAF et al. 1966).

5.32 Composition of formation waters

In most formation waters chloride forms more than 90 equivalent percent of the anions. However, especially in shallow and young formations, waters with high bicarbonate and correspondingly low chloride content occur. As examples, the anion compositions of relatively carbonate-rich formation waters from Tertiary beds of the Vienna Basin are plotted in fig. 5-13.

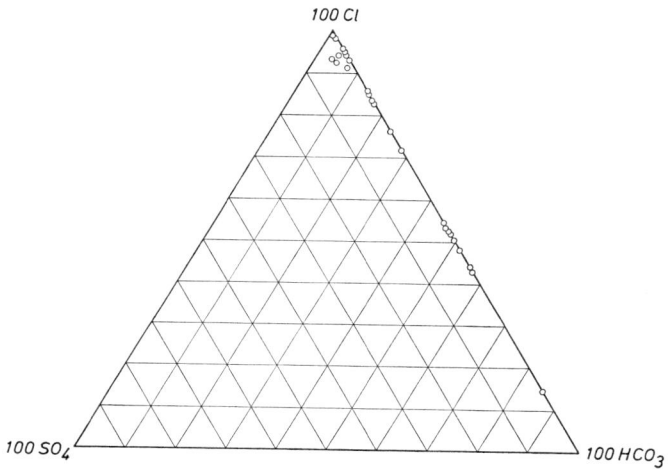

Fig. 5-13. Anion content (equivalent percent) in formation waters of Tertiary strata (Pannon, Sarmat, Torton) of the Gösting structure of the Vienna Basin (after FRIEDL 1956).

The total dissolved solids in formation waters vary over wide limits. Concentrations up to over 200 g/l occur. In many sedimentary basins penetrated by deep wells, the salt content of formation waters generally tends to increase with depth, as evidenced by electric well log measurements and by water analyses (see, for example, v. ENGELHARDT 1960, 1961; DICKEY 1969, and his literature citations).

According to a summary compilation by DICKEY (1969), in different formations in Oklahoma, Arkansas, Lousiana, and Texas, the concentration increase with depth is at first essentially linear. The gradient lies usually between 80 mg/l/m and 250 mg/l/m. In deep horizons the concentration often increases more slowly with depth, sometimes actually decreasing. For example in Gulf Coast sediments the amount of dissolved solids increases essentially linearly to a depth of 1200 m to about 80 g/l. Between 1200 and 2000 m depth the concentration remains about constant and then decreases down to 3000 m to about 40 g/l.

As an example for considering formation water concentrations in a large, relatively isolated sedimentary basin, one can consider in detail the Illinois Basin which has been penetrated by many deep wells. Many water analyses have been obtained from the strata in this basin (MEENTS et al. 1956; BREDEHOEFT et al. 1963; GRAF et al. 1966). The sedimentation began on the pre-Cambrian basement in late Cambrium time. Above the Cambrian follow a series of beds of Ordovician, Silurian, Devonian, lower and upper Mississippian, and Pennsylvanian ages. The total sedimentary sequence in the deepest part of the basin is more than 4000 m thick. The early Paleozic beds crop out around the basin 300 to 400 m above the present surface of the basin interior. Only in the southernmost part of the basin were the Paleozic strata covered by younger sediments. The structure of the basin is shown in fig. 5-14.

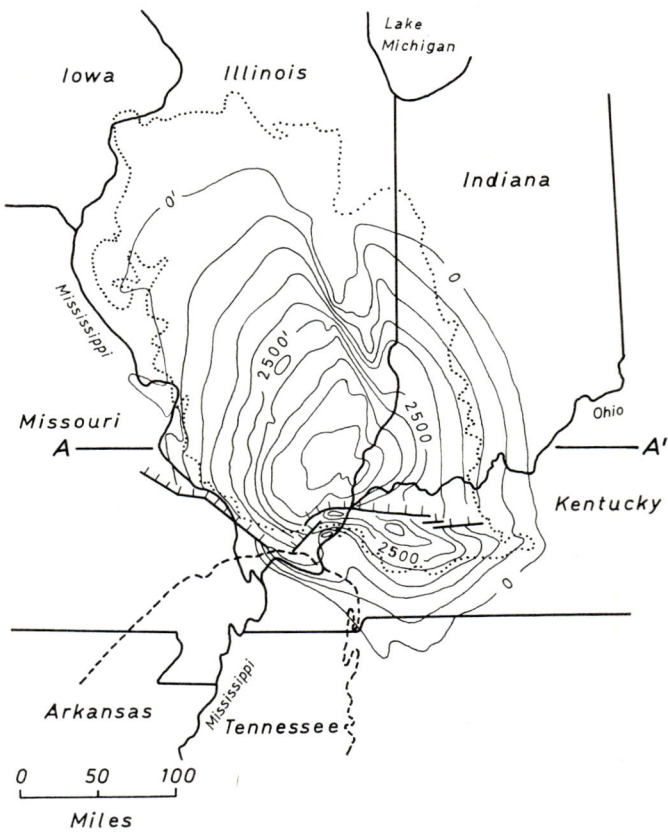

Fig. 5-14. Structure of the Illinois Basin. Indicated are the depth contours to the base of the Mississippian (in feet). Dotted line represents Pennsylvanian outcrops. A—A' is the line of the section in fig. 5-18.

The salt contents of waters obtained from individual formations from well holes in the central part of the basin (Marion County) are plotted in fig. 5-15. From the Pennsylvanian to the deeper Mississippian sediments the salinity increases quite regularly with depth. All salt contents lie above that of sea water (35 g/l). The lowest salt content was found to be 88 g/l in the Pennsylvanian, and the highest was 143 g/l in Mississippian sediments. This corresponds to a mean gradient of 70 mg/l/m. Devonian and Silurian rocks also contain saline solutions, which are more highly

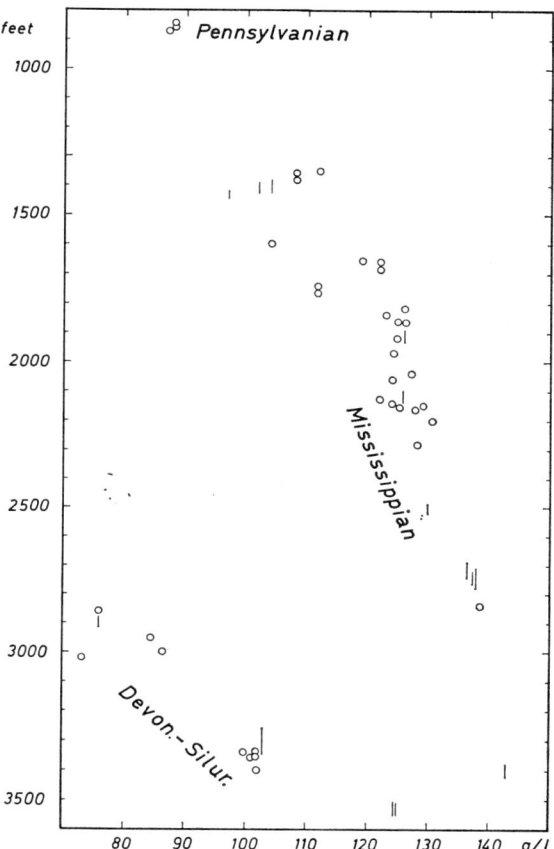

Fig. 5-15. Salinities of formation waters from deep bore holes in the central part of the Illinois Basin (Marion County) (after MEENTS et al. 1956).

concentrated than sea water. Apparently in this part of the basin no direct exchange exists between the Devonian and Silurian formation waters and those in the higher formations. In the Silurian-Devonian rocks the salinity of 75 g/l is lower than the overlying Mississippian rocks. This salinity then increases with a gradient of 128 mg/l/m up to about 280 g/l.

Fig. 5-16 shows the increase in salinity with depth within a given formation, the St. Genevieve limestone of lower Mississippian age. In this plot all water analyses for this formation from the entire Illinois Basin, both from MEENTS and co-workers (1956) and GRAF and co-workers (1966), were utilized. The values exhibit a broad scattering. Still an increase in concentration is clearly recognizable; the rate of rise in salinity decreases at greater depths where a value of 150 g/l is reached, that is, a concentration about 5 times that of sea water. The isocons (lines of equal

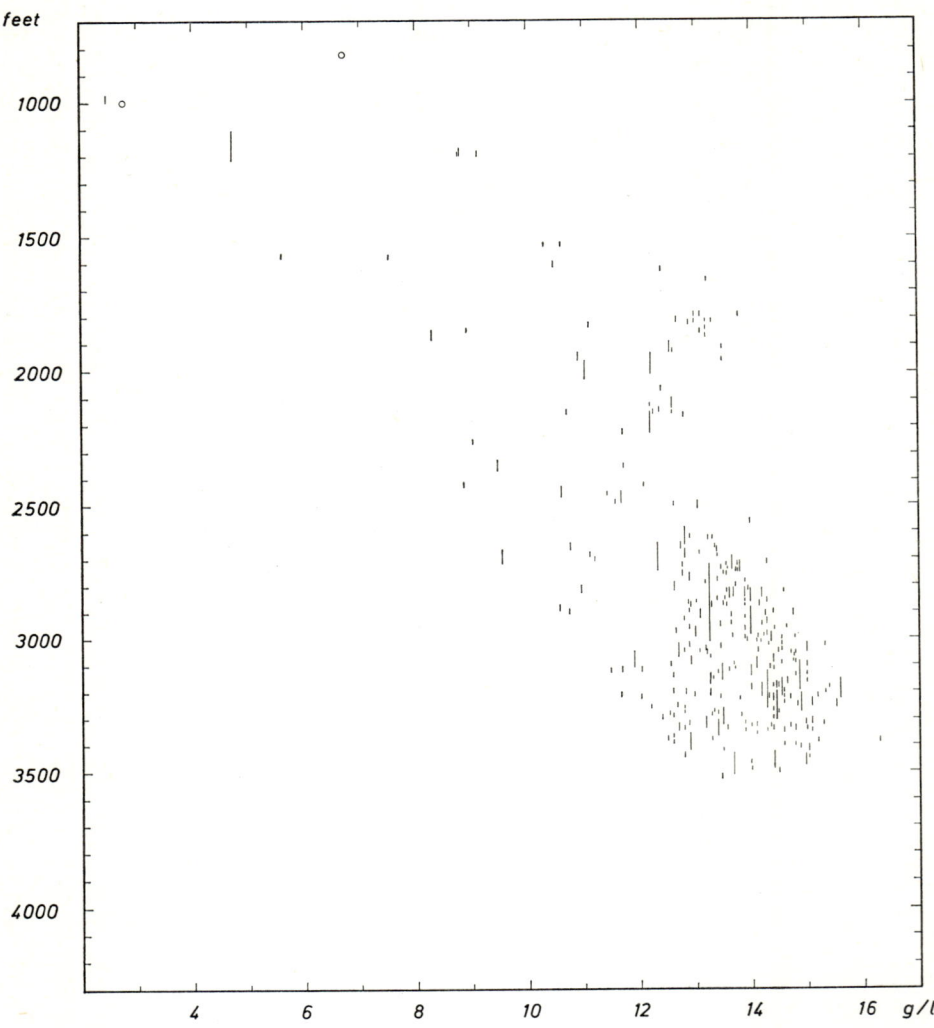

Fig. 5-16. Salinities of formation waters from the St. Genevieve limestone (lower Mississippian) of the Illinois Basin (after MEENTS et al. 1956).

formation water concentration) drawn in by MEENTS and co-workers for this limestone (fig. 5-17) conform approximately to the structure contours of the basin.

Fig. 5-18 shows in an east-west cross section the elevation of the base of the Mississippian and the concentration of formation waters in the St. Genevieve limestone along this section. The most concentrated water is found in the deepest part of the basin. Both east and west from the center the concentration falls off, although at different rates, as the limestone approaches the surface.

5.32 Composition of formation waters 221

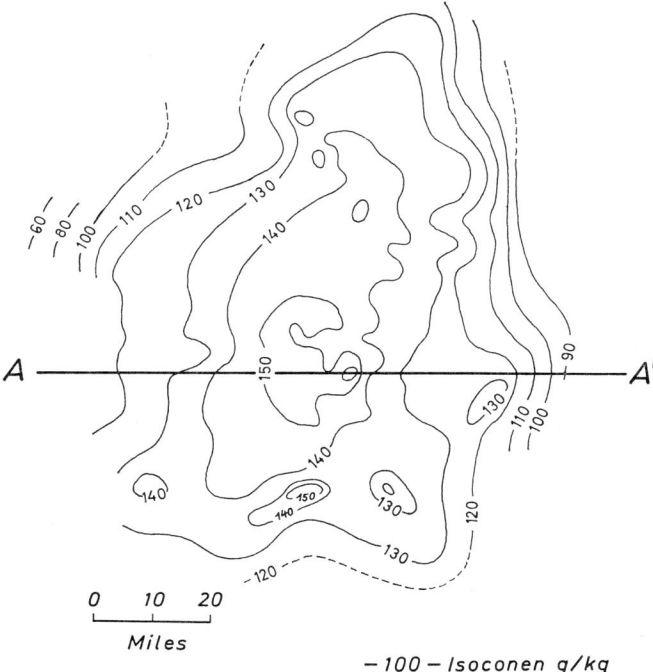

Fig. 5-17. Isocones (lines of equal salinity) of the formation waters of the St. Genevieve limestone in the Illinois Basin. A—A' is the line of section in fig. 5-18 (after MEENTS et al. 1956).

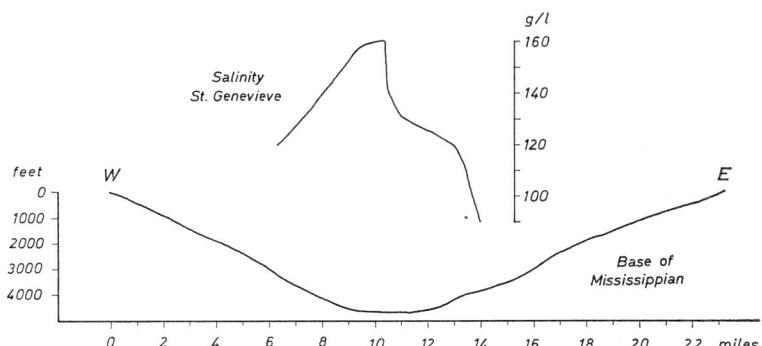

Fig. 5-18. Salinity of formation waters of the St. Genevieve limestone and depth to the base of the Mississippian along an E—W section (see figs. 5-17 and 5-14) through the central Illinois Basin (after MEENTS et al. 1956).

The composition of formation waters in the Illinois Basin is similar to waters of the West Canadian Basin (table 5-7). Here also the dominant anion is chloride. Sodium predominates among the cations, followed by calcium and magnesium.

Table 5-10. Mean composition of formation waters from deep bore holes in the Illinois Basin. 35 analyses of waters from the Pennsylvanian and Mississippian (GRAF et al. 1966). 15 analyses from Devonian and Silurian of the central basin (MEENTS et al. 1956).

Pennsylvanian and Mississippian:

	mg/l	mol/l	Equiv. %
Cl	52 205	1.472	98.94
SO_4	589	0.0061	0.83
HCO_3	216	0.0035	0.23
Na [1]	27 338	1.189	81.35
Ca	3 485	0.0870	11.90
Mg	1 199	0.0493	6.75
Dissolved solids	85 011	...	...

Silurian and Devonian:

	mg/l	mol/l	Equiv. %
Cl	50 612	1.428	99.42
SO_4	243	0.00252	0.35
HCO_3	200	0.00328	0.23
Na [1]	26 080	1.134	79.08
Ca	3 748	0.0935	13.04
Mg	1 373	0.0565	7.88
Dissolved solids	84 029	...	...

[1] Includes a small amount of K.

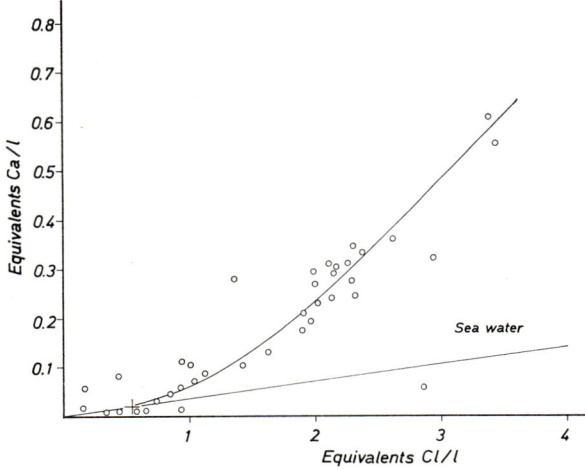

Fig. 5-19. Ca- and Cl-contents in formation waters of the Illinois Basin. + = sea water. The straight line corresponds to solutions resulting from the evaporation of sea water.

5.32 Composition of formation waters

The dependence of calcium and magnesium content on chloride content, which is plotted in figs. 5-19 and 5-20, shows that these waters could not be derived by the simple concentration of sea water. The equivalent ratio Ca:Cl = 0.037 in sea water. In the formation waters this ratio is higher and rises with increasing chloride content, that is, with increasing concentration. The reverse is true with respect to magnesium. In sea water the equivalent ratio Mg:Cl = 0.197. In the formation waters this ratio is lower and falls off with increasing concentration. Since concentration increases with depth and since deep lying waters represent the developing state of the pore water diagenesis, it can be concluded that diagenesis has resulted in a progressive reduction in the relative concentration of magnesium and a relative enrichment of calcium.

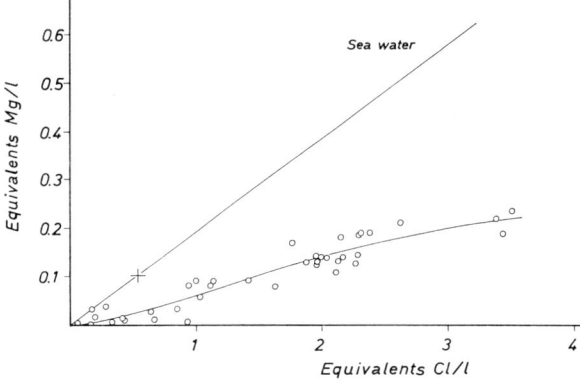

Fig. 5-20. Mg- and Cl-content in formation waters of the Illinois Basin. + = sea water. The straight line corresponds to solutions resulting from the evaporation of sea water.

CLAYTON and co-workers (1966) have determined the deuterium (D) and O^{18}-contents of the pore waters of the Illinois Basin. Taking the contents in present sea water as a basis, relative differences were determined:

$$\delta D\,(\%) = \frac{(D/H)_{\text{sample}} - (D/H)_{\text{sea water}}}{(D/H)_{\text{sea water}}} \times 100$$

$$\delta O^{18}\,(\%) = \frac{(O^{18}/O^{16})_{\text{sample}} - (O^{18}/O^{16})_{\text{sea water}}}{(O^{18}/O^{16})_{\text{sea water}}} \times 100$$

The results are reproduced in figs. 5-21 and 5-22. The deuterium contents of the formation waters are quite different from that of sea water. Waters of lowest salinity contain about as much deuterium as present surface water (Lake Michigan). With increasing salinity the deuterium content of formation waters increases. The O^{18}-contents behave similarly. In this case also low salinity formation waters contain about as much as O^{18} as surface waters. With increasing salinity O^{18}-content increases. The O^{18}-content is still low at a salinity corresponding to that of sea water.

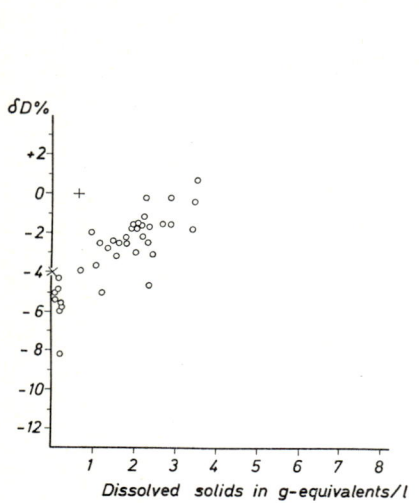
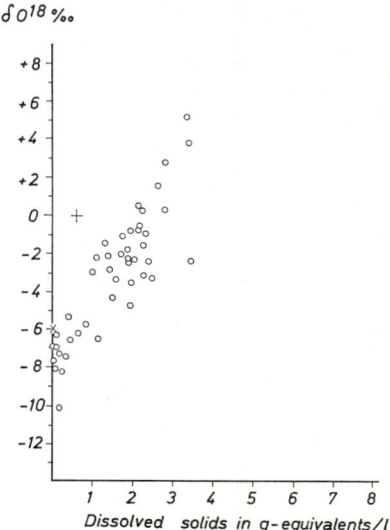

Fig. 5-21. Deuterium content of formation waters of the Illinois Basin. ✕ = Lake Michigan water; + = sea water (CLAYTON et al. 1966; GRAF et al. 1966).

Fig. 5-22. O^{18}-content in formation waters of the Illinois Basin. ✕ = Lake Michigan water; + = sea water (CLAYTON et al. 1966; GRAF et al. 1966).

The deuterium and O^{18}-contents show, as do the magnesium and calcium contents, that the formation waters of the basin could not have been derived by simple concentration of sea water. The low contents of heavy isotopes in the low salinity formation waters permit us to presume that these solutions contain infiltrated surface water. However, the formation waters could not have originated either by sample mixing of surface and sea water. This would require that plotted points for formation waters would lie on a line connecting the surface and sea water compositions. Therefore, it must be assumed that the sea water originally trapped in the sediments was on the one hand diluted by water penetrating from the surface, and on the other hand, that during the diagenetic alteration of pore solutions, enrichment of the heavy hydrogen and oxygen isotopes took place. The possible causes of such a change in isotope ratios are considered in the following section.

In the formation waters of the Michigan, Gulf Coast, and West Canadian basins, similar dependence of deuterium and O^{18}-content on salinity has been established. In these cases also the low salinity formation waters have the same heavy isotopes as the surface waters, and they increase with increasing depth or salinity (CLAYTON et al. 1966, HITCHON & FRIEDMAN 1969).

The fact that the formation waters of deep sedimentary basins contain heavy hydrogen and oxygen isotopes which originate at the surface, indicates that the flow of pore solutions within such basins has taken place, and still does, according to the model depicted in fig. 5-23 (BREDEHOEFT et al. 1963; CLAYTON et al. 1966; GRAF

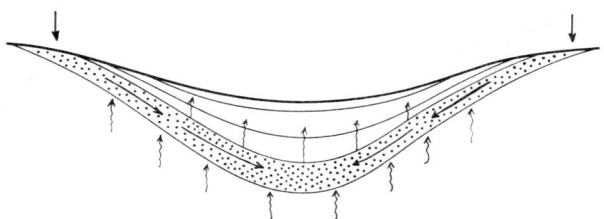

Fig. 5-23. Schematic diagram of flow of formation waters under the influence of compaction and artesian surface waters.

et al. 1966; HITCHON & FRIEDMAN 1969). During the course of geologic time considerable amounts of surface water have flowed into the deeper parts of the basin through permeable beds cropping out at sufficient elevation at the basin margins. This water ultimately enters the interior of the basin. The basin sediment fill becomes permeated not only by the fluids ejected by compaction, but also by an additional artesian flow. The amounts of solutions arising from compaction, previously calculated for the West Canadian basin, therefore, represent only a fraction of the total water which actually has flowed through the basin fill. Compaction flow will have been especially active in the early period of diagenesis, as sediments still were being compacted. In contrast the artesian flow played a greater and greater role, the deeper the basin subsided. Whenever the hydrologic conditions and rock permeability permit, the artesian waters can permeate over long periods of time even consolidated basin sediments.

The composition of formation waters, their alteration during the course of time and the diagenetic processes within the pore space of sediments are dependent not only on the chemistry of water entrapped during deposition. It depends also on the kind, amount, and duration of artesian supply.

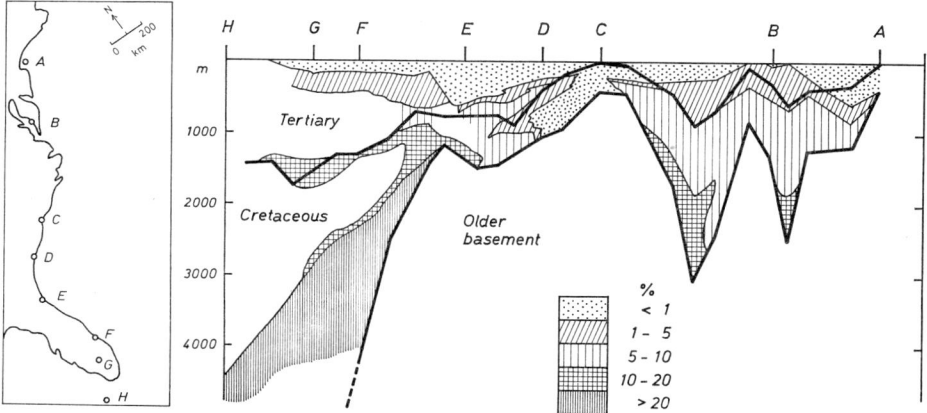

Fig. 5-24. Salinity of formation waters along a section of the North American Atlantic coast (after MANHEIM & HORN 1968).

The mobility of pore solutions in the conduits of permeable strata, the mixing of deep and surface waters and the dependence of these processes on the particular basin structure, its tectonic history, surface relief and temporal changes, all contribute to the differences in concentration and composition of formation waters in a given basin. Some special situations are illustrated in the following selected examples.

Concentrations of formation waters along a 1000 km section along the North American Atlantic coast are illustrated in fig. 5-24 (MANHEIM & HORN 1968). Electric well log measurements and chemical water analyses provided the data. Over the total profile, salinity increases with depth. In the Tertiary sediments low salinity waters are found (< 1 g/l); in the deep lying (2000—4000 m) Paleozoics highly saline solutions are obtained (> 200 g/l). The isocons are irregular, sometimes crossing formational boundaries. At some places more highly concentrated solutions are found above less saline ones. Fig. 5-25 shows a section perpendicular to the Atlantic coast of Florida. In off-shore boreholes in the Atlantic, Tertiary sediments filled with fresh water are found in rocks of the continental segment beneath a thin zone filled with sea water. The interface between salt and fresh water descends toward the open sea. The sloping boundary can be recognized by the fresh water flowing out to sea beneath the saline zone.

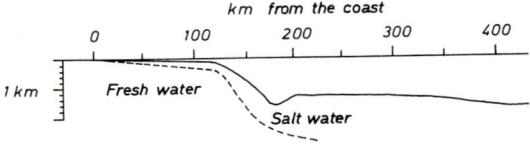

Fig. 5-25. Fresh and salt water distribution in Tertiary strata of the Florida coast (after MANHEIM & HORN 1968).

Another example illustrates how under special hydrologic conditions formation waters can flow outward from the interior of a basin toward its margin. LEMCKE & TUNN (1965) have described this situation for the Tertiary Molasse strata of southern Germany. Fig. 5-26 shows a section through the western part of the Tertiary Molasse basin north of the Alps. Above the southwardly (toward the Alps) dipping basement of karst-bearing upper Jurassic limestones, lies a series of Tertiary sands, marls, and limestones increasing in thickness to the south. Between the fluvial strata of the lower and upper fresh water Molasse lies the marine upper Molasse. In the formation water of the jointed Jurassic limestone a pressure prevails, as a result of connection to the Danube in the north, corresponding to the elevation at the level of the Danube. Since the Danube land surface rises to the south, the pressure in the Jurassic limestone is less than the hydrostatic pressure related to the land surface. In the Molasse strata above the Jurassic the normal hydrostatic pressure prevails instead. Under the influence of this pressure difference a downward flow of formation water results, across and through the Tertiary beds. As a result surface water must penetrate from above. The flow continues through the jointed Jurassic limestone into the Danube. As a result of this vertically downward and northerly directed flow, the salt water originally contained in the marine upper Molasse migrates downward. The boundary

5.32 Composition of formation waters

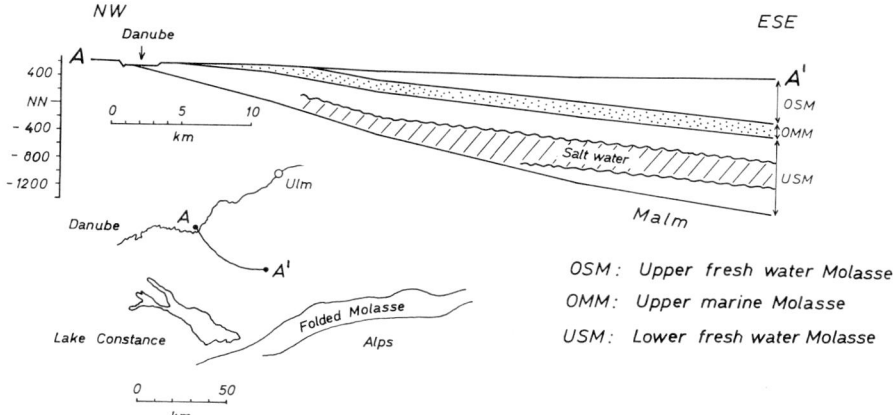

Fig. 5-26. Fresh and salt water distribution in Tertiary strata of the Molasse basin north of Lake Constance (after LEMCKE & TUNN 1956).

between fresh and saline pore water, from electric well log data, today extends below the lower margin of the marine strata in the middle of the lower fresh water Molasse. The boundary should be horizontal under static conditions. Its inclination to the south indicates the flow of the deeper water layers is directed toward the Danube.

In fig. 5-27 are plotted the concentrations and chloride contents of formation waters in the barely consolidated Tertiary strata of the Gösting structure of the Vienna Basin (after analyses by FRIEDL 1956, KREJCI-GRAF et al. 1957). The formation waters stem from sand layers interlayered with clays and marls. From faunal data the Pannon strata were deposited in a fresh-brackish water environment, the Sarmat sediments formed in brackish-saline water (16—30% salinity) and the Torton is actual marine (PAPP 1954). The waters are especially rich in carbonate (fig. 5-13). Among the cations, sodium predominates. The equivalent ratio Na:Ca varies in Pannon waters between 6 and 35, in the Sarmat between 30 and 170, and in the Torton between 25—45. The ratio is sometimes less and sometimes greater than that of sea water (23). The equivalent ratio Na:Mg in the Pannon is between 20 and 30, in the Sarmat between 50 and 100, and in the Torton between 50 and 80, and is, therefore, everywhere much higher than in sea water (4.4). Concentration and chloride content increase with increasing depth. A rapid increase occurs between 1000 and 1300 m, that is, from the middle of the Sarmat on (12th—17th sand horizons). Then at farther depth in the Torton a decrease in salinity is observed. The highest salinity in the deeper Sarmat and upper Torton almost reaches that of sea water. The salinity increase in the lower Sarmat may reflect the change from marine to brackish-limnitic environment between Torton and Sarmat times. It is also possible that a transition zone from saline to non-saline formation waters originally occurring at the boundary between Torton and Sarmat, has migrated upward as a result of the influx of artesian water.

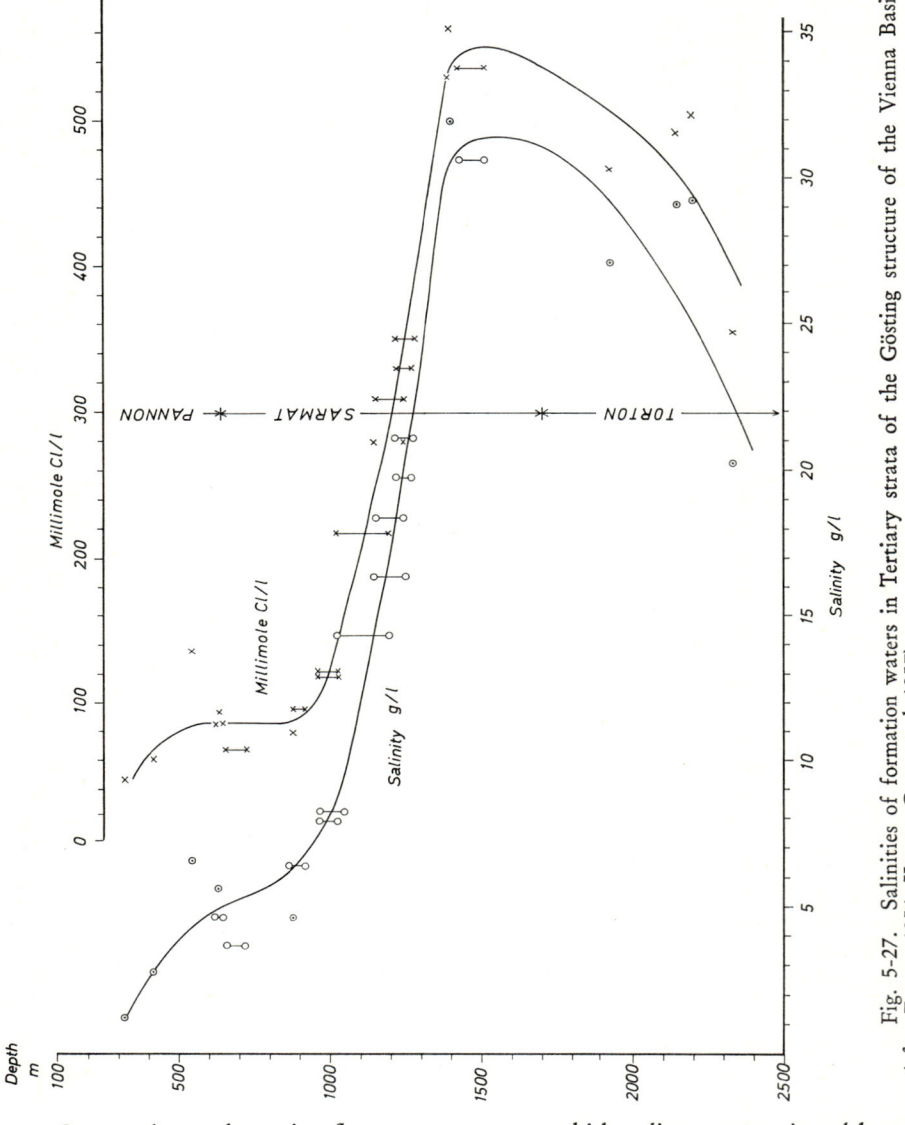

Fig. 5-27. Salinities of formation waters in Tertiary strata of the Gösting structure of the Vienna Basin (after FRIEDL 1956; KREJCI-GRAF et al. 1957).

Compaction and artesian flow permeate across thick sedimentary series, although very low-permeability argillaceous strata form volumetrically the greater part of the formations. Low permeability probably reduces the flow velocity of formation waters, but cannot stop their migration completely. However, where possible, formation waters will use permeable beds as preferred conduits, since in passing from a clay into a more permeable bed, the flow lines are strongly refracted in the perme-

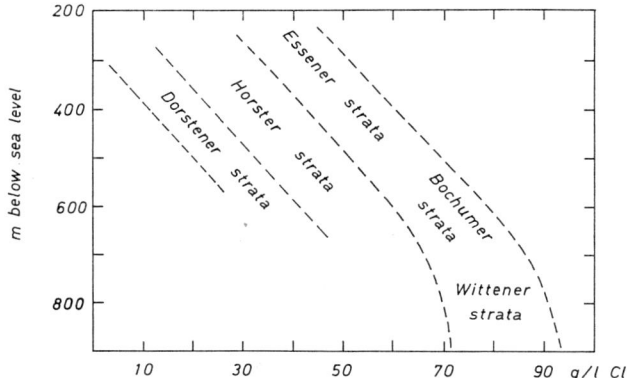

Fig. 5-28. Cl-content of formation waters in sandstones of the upper Carboniferous in the Ruhr region (MICHEL & RÜLLER 1964).

able bed, if the bedding plane is inclined only slightly to the isobars. Therefore, it is understandable that also adjacent to layers of high permeability separated by relatively less permeable strata, waters of different composition occur.

That such differences occur not only in young sedimentary series, but also in older, long-consolidated formations, has been shown by investigations of the compositions of waters in the upper Carboniferous of the Ruhr region (PUCHELT 1964;

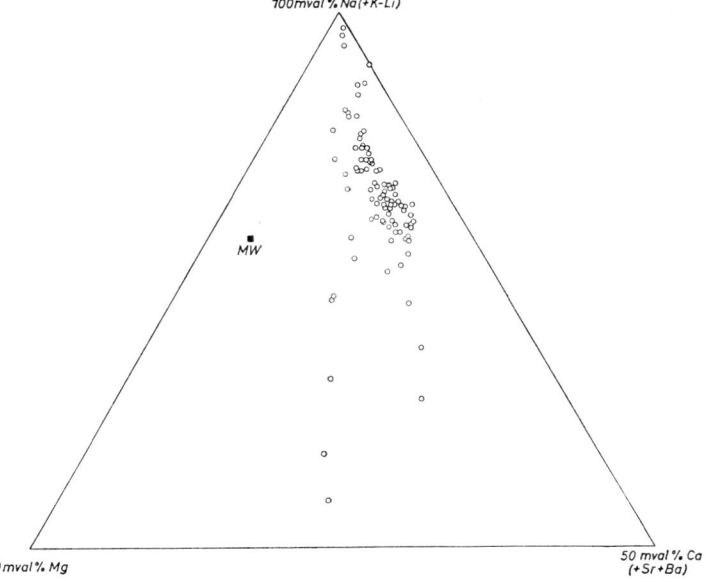

Fig. 5-29. Cation content (equivalent percent) of formation waters of the upper Carboniferous in the Ruhr region (PUCHELT 1964).

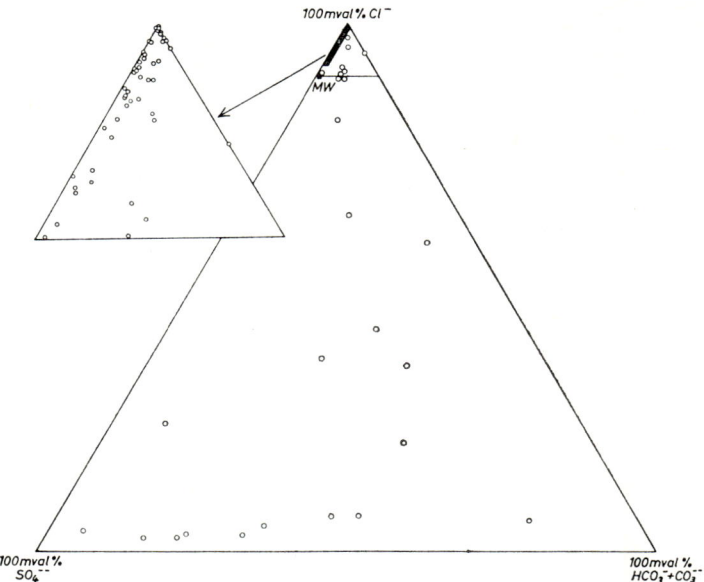

Fig. 5-30. Anion content (equivalent percent) of formation waters of the upper Carboniferous in the Ruhr region (PUCHELT 1964).

MICHEL & RÜLLER 1964). The upper Carboniferous is buried discordantly in the Ruhr region by Cenomanian. Below the Cretaceous the Carboniferous strata dip to the north. From older to younger the following are superimposed: Essener beds, Horster beds, and Dorstener beds. Fig. 5-28 shows the dependence on depth of the chloride content of formation waters from sandstones in these strata. The water samples were obtained from bituminous coal mines. In each stratum the chloride content increases with depth. At a given depth, however, the chloride content varies. It is higher in the Essener than in the Horster strata, which is in turn higher than in the Dorstener sandstones. Figs. 5-29 and 5-30 show the cation and anion relations, which correspond to those of other formation waters. The deep waters contain more than 90 equivalent % chloride. Carbonate- and sulfate-rich waters occur only near the surface. Fig. 5-31 shows along a NW—SE section in the Ruhr area, the spacial distribution of waters with predominantly carbonate, with high sulfate, and with predominate chloride without sulfate.

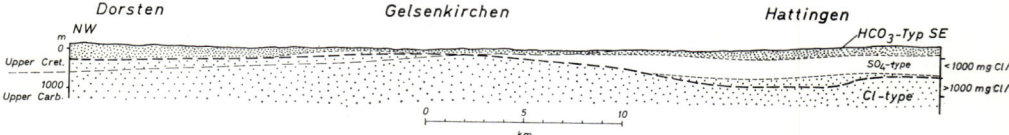

Fig. 5-31. Distribution of carbonate, sulfate, and chloride waters in Cretaceous and upper Carboniferous strata in the Ruhr region (MICHEL 1965).

5.33 Diagenesis of formation waters

Of all the changes which formation waters experience during diagenesis, the increase in total concentration is especially striking. As the specific examples have shown, pore solution concentrations increase quite regularly with depth, often exceeding many times over the salinity of sea water.

Saline solutions could occur as a result of solution of evaporite rocks, especially of halite. In fact there are observations of highly concentrated formation waters in the vicinity of salt bodies, which have originated by solution of solid salt. For example, MANHEIM & BISCHOF (1969) reported that two of six drill holes up to 300 m deep in the Gulf of Mexico (pelitic sediments, Pliocene to recent), drilled above salt domes, showed an increase in formation water salinity. In the other four wells water salinity did not change with depth. LÖHNERT (1967) investigated the distribution of saline waters around the salt body of Altona-Langenfeld in the metropolitan area of Hamburg. Fig. 5-32 shows some of the results of this study.

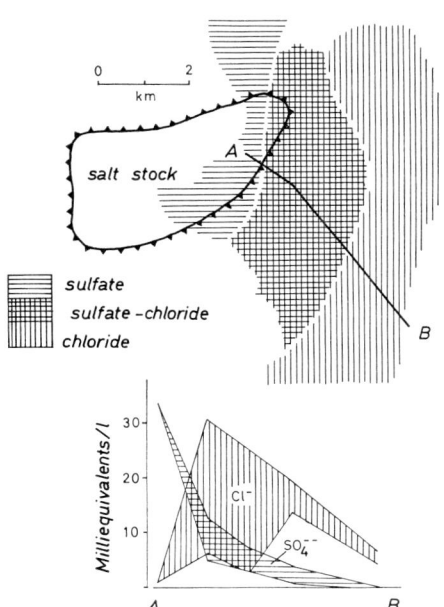

Fig. 5-32. Distribution of saline formation waters in Tertiary sands on the east flank of the salt stock of Altona-Langenfeld, Hamburg (after LÖHNERT 1967).

In Tertiary sands (100—200 m depth) sulfate-rich waters occur close at hand to the salt body, which is covered by a thick gypsum cap. Outward away from the body an approximately 5 km wide zone of saline water follows, with predominant chloride.

In this and similar cases, we are dealing with an aureole of limited extent, which cannot have produced the high salinities of the greater portion of formation waters. That solution processes take place within salt bodies is indicated by the gypsum and brecciated masses which envelop and cover salt plugs and which are interpreted as solution residues. These usually rather thick residual envelopes protect the salt body against further solution by the pore solutions from surrounding rocks. If salt solution in formation waters were to play a significant role, it would not be possible that salt bodies, by virtue of their buoyancy, could ascent practically intact through several thousand meters of sediment overburden. This has happened in many sedimentary basins as in northwest Germany and in the Gulf Coast during the course of geologic time. The actual existence of salt domes and salt plugs indeed speaks against the origin of the high salinities of formation waters by the solution of salt.

That other causes primarily are responsible for salt enrichment in deeper and older formation waters can be concluded from the fact that the increase in salinity with depth is a general phenomenon, observed in basins in which evaporite rocks are absent or occur in only very minor amounts. An example is the Illinois Basin which contains no salt beds and only small amounts of anhydrite. Also the Tertiary sediments of the Po Basin and Rhine Valley contain saline waters, although saline formations are absent or insignificant.

If the increase in salt concentration is not due to an introduction of salt, it must have been brought about by the removal of water. DE SITTER (1947) first advanced the idea that a pore solution, expelled from sediments by compaction, is depleted in ions as it migrates through argillaceous sediments. Non-saline water is expelled while saline formation water remains behind in the permeable strata.

That compressed clays act as ion filters results from their ion exchange capacities. The cation exchange of clays is related to the fact that individual clay particles carry negative electrical charges at their surfaces. These originate from the excess charge produced when hydrogen ions dissociate from OH-groups or from the dissociation of loosely bound cations between silicate layers. As a result a compacted clay forms a framework of negatively charged individual particles, each of which is surrounded by a layer of water containing the mobile cations. The negatively charged particle surface and its compensating cation cloud form a diffuse double layer. If a clay is compressed to the extent that the diffuse double layers contact or penetrate one another, neutral water molecules and cations involved in exchange are able to pass through them. A compacted clay, however, should be largely impermeable to anions, since the negatively charged particle surfaces prevent the passage of negative ions. If one forces an electrolyte solution through such a clay, it should act as an ion filter, and should allow the expulsion of pure water, since cations can not migrate without counter-anions.

Experimentally the effect of clays as semi-permeable membranes can be demonstrated. This is accomplished by determining the electrochemical potential between two salt solutions separated by a clay layer, and by actual filtration of salt solutions.

The electrochemical potential between two salt solutions of different concentration (c_1, c_2), without an intermediate semi-permeable layer, is determined by the mobilities of cations (u^+) and anions (u^-):

5.33 Diagenesis of formation waters

$$E = \frac{u^+ - u^-}{u^+ + u^-} \frac{RT}{F} \ln \frac{c_1}{c_2} \qquad (5.33)$$

(R = gas constant, F = Faraday constant, T = absolute temperature).

If the solutions are separated by semi-permeable membrane, which is permeable only for cations, not anions, then $u^- = 0$, and the above equation then gives the so-called Nernst potential:

$$E = \frac{RT}{F} \ln \frac{c_1}{c_2} \qquad (5.34)$$

If the concentrations differ by a factor of 10, the Nernst potential is 59 millivolts. With natural claystones and NaCl-solutions WYLLIE (1948, 1955) obtained the theoretical Nernst potential as long as concentrations were below 0.1 mol/l. At higher concentration the anion barrier became increasingly ineffective, so that low potentials resulted. Also HANSHAW (1962) obtained theoretical Nernst potential values with different strongly compacted illite and montmorillonite clays, again only with dilute NaCl-solutions (0.01—0.1 mol/l). The clays had to be compressed under pressures of at least 1000 to 2000 p.s.i. (70—140 kg/cm²). These experiments show that the effectiveness of clays as ion filters decreases with increasing salt concentration and requires a certain degree of compaction.

Different investigators have studied the changes in salinity as a result of passage through clays. MCKELVEY & MILNE (1962) and HANSHAW (1962) forced NaCl-solutions through clays and determined the salinity of the expressed solutions. MCKELVEY & MILNE investigated bentonite and powdered shale pastes, compressed under pressures up to 10,000 p.s.i. (700 kg/cm²). In all cases the concentration of the expelled solution was less than the initial concentration. The concentration of a 0.164 molar NaCl-solution was reduced by passage through bentonite (porosity 34—41%) by one-eight its original concentration. The effect decreases with increasing concentration and porosity. With a shale and a 0.129 molar solution a factor of 1.7 was attained. HANSHAW (1962) obtained, when filtering NaCl-solutions (0.001—0.08 molar) through illite and montmorillonite clays (compressed at 5000 p.s.i. = 350 kg/cm²) a reduction in filtrate concentration of $1/20$ to $1/80$. In this case also a decrease in filtration effect resulted with increasing oncentration.

Other authors (KRYUKOV & KOMAROVA 1956; v. ENGELHARDT & GAIDA 1963; RIEKE et al. 1964) have investigated the concentrations of solutions expelled from a clay mud during compaction. KRYUKOV & KOMAROVA found by compacting natural clays under increasing pressure that the salinity of the expelled solution becomes increasingly lower beyond a given degree of compaction (pressure). For a clay formed by the weathering of andesitic rocks ("askangel") the decrease in concentration began at a water content of 70% and a pressure of 600 kg/cm², for a bentonite at 50% water content and 1300 kg/cm², and for a kaolin at 10% water content and 3800 kg/cm². The mean salinity was 0.9 mol/l. RIEKE and co-workers found, by compressing a bentonite from an original 81 to 14% water content, a concentration decrease of about $1/5$ (pressure 40,000 p.s.i. = 2800 kg/cm²). v. ENGELHARDT & GAIDA determined the concentrations of solutions expelled from montmorillonite muds under pressures up to 3200 km/cm², and calculated the concentration of the pore solution remaining in the clay. The clay muds were made with solutions of known salinity. Down to compressed porosities of about 30% (about 1000 kg/cm²) the concentration of the solution remaining in the clay decreased slowly at first with increasing compaction, and then more rapidly to a few tenths of a mole/l (initial concentrations were: NaCl-solutions, 1 to 0.16 mol/l; $CaCl_2$-solutions, 1.2—0.6mol/l). Further decrease in porosity to 20% (3200 kg/cm²), resulted in a slight increase in pore solution concentration. These results suggest that with increasing compaction free saline water between the clay double layers is expelled, until the compressed clay almost contains only the non-saline water of the double layers. At this point the clay should have the best effect as an anion sieve. The increase in concentration at the highest pressure was interpreted to indicate that the clays studied were not quite homogeneously compacted. Therefore small droplets of free pore solution were entrapped, which upon further compaction, could release only water without salt, so that the mean salinity in the clay rose slightly.

Thus electrochemical and filtration experiments show that clays act as ion filters. If a salt solution flows through a clay the concentration at the side of entry increases, while less saline water exists on the other side. According to experiment the effectiveness of a clay as an ion filter requires a certain amount of compaction which increases with increasing ion exchange capacity and decreases with increasing salinity. Since the effectiveness of a clay as a filter for dissolved salts depends on the reduction of anion mobility, there is essentially no cation selectivity. According to the experiments thus far, the kind of anion seems to have no influence.

GRAF and co-workers (1965), as mentioned earlier, found in the formation waters of the Illinois and other basins, an increase in deuterium and O^{18}-content with increasing salinity or progressive diagenesis of the waters. They interpreted this effect as a diagenetic alteration of waters, which resulted chiefly from the original surface water. While the increase in O^{18}-content can be explained quantitatively by exchange with calcite at the higher temperatures at depth, the authors assume that the increased deuterium content resulted by filtration through clays. The experimental confirmation of such isotopic separation by clay layers has not yet been carried out.

As the individual cases have shown, during diagenesis formation waters experience, in addition to an increase in concentration, generally a reduction in magnesium and sulfate content, and frequently a lowering of the Na:Ca-ratio.

The activity of sulfate reducing bacteria such as *Desulfovibrio desulfuricans* is responsible for the reduction of sulfate content (ZOBELL 1958, 1963; KUZNETSOV, IVANOV & LYALIKOVA 1963; WALLHÄUSSER & PUCHELT 1966). Such bacteria reduce sulfate to sulfide with simultaneous oxidation of organic carbonaceous matter to carbonate. As a result precipitation of calcium carbonate and pyrite can occur. The net reaction can be written as follows:

$$CaSO_4 + 2\,C_{org.} + H_2O \rightarrow H_2S + CO_2 + CaCO_3$$

Sulfate reducing bacteria have been found to depths of several thousand meters in numerous formation waters. Their conditions of life cover a wide range. Certain forms occur in swamp waters with pH 3 and less than 100 ppm dissolved solids; other forms live in concentrated salt solutions at pH 10.4. Some forms are active at 0 °C; others grow above 100 °C and at pressures of 1000 atm. Sulfate reducing bacteria as a rule are tied to anaerobic conditions (Eh < -100 mV at pH 7).

Under favorable conditions sulfate reduction is initiated very early in the history of a sediment. KAPLAN et al. (1963) found within the upper 2 m, in recent marine sediments off the American Pacific coast, a decrease in sulfate content of the pore solutions to about $1/4$ the content of sea water (fig. 5-33). That the sulfate content does not decrease further with depth is related to the relatively low oxidizable organic content of these sediments. Sediments with more organic matter can reduce more sulfate. Therefore, the waters of petroleum deposits are generally sulfate free.

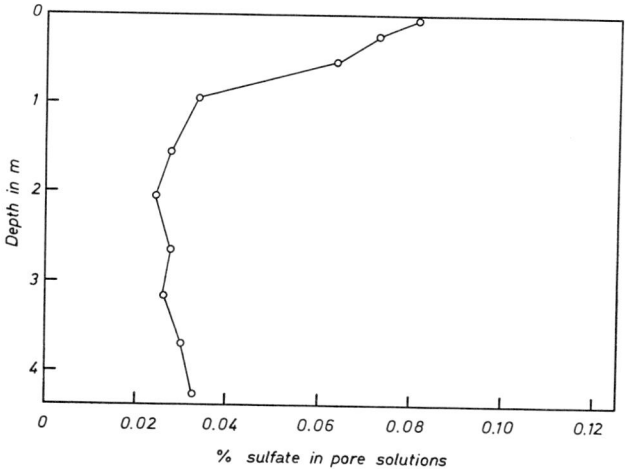

Fig. 5-33. Sulfate content in pore solutions in recent marine sediments from the Pacific off the California coast (Santa Barbara Basin) as a function of depth below the surface (KAPLAN et al. 1963).

The removal of sulfate from entrapped sea water displaces the formation water composition in the anion compositional triangle toward the Cl-HCO$_3$-side (fig. 5-34). Since carbonate forms during the bacterial reduction of sulfate, the water composition changes first along line A. Line B corresponds to the case where carbonate is tied up by precipitation of calcite or dolomite. Formation water compositions lying between lines A and B, like, for example, the mean value of waters from the West Canada Basin, could have resulted from bacterial sulfate reduction along with partial precipitation of the carbonate. Formation waters whose compositions fall below line A, as, for example, those from the Tertiary Vienna Basin, have an excess

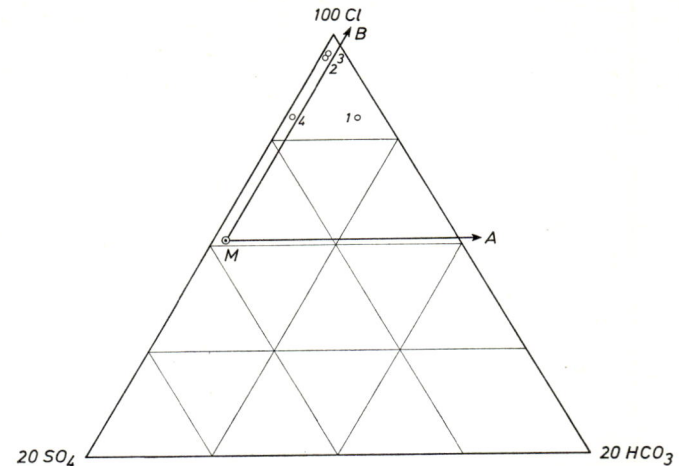

Fig. 5-34. Diagenetic changes in anion content of formation waters (explanation in text). M, 1—4, see fig. 5-35.

of carbonate, which cannot have been derived from sea water. Such waters do not originate directly from the marine environment. Carbonate must have been supplied to them from surface waters, and perhaps also by the assimilation of juvenile carbonic acid. Some compositions lie to the left of line B. In these cases more carbonate than that resulting from sulfate reduction has been removed.

Chloride does not significantly take part in chemical diagenetic processes. Therefore, it is enriched during the course of diagenesis both absolutely and also relative to the other anions. The almost pure chloride waters encountered in many deep zones of sedimentary basins, are the end products of pore water diagenesis.

The removal of magnesium from formation waters during diagenesis is a general and widespread phenomenon. Compared to the composition of sea water, the compositions of all formation waters are displaced in the Na—Ca—Na-composition triangle toward the Na—Ca-side (fig. 5-35). The diagenetic processes which remove magnesium from pore solutions are the formation of dolomite and secondarily, the formation of chlorite.

Chloritization requires exchange of magnesium for alkali or alkaline earth ions, that is, calcium or sodium, since potassium, because of its strong affinity for clays, does not remain in solution. Sodium and calcium may stem from decomposing feldspars. If exchange for calcium takes place, a pore solution derived from sea water would migrate along line A of fig. 5-35; exchange for sodium follows along line B.

If dolomite forms according to the reaction:

$$2\ CaCO_3 + Mg^{++} \rightarrow CaMg(CO_3)_2 + Ca^{++}$$

magnesium is exchanged for calcium, and the pore water composition changes along line A. Complete replacement of magnesium by calcium in solution would change the Na:Ca-ratio of 22.9 along the line B.

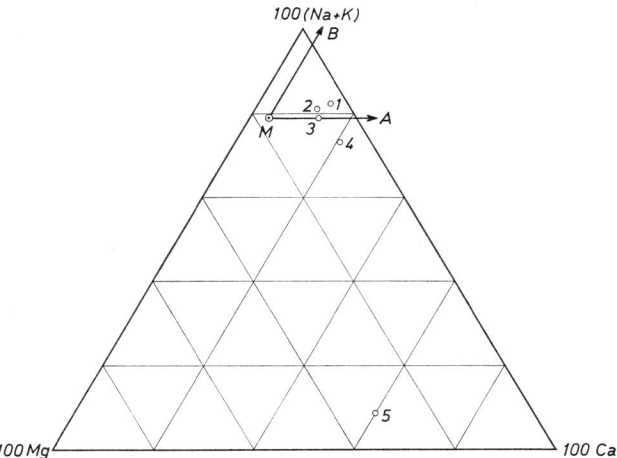

Fig. 5-35. Diagenetic changes in cation content of formation waters (explanation in text). M: sea water; 1: West Canada Basin, mean value; 2: Illinois Basin, Pennsylvania and Mississippian; 3: Illinois Basin, Silurian and Devonian; 4: Mean of 1860 formation waters from deep wells in the USA (KRAMER 1969); 5: Michigan Basin.

The numerous calcium-rich formation waters, whose compositions lie below line A, and whose Na:Ca-ratios are less than 3.71 (formation waters of the Michigan Basin, fig. 5-12), cannot be derived from sea water as a result of dolomitization or chloritization. Therefore, there must be diagenetic reactions which supply calcium to and remove sodium from these waters. Sodium depletion by albitization cannot be important since diagenetically formed albite is relatively rare. Since almost all calcium-rich formation waters contain over 90 equivalent percent chloride, relative to total anions, calcium addition cannot have resulted from the solution of calcium sulfates or calcium carbonate. This is indicated also by the fact that calcium-rich formation waters by no means occur chiefly in limestones or sulfate rocks. However, it is very frequently noted that the calcium content of formation waters increases with increasing total salinity. The data in fig. 5-19 show this. This fact makes it appear very probable that the Na:Ca-ratio of pore solutions is determined by cation exchange with clays. If a solution containing monovalent and divalent cations is in equilibrium with a clay of modest cation exchange capacity, the distribution of ions between the two phases, according to the law of mass action (see, for example, WALTON 1949) depends on the total concentration. The more concentrated the solution, the smaller the fraction of higher valence ions in the solid clay phase. With increasing concentration the fraction of higher valence ions in solution grows. The decreasing Na:Ca-ratio with increasing concentration could have been brought about by interaction of ion exchange clays with pore solutions made more and more concentrated during diagenesis.

In summary, it follows with regard to the cation content of formation waters derived from sea water: Waters whose compositions fall between lines A and B (fig. 5-35), can form from sea water as a result of diagenetic chloritization and dolomitization. In addition all waters with compositions above line A can be obtained by dolomitization and simultaneous removal of calcium by calcite precipitation. In particular the origin of all waters whose compositions lie above line B must be traced back to precipitation of calcite and the removal of magnesium by dolomitization and/or chloritization. Waters with compositions below line B (Na:Ca-ratio 3.71) can have been formed by cation exchange with clays. Sodium behaves like chloride. No significant diagenetic tying up of sodium takes place. Therefore, the sodium content of waters rises during diagenesis. NaCl-solutions of high concentration are the end products of formation water diagenesis.

In the foregoing discussion it was presumed that formation waters are derived from entrapped sea water. The data on heavy hydrogen and oxygen isotopes show, however, that at least in some, perhaps in many sedimentary basins, an addition of surface waters has been involved also (see section 5.32, also DEGENS et al. 1964). The compositions of surface waters are significantly different from that of sea water. As an example, the water of Lake Michigan is typical of the water likely to influx into the Illinois and Michigan Basins (table 5-11). It contains calcium and magnesium as its chief cations and carbonate as the primary anion. The addition of such water to formation waters could increase the carbonate, calcium and magnesium contents of pore solutions so much as to exceed the solubility products for calcite and dolomite.

Table 5-11. Composition of Lake Michigan water (GRAF et al. 1966).

	mg/l	mol/l	Equiv. % anions	Equiv. % cations
Cl	2.73	0.000077	3.54	
SO_4	7.26	0.000076	2.27	
HCO_3 as CO_3	58.35	0.00097	89.5	
Na	4.08	0.000177		8.14
K	0.66	0.0000169		0.76
Ca	26.21	0.000654		60.0
Mg	8.27	0.000340		31.2
Dissolved solids	107.27	. . .		. . .

The investigation of the interstitial pore solutions in young marine sediments indicates that diagenetic changes of the entrapped solutions begin very early. The differences, in comparison to sea water, tend to be noticeable only a few centimeters beneath the sediment surface. Only a very slight sediment overburden is sufficient to isolate the pore space from the influence of free sea water.

Studies of interstitial pore solutions in unconsolidated marine sediments by BERNER (1964), SIEVER, BECK & BERNER (1965) and FRIEDMAN et al. (1968) have resulted in the following picture.

The chloride content of interstitial solutions is higher by about 5% than that of the overlying sea water. While the Na:Cl-ratio of the interstitial water corresponds to that of sea water, the relative potassium content is clearly higher than in sea water, sometimes by more than a factor of 2. The (Ca + Mg):Cl-ratio is lower than in sea water, and especially the Mg:Cl-ratio is reduced. The SiO_2-content in all interstitial solutions studied is much higher than in sea water (pore solutions: 20—50 ppm, sea water: 0.1—4 ppm at the surface, 5—10 ppm at depth). In sediments of the Bay of California, sulfate content decreases with depth below the sediment surface, while the hydrogen sulfide content increases in pore solutions (see also fig. 5-33). Also in the gray clays of the northwest Pacific, but not in highly oxidized brown and red deep sea clays, the sulfate content of interstitial waters is lower than sea water. Th pH of the pore solutions is at 7.2—7.7, generally lower than for sea water (8.0—8.2). The pore solutions are, therefore, saturated or undersaturated with respect to $CaCO_3$ in contrast to supersaturated sea water.

From these data it would appear that the reactions which transform entrapped sea water into formation waters at depth begin shortly after deposition. These include removal of magnesium, probably by dolomitization, bacterial reduction of sulfate, and relative increase in chlorinity. The increase in potassium content may be related back to the decomposition of feldspar, perhaps also micas.

In addition it should be bourne in mind that the pore solutions occurring in the uppermost sediment layers originate in part from deeper layers, from which they ascend as compaction flow. The observed changes in chemical composition need not to have been produced in the very uppermost sediment layer.

5.4 Diagenesis of sands

5.41 Mechanical diagenesis

Deposition of sand from water or air leads to a relatively loose texture. Even homodisperse and well-sorted sands of spherical grains have porosities around 0.4 in freshly deposited sediment. In contrast densest packing of equal spheres, in which each sphere is in contact with 12 neighbors, leads to a porosity of 0.26. The total volume of a newly deposited, homodisperse sand, therefore, by simple rearrangement of grains, decreases about 20%. As a result 50% of the pore solutions originally in the sand would be expelled.

By shaking a quantity of a homodisperse sand in air or water the porosity can be reduced, but only to values just below 0.35, because the rearrangement into closest packing must overcome the frictional forces at the points of grain contact. Measurements of porosity of uncemented sands from deep boreholes show that these frictional forces have been overcome by the rock pressure. Also in such sands in which no reduction in pore space has resulted from chemical processes, either as a result of early saturation with petroleum or gas, or for other reasons, the porosity decreases with depth of burial to values of about 0.3 or less (see, for example, FÜCHTBAUER 1961, 1970, p. 105 f). This exhibits a distinct dependence on the grain size of the sand. While freshly deposited fine and coarse grain sands have the same porosities, mechanical compaction of coarse grain sand proceeds more rapidly with depth of burial than for fine grain sands. At the same depth of burial, clay-free, uncemented sands in coarse grain strata as a rule have lower porosities than fine grain ones. As an illustration of this relation, the porosities of some oil-filled, essentially uncemented lower Cretaceous sands (Valendis) from a depth of 1100 m are plotted in fig. 5-36.

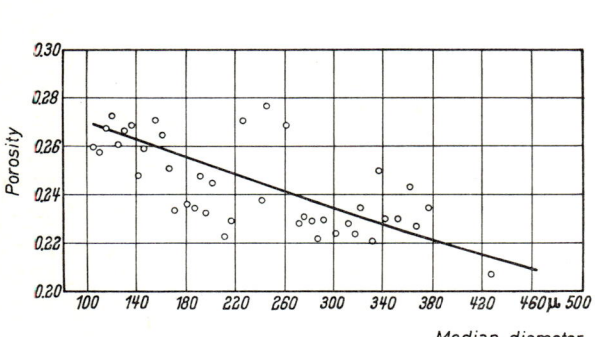
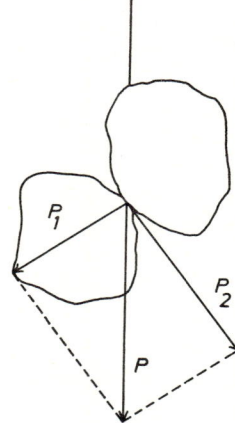

Fig. 5-36. Porosity of lower Cretaceous sandstones (Valendis) from the Scherhorn oil field, NW-Germany (1000 m depth) as a function of grain size.

Fig. 5-37. Friction at grain contacts (explanation in text).

As simple considerations show, this grain size dependence can be related to the fact that the friction at grain contacts is greater in fine grain than in coarse sands. The overburden pressure acting on a given volume of sand is distributed to the individual grains and transmitted to all of the points of grain contact. A coarse grain sand contains per unit volume fewer points of contact than a finer one. If d_1 and d_2 are the mean diameters of the coarse and fine sand respectively, for geometrically equivalent textural arrangements, the number of points of contact per unit volume for coarse and fine sand occur in the ratio of $1/d_1^3$ to $1/d_2^3$. At the same depth a higher pressure is transmitted to each grain contact point in a coarse sand than in a fine grain one. Let P be the vertical component of rock pressure, acting at a contact point (fig. 5-37). This pressure is resolved into the components P_1, directed normal to the contact surface, and a shear force P_2, acting at the surface. The latter force can affect movement of the contiguous grains, if it exceeds the frictional force R. It can be assumed that R is proportional to the normal pressure: $R = c \cdot P_1$. The frictional coefficient c depends on the shape and surface character of the grains. At equal depths the rock pressure P transmitted to a single point of contact is less in a fine grain sand (many contacts) than in a coarse grain sand (few contacts). But also the frictional force which must be overcome for the movement between two grains is smaller in the fine sand than in the coarse, because frictional forces depend on the normal pressure. Therefore, differences in the resistance to compaction can only arise because the frictional coefficient is lower for coarse than for fine sands. This is as a rule the case, since small and large sand grains experience markedly different stresses during transport. Smaller sand grains are transported chiefly in suspension and maintain their angularity and surface irregularities even after long transport distances. The coarser grains, which are moved along the bottom during transport become smoother and better rounded.

5.42 Chemical diagenesis

In contrast to metamorphism, all processes of chemical diagenesis are heterogeneous reactions, in which solutions filling the pore space of sediments are involved. The flow of formation waters, which gives rise to the constituents which form solid reaction products, and which remove components brought into solution, prevent equilibrium from being attained and cause diagenetic processes to proceed in effect, as long as connecting pore space is maintained.

Products or indicators of these chemical reactions can be recognized more or less distinctly in all sandstones. In their temporal sequence, these indicators reflect the diagenetic history of a sedimentary succession. Those sands are exceptions, in which pores were filled prematurely with petroleum or gases, or which for other reasons, were isolated from formation water flow. An example might be represented by sand lenses surrounded by impermeable clays.

Heterogeneous reactions within the pore space lead, on the one hand, to neoformation of solid phases, and on the other, to dissolution of solid phases in the sediments. One can distinguish formally the following cases.

1. **Dissolution without neoformation**: Rock constituents are dissolved by pore solutions without neoformation at the same time or place.
2. **Neoformation without dissolution**: Neoformation from the pore solution without dissolution of rock constituents at the same time or place.
3. **Neoformation with dissolution**: Rock constituents are dissolved and neoformation from pore solution occurs simultaneously, but not in the same space produced by dissolution.
4. **Alteration pseudomorphs**: Rock constituents are dissolved, and replaced simultaneously by minerals deposited from the pore solution. Dissolved and newly formed minerals have common chemical constituents. (Example: pseudomorphs of kaolinite after feldspar)
5. **Replacement pseudomorphs**: Rock constituents are dissolved and replaced simultaneously by newly formed minerals. Dissolved and newly formed minerals have no common chemical constituents. (Example: pseudomorphs of calcite after quartz).

As early as 1843, BLUM (see also NIGGLI 1920) distinguished between replacement and alteration pseudomorphs. In both cases the formation of the new phase takes place at the same time and place as the dissolution of a sedimentary mineral. Each and every particle of the dissolving mineral is replaced by a particle of the neoformation. In the case of alteration pseudomorphs certain components from the dissolving mineral react with the solution and form immediately a new phase, while the remaining constituents diffuse out and are carried away. In this way the alteration of K-feldspars to kaolinite takes place. First silica, aluminium, and potassium go into solution. A part of the dissolved silica and aluminum recombine forming kaolinite while potassium ions and the rest of the silica are carried away. A replacement pseudomorph forms when the solution at point A is undersaturated in phase B. An equilibrium is attained such that A goes into solution and the new phase B crystallizes at the same site. As an example, replacement of quartz by calcite is produced by a weakly alkaline solution supersaturated in calcium carbonate and undersaturated in SiO_2. As a result quartz is dissolved, calcite precipitated.

Whether or not the alteration or replacement pseudomorph involves the entire primary grain or only a portion of it, depends on the amount of essential matter supplied by the pore water flow.

In those cases where dissolution without neoformation takes place, solution phenomena resulting from differential pressure at the points of contact of sand grains, are especially significant. This so-called pressure solution results from the RIECKE principle, according to which the solubility of a solid body is increased by directed pressure. If a portion of the surface of a solid body is subjected to a higher pressure $\triangle P$ than the remaining surface, the ratio of the solubilities (saturation concentration) of the compressed surface (c) to that of the non-compressed surface (c_0) of the body is expressed as follows:

$$\ln \frac{c}{c_0} = \frac{V_0}{RT} \triangle P \qquad (5.35)$$

V_0 is the mole volume of the solid substance (see CORRENS & STEINBORN 1939; CORRENS 1949; BREHLER 1951). As WILLIAMSON (1917) showed (see also GORANSON 1940), the "free" surface, that is, that surface not perpendicular to the pressure direction also experiences an increase in solubility. The following expression applies for the ratio of solubility (c'), for surface parallel to the directed pressure, to (c_0) for a non-compressed portion of the crystal:

$$\ln \frac{c'}{c_0} = \frac{V_0}{RTE} (\triangle P)^2 \qquad (5.36)$$

E is the elastic modulus. Since for quartz, E is of the order of magnitude of 10^6 kg/cm², the solubility increase for the free surface is so low that it can be ignored.

The thermodynamic derivation of equation (5.35) (BECKER, in CORRENS 1949) shows, that in calculating the solubility increase, substitution of the total mole volume in the solid state is legitimate only if each molecule of dissolved matter in solution appears also as an independent particle. If this condition is not fulfilled, and α particles appear in solution per molecule (solid), then equation (5.35) must be modified:

$$\ln \frac{c}{c_0} = \frac{V_0}{\alpha RT} \triangle P \tag{5.37}$$

If $\alpha > 1$ (complete or partial dissociation) the calculated solubility increase is smaller than from equation (5.35); if $\alpha < 1$ (association) the increase is greater.

It is to be noted in addition that equations (5.35) and (5.37) are valid only for dilute solutions in which the activities can be represented by concentrations, and also for a pressure range over which V_0 can be assumed to be constant.

Solubility increases for some minerals, calculated from equation (5.35) are tabulated in table 5-12. On account of dissociation, according to equation (5.37), the values for feldspar and calcite may be too high.

Table 5-12. Solubility increase for some minerals as a result of differential pressure at 40 °C (calculated from equation 5.35).

Mineral	Mole volume	$C \triangle p : C_0$	
		$\triangle P = 10$ at	$\triangle P = 100$ at
Quartz	22.6	1.0088	1.0920
Calcite	37.0	1.0145	1.155
Orthoclase	108	1.043	1.523

In a sand, at the points of contact of individual grains, a component G of the rock pressure from the weight of overlying strata prevails; within the pore space the hydrostatic pressure p prevails. At the points of contact the grains are under higher pressure than where they are in contact with pore space. The pressure difference is:

$$\triangle P = G - p.$$

If a fluid film intervenes between contacting grains, according to equation (5.35) or (5.37), the saturation concentration (c) in the film becomes greater than that in the pore space (c_0). If the concentration difference leads to diffusion of dissolved components from the contact film into free pore space, pressure solution results, whose rate is controlled by the diffusion rate. The pressure solution rate depends, therefore, on the contact film thickness, the diffusion coefficients in the film, and on the saturation concentrations c and c_0. For a given mineral the latter depends on the differential pressure $\triangle P$.

Pressure solution by this mechanism is to be expected only when a fluid film of significant thickness occurs between the contacting grains and when the dissolved substance resulting from pressure solution does not produce supersaturation in the free pore space. The dissolved substance must be removed or precipitated elsewhere in the pore space.

As experiments by CORRENS & STEINBORN (1939) and by BREHLER (1951) on the behavior of differentially compressed alum crystals showed, the presence of a fluid film is of decisive importance. BREHLER (1951) noted pressure solution only when a layer of filter paper was inserted between two contacting alum crystals, so that the dissolved components could diffuse through it.

WEYL (1959) calculated a simple model for this qualitatively described pressure solution process. For the rate at which two cylinders of radius r, in contact on a section cut perpendicular to the cylinder axis, come together as a result of pressure solution, the following expression applies:

$$-\frac{dz}{dt} = 8 \, D \, m \, b \, \frac{\triangle P}{r^2} \tag{5.38}$$

For b, to a first approximation, the following applies.

$$b = \frac{c_0 \, V_0}{\alpha RT} \tag{5.39}$$

For this calculation it is assumed that the diffusion coefficient D in the contact film is constant and that the film thickness m is independent of pressure. $\triangle P$ is the prevailing pressure at the center of the surface of contact. The above equations are valid also when the film thickness decreases with increasing pressure, as long as this pressure is small with respect to the pressure necessary to reduce the film thickness to 1/e.

Applying this model to a simple cubic packing of spheres, WEYL determined that compaction resulting from pressure solution at grain contacts should proceed more rapidly in a fine grain sandstone than in a coarser grain rock.

Pressure solution of quartz is a very common phenomen in sandstones. This results in the interpenetration of adjacent quartz grains with narrow conical protrusions of one grain penetrating the other. In cross section the contact surfaces appear as a serrated or undulating boundary, referred to as a pressure suture. The dissolution rate of quartz grains evidently is controlled by crystal imperfections, so that at a given point the dissolution of one grain progresses rapidly and is penetrated by narrow protuberances of another grain. Nearby the second grain is more rapidly dissolved and penetrated by the first grain. Since quartz crystals dissolve more rapidly perpendicular than parallel to the c-axis (KENNEDY 1950; HEIMANN & WILLGALLIS 1969), one should expect a dependence of the relative amount of solution of the two adjacent quartz grains on crystallographic orientation. THOMPSON (1959) has established such a dependence at concave-convex grain contacts in a Silurian sandstone. Pressure solution is displayed mainly at those points of contact perpendicular of the prevailing orogenic pressure direction, therefore, generally along contacts parallel to the bedding. As a result rocks form with flattened quartz grains and sutured contacts parallel to the bedding (HEALD 1955; THOMPSON 1959). In rocks which contain grains of finely crystalline quartz rocks, such as chert, as well as quartz grains, it is established that the former are subject more rapidly to pressure solution (HEALD 1955; SLOSS & FERRAY 1948). Since the points of grain contact are enlarged with increasing pressure solution, so that the effective pressure decreases at an individual point of contact, the rate of pressure dissolution becomes smaller with increasing diagenesis, if the depth of subsidence does not increase correspondingly.

At the boundaries of mutually interpenetrating quartz grains dark accumulations of argillaceous, carbonaceous, or other substances are found. This was interpreted earlier as an insoluble residue. As a result of studies of different sandstones, it has been established, however, that pressure solution attacks primarily those contacts at which a fine film of clay or other fine-grained substances lies between the quartz grains. Therefore, in a sandstone body pressure solution is customarily most active in strata containing some clay, and is less effective in pure sandy strata (HEALD 1956a; THOMPSON 1959; LERBEKMO & PLATT 1962). HEALD (1956a) related this to a catalytic effect of the argillaceous substance. THOMPSON (1959) assumed that illitic clays at the grain boundary released potassium, giving rise to an alkaline reaction which increased the solubility of quartz. LERBEKMO & PLATT (1962) found pyrite, hematite, siderite, and carbonaceous matter at the contact boundaries of a Cretaceous sandstone of Alberta and presumed that reduction of hematite in the presence of sulfur-bearing organic matter led to the formation of pyrite and increase in quartz solubility by the resulting alkaline reaction.

Since pressure solution presupposes the presence of a fluid film at the contact point, it is more likely that the promotion of pressure solution by films of clay or other fine-grained substances results from the ability of these films to allow diffusion away of the dissolved substances (WEYL 1959). The films at the points of contact play the same role as the filter paper in the earlier-mentioned experiments on alum dissolution by BREHLER (1951).

For significant pressure solution of quartz, apparently modest pressure is necessary. Observations to date indicate that pressure solution of quartz begins to be an important factor in chemical diagenesis at depths of burial of about 1000 to 1500 m. Not only the weight of the overlying strata, but also tectonically caused pressure can lead to solution of quartz. This is shown by observations of SIEVER (1959). He noted in the Appalachian region that strongly folded Pennsylvanian sandstones are much more strongly lithified by pressure solution than just as deeply buried, but weakly folded sandstones of the same formation. Pressure solution of quartz by tectonic stresses is verified also by observations of HOEPPENER (1956) from the Rhenish Schiefergebirge, where quartz-rich layers occur in shales and are partially dissolved parallel to the schistosity.

In some sandstones continuous stylolite seams are observed which run long distances in the rock parallel to the bedding (HEALD 1955, 1959). Stylolites always contain a thin layer of "clay" material which may actually be different clay minerals, but may also consist of carbonaceous matter, iron oxides, pyrite, and other fine-grained mixtures. The seams develop from thin layers of fine-grained material interlayered as primary deposits in homogeneous sand bodies. The prevalent pressure solution along such layers has been related to an increased movement of water in these planes (STOCKDALE 1926). That this is not the essential factor in stylolite formation follows from observations by HEALD (1959) of very permeable sandstones of the United States. The high permeability allows water movement through the entire layer complex of these rocks. If flow of pore solutions were alone decisive, then there would be no reason for stylolites to form just at the location of the occasional clay layers. It must be assumed more likely that, as in the case of individual grain boundaries, clays and other fine materials bring about or promote the pressure solution. Therefore, pressure sutures develop along thin layers of fine material which are primary interlayers in the sandstone. Pressure solution along stylolitic seams is a long lasting process, by means of which a considerable amount of quartz is dissolved and can be brought into the pore solutions. In some cases the amount of dissolution can be estimated from the enrichment of relatively insoluble heavy minerals in the seams. The often observed truncation by stylolites of secondary quartz overgrowths shows that pressure solution was active even after quartz had been deposited on the pore space (HEALD 1959).

Calcite (and probably also dolomite) is more susceptible to pressure effects than quartz. Pressure solution appears especially significant in clastic sediments with carbonate components. From different regions young conglomerates have been recognized which contain "indented" limestone and dolomite pebbles (pitted pebbles) which result from pressure solution. Observations by MORAWIETZ (1958) on the Tertiary Jura gompholites (coarse-grained clastic sediments of mainly carbonate material) of the Swabian Alb, showed that pressure solution of carbonates, which leads to the formation of sutured grain boundaries and styolites, takes place after burial of only a few meters. This indicates appreciably lower pressures than in the case of quartz. In sands which contain large carbonate fossil remains as well as quartz, the former can be destroyed completely by pressure solution during diagenesis. SEILACHER (1968) described an example of pressure dissolved calcarous shells from a Devonian sandstone of the Appalachians. Within this rock the intensity of pressure solution varies among layers separated by only a few centimeters, perhaps because of different rates of removal and transport of dissolved substances. Tectonic stress also induces pressure solution of carbonates. Calcareous fossils and ooides in sandstones, which were partially dissolved predominantly parallel to the schistosity and "jointing" were described by PLESSMANN (1966) from the Rhenish Schiefergebirge. In addition to pressure effects, in this case the removal of dissolved material and the introduction of fresh solvent along the schistosity may have played a role.

Feldspar grains show the effects of pressure solution in some sandstones, although to a much lesser extent than quartz (GILBERT 1949; HEALD 1955, 1956 a). Even less distinct is the pressure solution of other minerals in sandstones. According to the observations by HEALD (1955) of Cambrian to Triassic sandstones, the following series represents decreasing susceptibility to pressure solution: calcite — quartz — feldspar — sphene and tourmaline — zircon and pyrite.

The great difference in behavior of individual minerals is noteworthy. As equation (5.38) shows, the rate of pressure solution depends on the normal solubility, on the diffusion coefficients of the dissolved components with the contact film, and on the factor $V_0/\alpha RT$. Of special importance is the influence of solubility. Relatively soluble minerals will show especially strong effects. This is probably the major reason for the high pressure sensitivity of carbonates.

While pressure solution results from physical causes, chemical processes give rise to varied dissolution, recrystallization, and neoformation in the pore space of sandstones. The most important of these will be discussed in the following sections.

Neoformation of quartz (WALDSCHMIDT 1941; DAPPLES 1949, 1967; SIEVER 1959; CAROZZI 1960; FÜCHTBAUER 1961; MEISL 1970). Within the pore space of sandstones newly formed quartz is exceptionally widespread. Usually it is represented as overgrowths added so as to maintain the same crystallographic orientation as the detrital grain. In thin section these newly formed overgrowths are distinguished from their detrital cores by their greater clarity and lack of inclusions. Frequently a ring of dark inclusions designates the original grain surface, so that it is possible in thin section to estimate the porosity of the rock before secondary silicification and also the amount of quartz deposited. Crystal faces are observed often on the overgrowths, if the pore space has not been filled completely. Growth of such single crystals can only take place if a small number of nuclei develop and at low growth rates. They formed, therefore, from weakly supersaturated solutions. With greater nucleation and more rapid growth rates, that is, from strongly supersaturated solution, polycrystalline quartz forms. This occurs as pore filling in some sandstones which contain relatively soluble silica, such as biogenic silicious skeletons or volcanic glass.

In a continuous sandstone sequence the neoformation of quartz is never uniformly distributed. Newly formed quartz rather is concentrated within limited zones or layers. On a small scale adjacent grains occur, some with and some without secondary quartz overgrowths. Supersaturated silica solutions can be maintained for a long time and can migrate long distances. As the precipitation of oriented overgrowths shows, supersaturation is alleviated only where the work of nucleation is reduced by the possibility of further growth on appropriate quartz surfaces. The character of the quartz grain surface is decisive in determining the locus of secondary quartz precipitation. Because a clay coating prevents or hinders oriented overgrowth, argillaceous sandstones are usually free of quartz neoformations (SIEVER 1959; HEALD & ANDEREGG 1960; HORN 1965). Especially abundant secondary quartz is found in layers of pure quartz sandstones. Frequently greater quartz precipitation is observed in fine-grained layers than in coarse-grained ones. This phenomenon may originate from the greater angularity and roughness of smaller sand grains, which increases the probability of nucleation. The observed small scale differences in quartz growth are apparently related to differences in surface quality, such as adsorption films or other surface layers not seen under the microscope.

In a uniform sedimentary basin diagenetic quartz neoformation increases with depth of burial. In fig. 5-38 this is shown for the Dogger sandstone of the Gifhorn trough of northwest Germany, after investigations of PHILIPP, DRONG, FÜCHTBAUER, HADDENHORST & JANKOWSKY (1963). Quartz neoformation is represented as that portion in percent of the total quartz grains which exhibits overgrowths. This is plotted as a function of maximum depth of burial. The circles correspond to water-

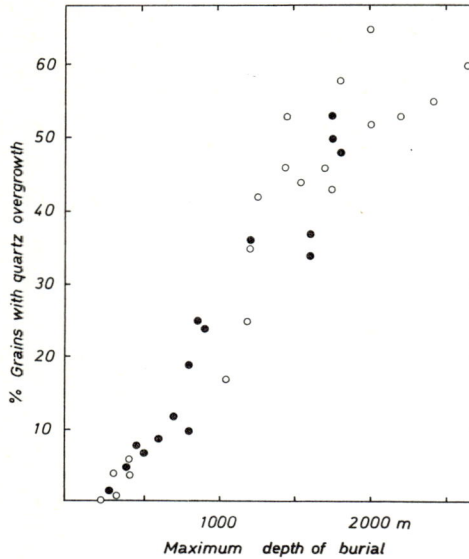

Fig. 5-38. Quartz grains with overgrowths of diagenetic quartz in % of total quartz grains. Middle Jurassic sandstones from deep wells in NW-Germany. Maximum depth of burial related to the oilfilled sandstones (•) at the time before oil migration (after Philipp et al. 1963).

filled and the points to oilfilled sandstones. For the latter, the maximum depths of burial before migration of oil into the sandstones were estimated from geologic considerations.

The silica deposited diagenetically as quartz can stem from different sources. Quartzites formed near the surface, whose pore space is filled with quartz originating from weathering solutions, represent a very early stage of diagenesis. Examples of these are the Tertiary quartzites of central Europe, whose silica stems from the intensive kaolinitic weathering beneath coal bogs. Also desert quartzites with opal and chalcedony cements represent transition between weathering and diagenesis.

An important source of silica in deeply buried sandstones is pressure solution. One would expect that the silica dissolved at the points of grain contact would be redeposited in the immediate vicinity on the non-stressed surfaces of sand grains. However, very frequently this is not the case, and the greatest precipitation of secondary quartz is not observed in the same strata which show the strongest evidence of pressure solution. The reason is the mentioned stability of supersaturated silica solutions. These only precipitate quartz where suitable quartz grain surfaces are available. Since clay content promotes pressure solution and clay coatings on the sand grain inhibit quartz deposition, the silica resulting from pressure solution migrates from argillaceous layers into regions where quartz neoformation takes place more easily, that is, into clay-free and finer-grain layers.

Another important source of silica are the organic shells and exoskeletons embedded in sands, which consist of amorphous SiO_2. Since the solubility of amorphous SiO_2 is about ten times greater than that of quartz (section 2.331), the dissolution of organogenic SiO_2 results in solutions which are strongly supersaturated with respect to quartz. During the course of diagenesis amorphous silica must gradually disappear simultaneously with the neoformation of quartz.

Additional SiO_2 is supplied from the dissolution of various silicates. This is especially so for the transformation of feldspar to kaolinite, which will be treated in detail later. Finally, it is likely that SiO_2 is derived from the diagenesis of clays and is introduced into sands by the compaction flow of fluids.

Since silica solubility increases significantly with temperature, solutions which are saturated at greater depths and higher temperatures, become supersaturated when they ascend and cool off during compaction flow. In this way SiO_2 can be transported upward from greater depths. Solutions arising from sediment compaction bring SiO_2 into the pore space of sandstones where it remains, because the solutions, passing again into another clay layer, leave their dissolved silica in the sandstone as a result of ion filtration. An example of this phenomenon was described by FÜCHTBAUER (1961) from the Dogger of the Gifhorn trough, where zones of intensive quartz neoformation occur in oilfilled sandstones adjacent to sub- and superjacent clay layers.

Also the dependence of silica solubility on pH can lead to supersaturation and neoformation of quartz. Above pH 8 its solubility increases markedly with pH (section 2.331). Neutralization of alkaline solutions can evoke quartz precipitation.

Neoformation and dissolution of carbonates. Diagenetic neoformation of carbonates is widespread in porous sandstones. In part this involves reprecipitation of carbonate dissolved by pressure solution, the carbonate having been admixed with the sand as detritus (carbonate grains or organic remains). In this way a fine carbonate cement forms in the free pore space.

From the carbonate dissolved in the pore solution calcite or dolomite can precipitate, if the temperature, pH, or CO_2-pressure changes. Heating the pore solution saturated in carbonate at greater depth must lead to calcite or dolomite deposition. This can happen by the subsidence of sediments or by introduction at depth of carbonate saturated surface waters into the sandstone layer. An increase in pH and reduction of CO_2-pressure, for example, by the dissolution of alkali-bearing silicates or degassing of pore solutions, can result in carbonate deposition.

In the compaction flow, ascending formation waters can experience supersaturation by ion filtration, when they pass from sandstone into clay layers. In this way the occurrence of calcified zones is explained, which frequently are encountered in sandstones near the contact of clay layers (FOTHERGILL 1955). An example of such an occurrence is described by WERNER (1961) from a Dogger sandstone of southern Germany. There a calcified zone is found beneath the overlying shale, while in the underlying sandstone calcareous fossil remains have been dissolved. In other cases calcified margins are noted at both the upper and lower sandstone contacts. This is true also of sandstones which were filled with oil at an early stage. Here also diffusion of ions from fine-grained sediments into the sandstone could have caused supersaturation in carbonate (FÜCHTBAUER 1970).

For the precipitation of dolomite in pore spaces and the frequently observed transformation of primary or diagenetic calcite into dolomite, the temperature dependence of calcite and dolomite stability (shown in fig. 4-7) could be of importance. The Ca:Mg-ratio of solutions in equilibrium with dolomite changes to higher calcium concentrations with increasing temperature. A given solution at equilibrium with calcite at a lower temperature is in equilibrium with dolomite at higher temperatures. As the temperature increases calcite should dissolve and dolomite be deposited.

Dissolution of carbonates results from a reduction of temperature or pH and an increase in CO_2-pressure in the pore solution. The commonly observed phenomenon that calcareous sandstones are decalcified, where permeated by cool, acidic, and CO_2-bearing surface water is explained in this way.

Replacement of quartz and feldspar by carbonates. In some sandstones diagenetic replacement of quartz and feldspars by calcite, and sometimes dolomite, is observed (WALKER 1957, 1960, 1962; DAPPLES 1959, 1967; CAROZZI 1960; ZIMMERLE 1963; HELING 1963, 1970a; SHARMA 1965; FÜCHTBAUER 1967b; KULKE 1969; MEISL 1970).

Quartz grains are attacked from the surface inward by newly formed carbonate. In progressive stages only skeletons or aggregates of quartz fragments remain, whose optical continuity still indicate the outline of the original grain. In this way rocks can form finally, which consist of a dense calcite mass, in which individual, usually clearly corroded quartz grains are floating.

Replacement of quartz by carbonate must have been brought about by a change in the pore solution, which increased the quartz solubility, while simultaneously depositing carbonate. Quartz solubility is augmented by an increase in temperature and pH, both of which decrease the solubility of calcite. Therefore the replacement reactions depend upon either or both of these factors. CORRENS (1950) first suggested the effect of pH. A pH-increase can result if the CO_2-content of the pore solution is decreased by separation of a gas phase. A solution saturated in $CaCO_3$ has at a negligible CO_2-pressure a pH of almost 10 and contains only 13 mg/l $CaCO_3$, while a saturated solution in equilibrium with atmospheric CO_2 has a pH of 8.4 and contains 50 mg/l $CaCO_3$ (both at 25 °C). Removal of CO_2 from a formation water can, therefore, lead to calcite deposition and simultaneously increase the pH sufficiently so that SiO_2 dissolves. Temperature increase works in the same direction. Also at fixed CO_2-pressure the solubility of calcite decreases with increasing temperature. For example, at a given CO_2-pressure, at 0 °C a saturated solution contains 2.4 times more $CaCO_3$ than at 40 °C. The same temperature rise increases the solubility of quartz from around 5 ppm to 16 ppm (see sections 2.331 and 4.2).

It can be assumed that the frequently observed replacement of quartz by carbonate results from reduction of the CO_2-content of the pore solution and by a rise in temperature or by both effects working in concert. In addition to quartz, feldspars are replaced also by calcite or dolomite. Replacement proceeds along cleavage cracks, twin lamina, or perthite spindles, and leads finally to more or less complete pseudomorphs. According to observations so far, potassium feldspar appears to be more readily and rapidly replaced than plagioclase. Since the solubility of feldspar is least in neutral solutions (pH 7) and increases with increasing pH, replacement of feldspar by carbonates can be promoted by a pH-increase in the pore solution, in this case also resulting from a decrease in CO_2-content.

Carbonate replacement takes place as a rule at later stages of diagenesis, that is, after considerable subsidence.

Replacement of carbonates by quartz. Replacement of carbonate minerals by chalcedony or coarsely crystalline quartz is widespread in carbonate rocks and leads then to the formation of chert or flint concretions. The same replacement is observed in sandstones. This shows that in the system quartz-carbonate-pore solution, reversible equilibrium exists which depends on temperature and CO_2-pressure or hydrogen ion concentration. Increase in temperature and reduction in CO_2-pressure (elevation of pH) lead to the replacement of quartz by carbonate; reversal of these factors can result in replacement of carbonate by quartz. In some cases the rock reflects a repeated change in these relations in the pore solution: WALKER (1962) observed the following sequence of events in a silicified calcareous oolite from Wisconsin (Ordovician):

1. Deposition of a calcareous oolite.
2. Partial replacement of carbonate by chalcedony.
3. Partial replacement of chalcedony by dolomite.
4. Partial replacement of dolomite by chalcedony.

Replacement of carbonates by quartz in sandstones is less common than replacement of quartz by carbonate.

Neoformation of alkali feldspars. In many sandstones diagenetically formed alkali feldspars are found. As a rule only the pure end members (albite with less than 3 mol% K-feldspar, K-feldspar with less than 2 mol% albite) occur (BASKIN 1956). This would be expected by their low temperature of formation. Nevertheless HELING (1965) found neoformed K-feldspars with up to 30 mol% albite in a Keuper sandstone. These feldspars formed either overgrowths on different detrital feldspars or crystallized in the pore space in the rock. The overgrowths have the same optical orientation as the core, when they have corresponding compositions. Secondary albite is often polysynthetically twinned, although the sharpness of the twin lamella decreases adjacent to the core (HEALD 1956 b). The newly formed feldspars as a rule have crystal faces. Newly formed albites are tabular after (001) or (010) and are twinned often as quadruplets according to the Rock Tournée law. Neoformed K-feldspars often have the adularia habit. According to X-ray and optical studies microcline, "orthoclase" and sanidine occur as neoformations. BASKIN (1956), SAENZ (1963), and GLOVER & HOSEMANN (1967) found in a Devonian sandstone overgrowths on detrital K-feldspar which consist on (001) of high sanadine, on (010) of orthoclase and low sanadine ond on (110) and (1$\bar{1}$0) of all three modifications. Symmetry and optics of neoformed feldspars probably depends on the growth rate, that is, on the degree of supersaturation. The low symmetry, well-ordered phase may result by slow growth from very dilute solutions.

As was the case with quartz, neoformed feldspars precipitate preferably where surfaces of detrital feldspar promote nucleation. Therefore, neoformed feldspars occur predominantly in arkoses and graywackes. Füchtbauer (1967 a) found albite in brackish-marine middle Buntsandstone and K-feldspars in fluviatile rocks (see also Meisl 1970), indicating the influence of the primary entrapped pore solutions.

Neoformation of feldspars infers a certain amount of burial and is promoted by higher temperatures. Pore solutions precipitating feldspars must have been weakly alkaline.

Alteration pseudomorphs of kaolinite after feldspars and neoformation of kaolinite with simultaneous dissolution of feldspar. The well known dissolution of feldspar during weathering, accompanied simultaneously by kaolinitization, is a reaction observed also in many feldspathic sandstones. This is a phenomenon of early diagenesis at low subsidence depths. A summary concerning sandstones containing newly formed kaolinite has been published by Shelton (1964). Detailed descriptions of specific examples are found in the works of Fothergill (1955), Füchtbauer & Goldschmidt (1963), Heling (1963), Carrigy & Mellon (1964) and Kulke (1969). The newly formed kaolinite is deposited either as vermicular aggregates in the free pore space, or it occupies the space of dissolved feldspar, forming alteration pseudomorphs. The dissolution of the incompletely replaced feldspar is recognized by corrosion of the grain borders and internal cavities. Their delicate forms distinguish these newly formed kaolinite aggregates from detrital kaolinite deposited along with the sand. Another way of distinguishing diagenetically formed from detrital kaolinite involves mineralogical comparison with interstratified clay beds. The detrital clay mineral compositions of the sands and clays should be approximately the same. Sandstones with newly formed kaolinite contain much more of this mineral than do the accompanying clay beds (Glass, Potter & Siever 1956; Füchtbauer & Goldschmidt 1963).

Kaolinite is deposited first as the disordered "fireclay" type, which gradually transforms to the well-ordered variety. The disordered modification may remain in early oil-filled sandstones, indicating that the re-ordering does not develop in the solid state, but results from interaction with pore solutions (Füchtbauer & Goldschmidt 1963).

As illustrated in section 2.333, kaolinite forms from weakly acid solutions (about pH 5). Dissolution of feldspar into potassium and aluminum ions and silica, as well as the formation of kaolinite from these constituents, can be described by the following equation:

$$2 KAlSi_3O_8 + 16 H_2O \rightarrow 2 K^+ + 2 Al^{3+} + 8 OH^- + 6 H_4SiO_4$$
$$\rightarrow Al_2(OH)_4Si_2O_5 + 4 SiO_2 + 2 K^+ + 2 OH^- + 13 H_2O$$

The process leads, therefore, to the liberation of SiO_2 and potassium and hydroxyl ions as well as the formation of kaolinite. For feldspar dissolution and kaolinitization to preceed, SiO_2 must be precipitated, and the flow of formation waters must suppy hydrogen ions to neutralize the OH-ions and remove the K-ions. In all sandstones containing diagenetic kaolinite, newly formed quartz is found also, as overgrowths on detrital grains. As to the whereabouts of the potassium there is no direct evidence. It may be assumed that potassium is adsorbed in the clay layers or fixed within previously degraded micas.

Neoformation and transformation of clay minerals. Depending on the primary mineral composition of the sand and the composition, temperature and hydrogen ion concentration of the pore solutions, different clay minerals form by diagenetic alteration and neoformation in the pore space of sandstones. These reactions continue processes initiated during weathering on the one hand, and on the other may be processes also observed in the realm of metamorphism.

The neoformation of kaolinite in a weakly acid environment is observed also in sandstones without noticeable feldspar decomposition. In these cases aluminum and silica were not mobilized in situ, but introduced by the flow of formation waters.

Detrital dioctahedral micas by means of acid pore solutions can be degraded to potassium deficient mixed-layer minerals and finally to montmorillonite.

However, if pore solutions supply potassium and aluminum, which may stem from the dissolution of feldspars, degraded detrital micas (illite) may be transformed more or less completely to muscovite. From the fine grain primary clay matrix of sandstones, coarse grain muscovite (sericite), and, less frequently, biotite can form (DAPPLES 1967). The crystallization of biotite is possible only when the pore solution contains divalent iron, and thus has a very low oxidation potential. From primary glauconite, mixtures of muscovite and biotite form (DAPPLES 1967).

At greater depths muscovite forms at the expense of K-feldspar also. K-feldspar-muscovite equilibrium depends on the amount of dissolved silica in the pore solution (GARRELS & CHRIST 1965, p. 361). At a given potassium concentration K-feldspar is formed at higher H_4SiO_4-concentrations, and muscovite at lower concentrations.

Trioctahedral and dioctrahedral chlorites are characteristic neoformations in many sandstones. Trioctahedral, usually rather iron-rich chlorites occur in some sandstones as fine crystalline pod-like rinds coating the sand grains (HEALD 1950; v. ENGELHARDT 1960; CARRIGY & MELLON 1964; HORN 1965). HORN (1965) found as a crystal coating in the pore space of a Dogger sandstone, an iron-rich "7 Å"-chlorite = chamosite, a two-layer modification with the kaolinite structure, belonging to the berthierine-cronstedite group. The neoformed chlorite in an upper Carboniferous sandstone studied in detail by HELING (1970) is a normal chlorite with $Si_{4.4}Al_{3.6}$ in tetrahedral sites, and with sites of both octahedral layers half occupied by iron. In some sandstones the replacement of detrital biotite by chlorite can be observed (SCHERP 1963; HELING 1965, 1970a). In other sandstones chlorite forms at the expense of primary glauconite (DAPPLES 1967). The alteration pseudomorphs of chlorite after biotite or glauconite were produced by a pore solution which was poor in potassium and contained either much iron and magnesium or little silica. In each case the diagenetic formation of trioctahedral chlorite presupposes pore solutions with a pH above 7 (GARRELS & CHRIST 1965, p. 364) and a very low oxidation potential, permitting the existence of divalent iron ions in solution.

Dioctahedral chlorites, and indeed the aluminum-chlorite, sudoite, as well as irregular mixed species (tosudite), are probably more widely distributed in sandstones than is generally realized. They were found by KULKE (1969) as fine feathery, fan-shaped, and vermicular aggregates in the pores of a south German Keuper sandstone, where they occurred as alteration pseudomorphs from previously diagenetically formed kaolinite. The factors which differentiate the stabilities of kaolinite and sudoite are not clear. In some known occurrences sudoite and mixed-layer phases occur at greater depths in place of kaolinite. The effect of higher temperature may be an important factor. It is more likely that dioctahedral, like trioctahedral chlorites are stable in alkaline solutions, whereas kaolinite forms in acid solutions.

Neoformation of sulfates. In some sandstones anhydrite occurs as a neoformation of late diagenesis (WALDSCHMIDT 1941; HELING 1965; FÜCHTBAUER 1967 a, b; MEISL 1970). It is found especially in the vicinity of evaporite sediments. The anhydrite occurs as a pore filling or forms replacement pseudomorphs after quartz, feldspar, and clay minerals. Since, in contrast to the replaced minerals, the solubility of anhydrite decreases with increasing temperature (fig. 4-10), precipitation of anhydrite, as well as the observed replacement, could result from the heating of a $CaSO_4$-saturated pore solution.

Alteration pseudomorphs of calcite after diagenetically formed anhydrite, as observed by HELING (1965) in the Schilf sandstone (Keuper) of the Rhine graben, show that these sandstones after anhydrite deposition were permeated by pore solutions in which the concentration of carbonate/sulfate was so great, that alteration of anhydrite to calcite took place.

Diagenetically formed barite (less frequently baritocelestite and celestite) is not rare in sandstones. It is found as pore fillings, often in the form of idiomorphic crystals. Also replacement of quartz and feldspars has been described (FÜCHTBAUER 1967 b; MEISL 1970). The barium may originate from the dissolution of K-feldspar. Because of the very low solubility of $BaSO_4$, very small amounts of barium and sulfate in the pore solution are sufficient to lead to precipitation of barite.

Neoformation and alteration of zeolites. Zeolite-bearing rocks occur as deposits of saline alkaline lakes, and in most thick sequences of tuffs and tuffaceous sandstones which contain volcanic glass, as well as some sandstones containing feldspar and rock fragments. Such zeolite-rich sandstones are widely distributed in deeply buried sediment series in the geosynclines of western North America, New Zealand and the Russian Platform. DEFFEYES (1959) and HAY (1966) have summarized the occurrences of zeolites in sediments.

Zeolites in these rocks either fill the pore space or form pseudomorphs after clastic components such as feldspar or glass. They are formed either as a result of sublacustrian and submarine weathering or during diagenesis.

Table 5-13 indicates the compositions of the most important zeolites that occur in sediments. According to the results of synthesis studies (summarized by COOMBS et al. 1959; FISCHER & MEIER 1965) zeolites form from alkaline solutions (high cation: hydrogen ion ratios) containing much dissolved silica, at temperatures up to about 450 °C. Experimentally produced zeolites form as a rule as metastable phases. Therefore, their exact equilibrium conditions are still not known.

From the natural occurrences one can conclude that chabasite, offretite (erionite), natrolite, and phillipsite form predominantly through subaquatic weathering on the

Table 5-13. Most important sedimentary zeolites.

Name	Ideal formula	Compositional range (Coombs et al. 1959)	Si : Al
1. Formed primarily by subaquatic weathering			
Chabasite	$(Ca_{1/2},Na,K)_2Al_2Si_4O_{10} \cdot 6 H_2O$	$(Ca_{1/2},Na,K)_{10}Al_{10}Si_{26}O_{72} - (Ca_{1/2},Na,K)_{13.5}Al_{13.5}Si_{22.5}O_{72} \cdot 36 H_2O$	2.6—1.7
Offretite (Erionite)	$(Ca_{1/2},Mg_{1/2},Na,K)_3Al_3Si_9O_{24} \cdot 9 O_2O$		3.0
Natrolite	$Na_2Al_2Si_3O_{10} \cdot 2 H_2O$		1.5
Phillipsite	$KCaAl_3Si_5O_{16} \cdot 6 H_2O$	$(Ca_{1/2},K)_5Al_5Si_{11}O_{32} - (Ca_{1/2},K)_7Al_7Si_9O_{32} \cdot 12 H_2O$	1.7
2. Formed primarily in the sediment at low temperatures			
Heulandite	$CaAl_2Si_7O_{18} \cdot 6 H_2O$	$(Ca_{1/2},Na)_{9.6}Al_{9.6}Si_{26.4}O_{72} - (Ca_{1/2},Na)_8Al_8Si_{28}O_{72} \cdot 24 H_2O$	2.8—3.5
		Clinoptilolite: $Ca(Na,K)_4Al_6Si_{30}O_{72} \cdot 24 H_2O$	5.0
Analcite	$NaAlSi_2O_6 \cdot H_2O$	$Na_{17}Al_{17}Si_{31}O_{96} - Na_{14}Al_{14}Si_{34}O_{96} \cdot 16 H_2O$	1.8—2.4
Wairakite	$CaAl_2Si_4O_{12} \cdot 2 H_2O$		2.0
Mordenite	$(Ca_{1/2},Na,K)_2Al_2Si_{10}O_{24} \cdot 7 H_2O$		5.0
3. Formed in sediments at temperatures above 200 °C			
Laumontite	$CaAl_2Si_4O_{12} \cdot 4 H_2O$	$Ca_{4.25}Al_{8.5}Si_{15.5}O_{48} - Ca_{3.75}Al_{7.5}Si_{16.5}O_{48} \cdot 16 H_2O$	1.8—2.2

sea floor or especially in salt lakes with high pH and high alkali contents (soda lakes, borate lakes).

At shallow depths chiefly analcite, wairakite, and clinoptilolite, the SiO_2-rich variety of heulandite, are formed diagenetically. Prerequisite for diagenetic zeolitization is a pore solution high in sodium, calcium, SiO_2, and alkalinity. Zeolites form chiefly in the pore space of rocks which contain volcanic glass, amorphous SiO_2 or appreciable feldspar, and, therefore, contain pore solutions which are saturated in silica with respect to quartz. A further condition for diagenetic zeolite formation may be a low concentration of carbonate ions, so that no carbonate minerals can form, which are otherwise the most common precipitates from alkaline solutions.

As indicated in table 5-13, the composition of individual zeolites varies considerably. The general formula for all zeolites is $(M^I, M_{1/2}^{II})_x \cdot n\, AlO_2 \cdot y\, SiO_2 \cdot z\, H_2O$, with M^I = Li, Na, K and M^{II} = Mg, Ca, Sr, Ba. Among some types, x, y, and z vary. Since the cations are mobile within the structure and can be exchanged, their relative amounts are variable. These variations in composition reflect the composition of the parent materials and the pore solutions and are dependent also on pressure and temperature.

COOMBS & WHETTEN (1967) differentiate three groups (A, B, C) of analcites. Si-rich A-analcite ($Na_{13}Al_{13}Si_{35}O_{96} - Na_{14}Al_{14}Si_{34}O_{96} \cdot z\, H_2O$) forms near the surface in solutions high in cations and silica. B-analcite with the composition $Na_{14}Al_{14}Si_{34}O_{96} \cdot z\, H_2O$ forms diagenetically during burial to greater depths. Si-deficient C-analcite ($Na_{15}Al_{15}Si_{33}O_{96} \cdot z\, H_2O - NaAlSi_2O_6 \cdot H_2O$) often occurs together with neoformed dolomite. It does not form from volcanic glass, but directly from alkaline solutions or by reaction of such solutions with clay and other sedimentary minerals. According to experiments by FYFE (COOMBS et el. 1959) increasing pH reduces the SiO_2-content of synthetic zeolite and promotes the precipitation of quartz.

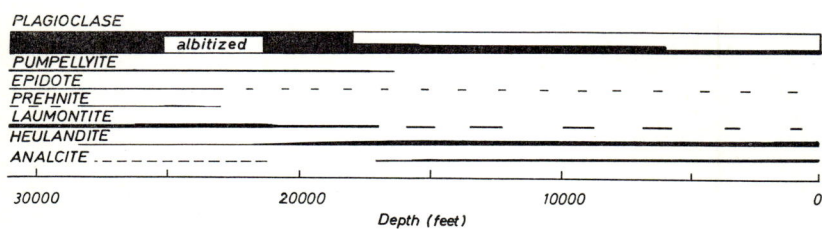

Fig. 5-39. Diagenetic or metamorphically formed minerals in Triassic graywackes, siltstones, tuffites, and tuffs of Taringatura, South Island New Zealand (after COOMBS et al. 1959).

In deeply buried rocks analcite, heulandite, and wairakite disappear and laumontite, albite, and quartz appear in their place.

An example of this depth relationship is shown in fig. 5-39, which summarizes the occurrence of zeolites in a 10,000 m series of Triassic tuffaceous graywackes, siltstones and tuffs of the south island of New Zealand (after COOMBS et el. 1959).

With increasing temperature albite forms from analcite and quartz, according to the following reaction:

$$\underset{\text{analcite}}{NaAlSi_2O_6 \cdot H_2O} + \underset{\text{quartz}}{SiO_2} \rightarrow \underset{\text{albite}}{NaAlSi_3O_8} + H_2O$$

Correspondingly in the section illustrated in fig. 5-39, marked albitization of plagioclase accompanies the disappearance of analcite. Because of the low reaction rate, the equilibrium temperature of this reaction has not yet been determined accurately. According to available data, analcite may react with quartz to form albite at temperatures around 200 °C, at water vapor pressures below 2000 bars (COOMBS et al. 1959; WINKLER 1967).

In contrast to the other zeolites laumontite occurs only in tuffs and sediments which have been buried to a certain depth. In sections of homogeneous sedimentary series, such as shown in fig. 5-39, the beginning of laumontite formation starts abruptly, apparently related to temperature. At the same boundary analcite and heulandite disappear. Laumontite is formed at the expense of heulandite or the anorthite component of plagioclase, according to the following reactions:

$$\underset{\text{heulandite}}{CaAl_2Si_7O_{18} \cdot 6\ H_2O} \rightarrow \underset{\text{laumontite}}{CaAl_2Si_4O_{12} \cdot 4\ H_2O} + \underset{\text{quartz}}{3\ SiO_2} + 2\ H_2O$$

$$\underset{\text{anorthite}}{CaAl_2Si_2O_8} + \underset{\text{quartz}}{SiO_2} + 4\ H_2O \longrightarrow \underset{\text{laumontite}}{CaAl_2Si_4O_{12} \cdot 4\ H_2O}$$

The formation of laumontite by these reactions in the laboratory has not yet been successful. According to observations in the fumerole region of Wairaki, New Zealand, COOMBS et al. (1959) concluded that laumontite formation, like albitization from analcite, begins at about 200 °C. The occurrence of laumontite in some sandstones which were not deeply buried, and apparently not subjected to temperatures greater than about 100 °C, makes it appear likely that laumontite can form also at lower temperatures from appropriate pore solutions (alkaline, much SiO_2 and Ca, no carbonate).

In an attempt to define the boundary between diagenesis and metamorphism, the beginning of laumontite formation has been used (TURNER & VERHOOGEN 1960; WINKLER 1967). Above this boundary, that is, at temperatures above 200°, the laumontite and prehnite should form and the zeolite- (TURNER & VERHOOGEN 1960) or laumonite-prehnite metamorphic facies (WINKLER 1967), should begin.

Dissolution of heavy minerals (intrastratal solution). Like the major constituents of sandstones, the accessory heavy minerals are dissolved also by permeating pore solutions. This dissolution does not differ in principle from the corresponding processes during weathering. Since heavy minerals have received special attention for purposes of correlation and in determining transport direction, their dissolution during diagenesis has been of interest for some time. This phenomenon was designated as intrastratal solution by PETTIJOHN (1941, 1957).

From a statistical study of heavy minerals described from formations of different ages, PETTIJOHN formulated a persistance series. At the head of the series stand minerals concentrated in the oldest rocks, such as rutile, zircon, and tourmaline. These are judged to have the greatest stability. At the bottom of the series are the most abundant accessories of the youngest rocks, such as olivine and pyroxene. These are considered to be especially unstable.

Observations of different authors, summarized by FÜCHTBAUER (1970) have essentially substantiated this stability series. As is the case during weathering the influence of carbonate content and pH is particularly noticeable. Dissolution is especially intensive in sandstones in which kaolinitization of feldspars has taken place, indicating the effect of acid pore solutions. In calcareous sandstones heavy mineral dissolution is less intense, because of less acid pore solutions, as well as early isolation of pore space by diagenetically deposited carbonate. The latter inhibits movement of pore solutions. Likewise, dissolution of heavy minerals comes to a standstill in oilfilled sandstones, while it continues in adjacent water-saturated beds.

Dissolution of heavy minerals in a continuous series of strata increases with increasing depth, as numerous examples have demonstrated. WIESENEDER (1952, 1958) found an increasing loss of epidote, hornblende, garnet, and staurolite with depth of burial (0—2500 m) in the Miocene sands of the Vienna Basin. The formation waters of these strata (see section 5.32) contain carbonate and free carbonic acid, whose content increases with depth. In Dogger sandstones of northwest Germany DRONG (FÜCHTBAUER 1970) established a decrease in kyanite, staurolite and garnet with increasing depth of burial (500—2000 m).

The following is the stability series for heavy minerals with respect to diagenetic dissolution, as indicated by FÜCHTBAUER (1970), based on observations from many different sandstones:

Amphibole, pyroxene, olivine	unstable
Epidote	↑
Kyanite	
Staurolite	
Apatite, garnet	↓
Tourmaline, zircon, rutile	stable

The series corresponds approximately to the statistically determined persistence series of PETTIJOHN. It represents apparently the stabilities of different minerals in the most common pore solutions in sandstones. In particular one should expect during diagenetic dissolution, as with weathering, a dependence of the stability series on the particular chemistry of the pore solutions. In contrast to observations in most soils, the diagenetic series is different, in that apatite and garnet appear as relatively stable minerals. This is because pore solutions of most sandstones during the course of diagenesis usually are not strongly acidic and probably also contain relatively large amounts of calcium.

Neoformation of heavy minerals. Some heavy minerals which normally form at high temperatures from magmas or hydrothermal solutions can form at low temperatures during the course of diagenesis. Their exact conditions of formation are not known. Diagenetically formed tourmalines have been described from different sedimentary rocks, including sandstones (BOSWELL 1933; KRYNINE 1946; PETTIJOHN 1957; PUSTOWALOFF 1955; AWASTHI 1961). As a rule these are clear euhedral overgrowths over detrital grains. In addition fine, newly formed tourmaline needles may occur in the same rock (VALETON 1955). Similarly zircon grains with diagenetic overgrowths have been described from various sandstones (PETTIJOHN 1957; VALETON 1955 b; AWASTHI 1961). On zircons from the Buntsandstone VALETON observed preferential growth on the prism surfaces of turbid or zoned zircon, with development of (111) planes, while on the same grain other parts of its surface appear badly corroded. It is possible that what is involved in this case, as with some tourmalines, is material transfer by surface diffusion, rather than deposition from pore solution. Anatase and brookite, distinguished from detrital grains by euhedral growth, are common neoformations in sandstones (BOSWELL 1933; PETTIJOHN 1957). The titanium may originate from the decomposition of biotites. Apatite has been described also as a neoformation from some sandstones. Examples are overgrowths over detrital apatites in the Buntsandstone (VALETON 1953).

Neoformation of sulfides. Diagenetically formed iron sulfides (marcasite and pyrite) occur in many sandstones. Sulfide is fixed in these minerals as a result of early diagenetic reduction of sulfate in pore solutions.

The majority of the diagenetic processes occurring in the pore space of sandstones cannot be explained alone by reactions between detrital sand constituents and entrapped pore solution. The quantitatively important transformations which result, for example, in kaolinitization of feldspathic sands, dissolution of carbonates, recrystallization or neoformation of phyllosilicates, and replacement of quartz and feldspar by carbonates, all necessitate an influx of fresh solution and removal of dissolved matter over a long period of time. The same applies to the diagenetic dissolution of hard to dissolve minerals, such as the heavy minerals. As the studies of formation waters have shown, the flow is fed on the one hand by water released by sediment compaction, and on the other by an influx of surface water.

The compaction flow arises from compaction of subsiding sediments. The major amount of ascending solutions is supplied by the compaction of clays (to be considered in the next chapter). These constitute the main portion of most basin fill and experience the greates volume reduction upon burial. The expelled solutions migrate upward under the influence of the vertically directed pressure from overlying strata. However, the flow in a sedimentary sequence consisting of different strata, does not proceed vertically to the surface. It takes place predominantly in the permeable sandstone layers. As detailed considerations have shown (see section 5.23), the flow lines of water flowing from impermeable into permeable strata are refracted strongly in the permeable bed, when the interface is inclined only slightly from the horizontal. Since bedding planes are rarely exactly horizontal, the compaction flow from a compacting basin uses mainly sandstone layers as conduits for flow. These layers during the course of diagenesis are permeated by much larger volumes of solution, than the compaction volume of underlying clay layers, since the sandstones drain a larger volume of compacting basin sediment (v. ENGELHARDT 1967).

Since the flow of formation waters takes place essentially in sandstone strata during diagenesis, under certain circumstances different chemical conditions can prevail within adjacent sandstone layers, if they are separated from one another by relatively impermeable shales. An example of this are the Schilf- and Stubensandstones of middle Keuper age in southern Germany, which are separated by a 60 to 70 m thick marl (HELING 1963, 1965). While in the Stubensandstone feldspar is first dissolved and quartz and kaolinite newly formed and dolomite and calcite later deposited, in the underlying Schilfsandstone diagenetic neoformation of alkali feldspar and chlorite takes place.

Only a small portion of the compaction flow passes from the sandstone layers into the overlying shales. As a result suspended gas bubbles and oil droplets, as well as dissolved ions, are filtered out, remaining in the sandstone, the gas and oil, because of the capillary pressure, and ions by ion filtration effects.

In this way gas and oil, as well as dissolved matter, are concentrated in permeable strata. In this way, near the sand-shale boundary, saturation of certain minerals, for example, quartz or calcite, can be exceeded, leading to deposition there.

Special situations result in anticlinal and similar structures in sands. The compaction flow converges on the crest of the structure from all sides and from there is forced into the overlying shales. Here accumulations of gas and oil or diagenetic minerals can form.

The compaction flow will flow in dipping sandstone strata as a rule up dip. It can flow differently, however, if the permeability of the overlying strata varies or if at certain places it is greater, because of changes in rock characteristics or deformation and faulting. Then the flow in the sandstone may be directed away from these places, independent of dip in certain instances.

As analyses of formation waters have shown, their movement within sediments is not determined always by the compaction flow alone. Under certain conditions large amounts of surface waters can penetrate deep strata. This influx of surface waters can be expected especially in older, consolidated sedimentary basins, in which subsidence is taking place no longer. Therefore, the compaction flow has ceased. Also a later uplift of sediments compacted at greater depth can bring these into the sphere of influence of surface waters.

In sandstones of consolidated sedimentary basins, two epochs of diagenetic reactions, caused by flow of formation waters, should be differentiated. In a first period of subsidence, dissolution, recrystallization and neoformation, with simultaneous increase in pressure and temperature, are produced by the compaction stream ascending from depth. After consolidation of the basin, a second period can follow, during which surface waters penetrating from above bring about changes within the pore space of the rocks.

FÜCHTBAUER (1970) has given descriptions of the temporal sequence of chemical events during diagenesis, to which reference should be made. From these examples, it is concluded that the sequence of diagenetic dissolution, recrystallization, and neoformation, as expected, is very different in individual cases, since they depend on many factors. Therefore, there is no generally applicable scheme representing the course of chemical diagenesis. The most important factors, apart from the compositions of the sands and primary pore solutions are the subsidence history, the effects of later uplift, and changes in the chemistry of formation waters which penetrate a sand during diagenesis.

Of special importance to the diagenetic process is the hydrogen ion/cation ratio of the pore solution. In the early stages of diagenesis this ratio is determined essentially by the nature of the depositional environment. As a rule the ratio is higher in deposits of the continental environment than in marine sediments.

For the early diagenesis of many continental sandstones, reactions are typical which take place in an acidic milieu. Examples are the dissolution of feldspars with simultaneous formation of kaolinite and quartz, the preferential dissolution of potassium from detrital micas, the dissolution of carbonates and their replacement by quartz, and the dissolution of heavy minerals, especially apatite, garnet, and hornblende.

In marine sandstones and those continental deposits, which did not entrap acidic weathering solutions, these reactions do not take place. Here rather early neoformation of feldspars, deposition of carbonates and neoformation of chlorites occur. Primary or secondary calcite is transformed later into dolomite, which, like chlorite, removes magnesium from the pore solution. At especially high silica and cation concentrations, promoted by the presence of much feldspar and volcanic glass, zeolites are formed, especially analcite and heulandite.

If, within a formation, a regional facies change is present, corresponding regional differences in early diagenesis are found. The diagenetic neoformations in the middle Buntsandstone of Germany are determined by different depositional environments (FÜCHTBAUER 1967 a). In the southern region fluviatile sandstones formed K-feldspar in addition to quartz and then dolomite in a later phase. In contrast in the northern brackish-marine region, analcite, albite and quartz formed and later anhydrite, barite, baritocelestite, celestite and occasionally halite, in addition to calcite.

Pressure solution of detrital constituents is a general diagenetic reaction in sandstones of all environments. Following very slight burial, pressure solution of detrital carbonates begins, with redeposition of a finely crystalline form in the pore space of sediments. Pressure solution of quartz (to a lesser extent also feldspar) begins to take effect at subsidence depths of about 1000 m. Neoformation of quartz, observed in almost all sandstones, and initiated after slight subsidence, depends to a degree on pressure solution. The additional SiO_2 originates from the dissolution of readily soluble organogenic forms of SiO_2, from the dissolution of feldspar and probably also from the cooling of pore solutions ascending from greater depths.

At greater depths diagenetic processes are determined by elevated temperature and pressure, and their characteristics can be understood from the changes in formation water composition experienced during diagenesis and subsidence. The generally observed increase in cation content in formation waters with depth, that is, the decrease in hydrogen ion/cation ratio, corresponds to a diminution of reactions taking place in an acid environment. The differences due to depositional environment, still preserved at shallow depths, are reduced or completely obliterated at greater depths. Evidence of kaolinitization during early diagenesis is lost at depth with the formation of dioctahedral and trioctahedral chlorites. From alkali-rich solutions alkali feldspar is formed. Muscovite forms from degraded micas or in place of or at the expense of K-feldspar, if the solution contains much silica. From carbonate-bearing solutions, with the increase in temperature and alkalinity at greater depths, carbonate precipitates. This frequently replaces quartz and feldspar. Analcite and heulandite, formed at the beginning of diagenesis, are transformed at greater depths into albite, quartz, and laumontite. Anhydrite and barite appear as a rule to belong to a late stage of diagenesis.

Diverse are the diagenetic reactions which can be induced by mixing formation waters with penetrating surface waters. Different formations and transformations of late diagenesis may be traced back to the influence of surface waters. As already mentioned, the high Ca-content of formation waters of the Michigan Basin probably originates from surface waters. The heating of carbonate-bearing Ca- and magnesium-rich surface waters at depth could have caused the late precipitation of carbonates observed in many sandstones. Considering the usually low carbonate content of formation waters, it is difficult to explain this otherwise. The same applies to the late deposition of barite and anhydrite. Since deep formation waters contain very little sulfate, sulfate-bearing solutions must have been introduced. Where they could not have originated from evaporites, they could have been supplied from the surface.

The chemical processes of diagenesis are interrupted, when the pore space becomes filled with oil or gas. In oil- and gasfilled sandstones, that stage of diagenesis is preserved, which was attained at the time these phases migrated into the rock. This is the basis for the established fact, concerning many oil and gas deposits, that sands filled with oil or gas have a higher porosity than where they are water-filled. The boundary between intensive and weak diagenesis, established, for example, by the amount of diagenetic quartz, seldom coincides exactly with the present oil or gas/water boundary. From this deviation, one can infer conclusions concerning the history of oil or gas migration into the reservoir as well as structural history (FÜCHTBAUER 1961; PHILIPP et al. 1963).

5.5 Diagenesis of clays

5.51 Mechanical diagenesis (compaction)

Newly deposited clay sediments have porosities in the range between 0.7 and 0.9. From the pressure of overlying strata this porous texture can experience considerable compaction by rearrangement and deformation or fracture of individual particles, without involvement of chemical processes. This compaction is much more prevalent for argillaceous sediments than it is for sands. This is because the primary porosity of clays is significantly greater, and also because platy clay particles can pack geometrically more densely than more equidimensional sand grains.

Compaction of clays involves two kinds of textural alteration. First of all a reduction in pore space takes place as a result of expulsion of pore solution. Secondly the primary particle arrangement is changed, as the primary aggregates or floccules are destroyed (homogenization), resulting in a texture in which platy particles arrange themselves parallel to one another and perpendicular to the directed pressure.

Related to the practical problems of soil mechanics, the compaction of clay muds has been investigated in many laboratories. In such compaction experiments a clay sample, confined in a cylinder, is compressed by means of a piston. The expelled water is allowed to escape through the substrate or through the piston. The volume assumed by the clay sample under a given load is measured. If V_0, ε_0, and E_0 are the initial volume, porosity, and relative pore volume of the sample, and V, ε, and E the same quantities after compaction, the following equations apply for the relation between relative volume decreases, $\triangle V$ and porosity or relative pore space:

$$\triangle V = \frac{V_0 - V}{V_0} = \frac{\varepsilon_0 - \varepsilon}{1 - \varepsilon} = \frac{E_0 - E}{1 + E_0} \tag{5.40}$$

$$\varepsilon = \frac{\varepsilon_0 - \triangle V/V_0}{1 - \triangle V/V_0} \tag{5.41}$$

$$E = E_0 - (1 + E_0) \triangle V/V_0 \tag{5.42}$$

Fig. 5.40 illustrates some examples of experimentally determined compaction curves for some different argillaceous sediments at pressures up to about 100 km/cm². If one plots the relative pore volume against the log of the pressure, the resultant curve in many cases is almost linear. The compaction can be described to a good approximation by the following formula suggested by TERZAGHI (1925):

$$E = E_1 - \beta \lg P/P_1 \tag{5.43}$$

E_1 is the relative pore volume at pressure $P_1 = 1$ kg/cm². The compression index β is equal to the decrease in relative pore space produced by an increase in pressure from P to $10 P$. As table 5-14 and figs. 5-41 and 5-42 show, E_1 and β increase with increasing clay content of the sediment. Fig. 5-43 shows the relation between E_1 and β. The higher the initial porosity (determined by grain size), the greater the compression index. Fine grain clays, which form very voluminous sediments with water at normal pressure, experience at a given pressure relatively greater compression than argillaceous sediments with coarser constituents. For this reason the compaction curves of the individual sediments in fig. 5-40 converge with increasing pressure.

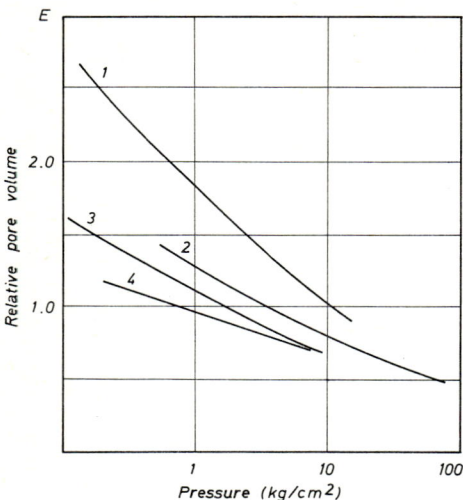

Fig. 5-40. Experimental compaction of different clay sediments at pressures below 100 kg/cm². 1: Plastic clay, marine (Eocene), Paris Basin; 2. London blue clay, marine (Eocene); 3. Gosport clay, estuarian (Recent); 4. Oxford clay, marine (Jurassic) (after SKEMPTON 1944).

Compaction experiments show that argillaceous sediments, subjected to pressures of about 50 kg/cm², are compressed to relative pore volumes of about 0.5 or porosities of about 0.3. Thus with an initial porosity of 0.7, the original volume is reduced to less than one-half. A pressure of 50 kg/cm² corresponds in nature to burial depths of only a few hundred meters. To understand diagenesis of clays at the especially interesting greater depths of burial, compaction experiments at higher

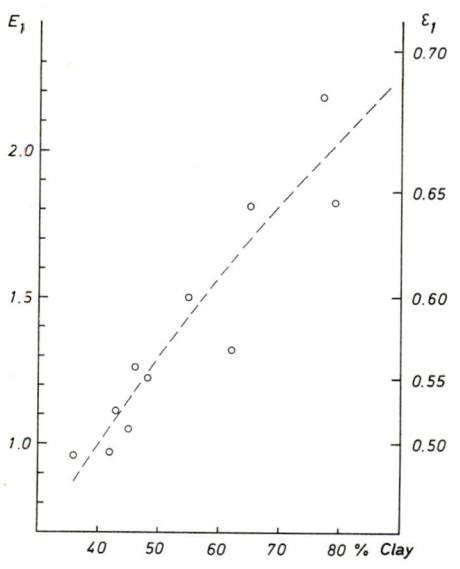

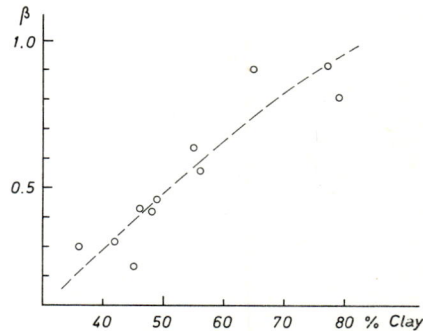

Fig. 5-42. Dependence of compression index β of the sediments in table 5-14 on clay content.

◄ Fig. 5-41. Dependence of initial pore volume E_1 of sediments in table 5-14 on the clay content.

5.51 Mechanical diagenesis (compaction)

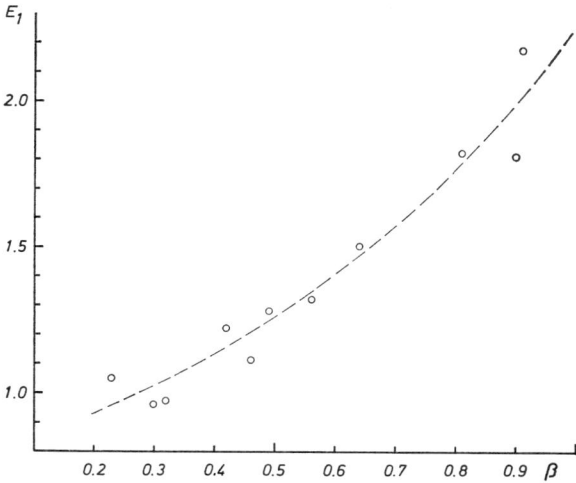

Fig. 5-43. Relation between initial pore volume E_1 and compression index β for the sediments of table 5-14.

pressures are important. Fig. 5-44 shows the results of experiments by PARASNIS (1960) with different natural argillaceous sediments. Fig. 5-45 reproduces experimental data by v. ENGELHARDT & GAIDA (1963) from pure clay minerals. Similar results were obtained by CHILINGAR & KNIGHT (1960) with montmorillonite, illite, and kaolinite clays. In all of these experiments the clays were compacted by pistons in compression cylinders.

The linear relation between relative pore volume and the log of the applied pressure apparently applies less adequately at these higher pressures. The linear correlation can be used as a rule only over certain pressure ranges, but not generally. Especially for fine grain clays does the porosity decrease more slowly with increasing pressure. At higher pressures also, the compaction curves for different clays converge with increasing pressure. At the highest pressures, the initial porosity differences of coarse and fine grain clays disappear.

The complex processes involved in the compaction of clay muds can be understood in their main features by the following model (see VAN OLPHEN 1963, 1964). The platy clay mineral particles, based on electrophoritic relations and cation exchange properties, are negatively charged overall. The negative charge originates from the isomorphous substitution of Al for Si and Mg for Al and is compensated by the adsorption of cations on the particle surfaces. In the presence of water the cations are distributed in a diffuse double layer about the particles. The double layer thickness depends on the electrolyte concentration of the pore solution. It is large in dilute solutions and decreases with increasing salinity. If two clay particles line up opposite one another with basal surfaces parallel, the double layers repel one another and only van der Waals forces provide a weak attraction. It is different if the edge of a clay platelet contacts the basal surface of another platelet. At the

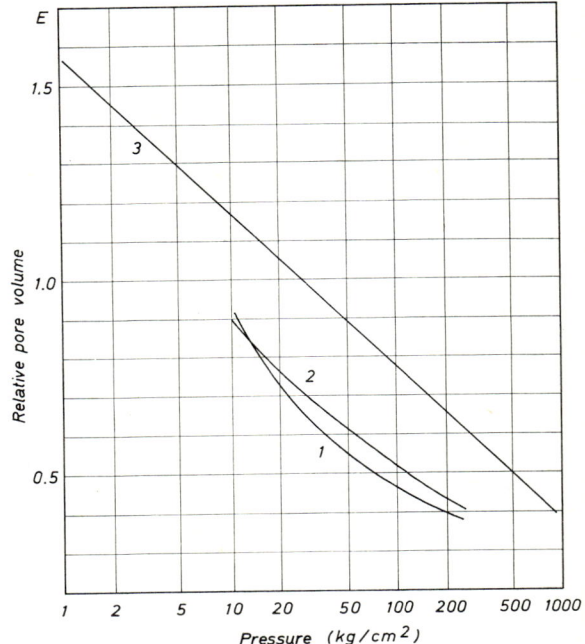

Fig. 5-44. Experimental compaction of different clay sediments at pressures up to 1000 kg/cm², after PARASNIS (1960). 1: Gault clay, Cambridge, England; 2: Oxford clay, Bedford, England; 3: Deep sea clay (4700 m), Atlantic (45° 26' N, 9° 20' W).

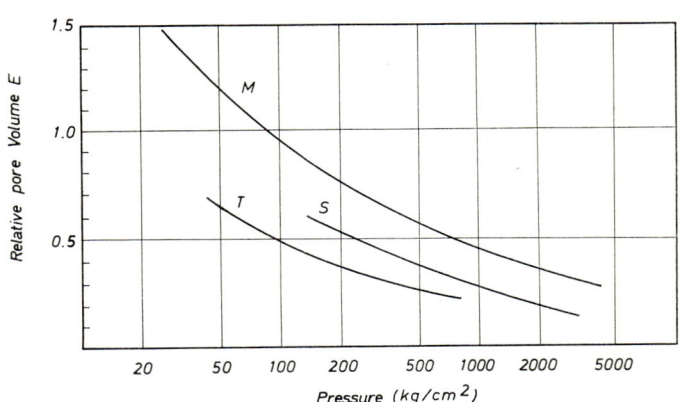

Fig. 5-45. Experimental compaction of kaolinite and montmorillonite muds at pressures up to 3000 kg/cm². M: Montmorillonite (Tixoton); S: Kaolinite, fine grain (Commercial product); T: Kaolinite, coarse grain (Tirschenreuth) (v. ENGELHARDT & GAIDA 1963).

5.51 Mechanical diagenesis (compaction)

Table 5-14. Compaction of argillaceous sediments at pressures up to 70 kg/cm² (after SKEMPTON 1944).

	% Clay	E_1	ε_1	β	Max. pressure kg/cm²
1) Oxford clay Jurassic, marine Peterborough	36	0.96	0.49	0.30	8.6
2) Gosport clay estuarian, Recent	43	1.11	0.53	0.46	8.9
3) Kaolin Cornwall	45	1.05	0.51	0.23	8.9
4) Blue marine clay Bosporus	42	0.97	0.49	0.32	18
5) London blue clay Eocene, marine	46	1.28	0.56	0.49	70
6) Ganges Delta clay Calcutta, Recent	48	1.22	0.55	0.42	9.1
7) Mississippi clay	ca. 55	1.50	0.60	0.64	3.2
8) London brown clay Eocene, marine weathered	62	1.32	0.57	0.56	8.6
9) Yaguajay clay Cuba	ca. 65	1.81	0.65	0.90	3.1
10) Kleinbelt clay Eocene, marine Denmark	77	2.18	0.69	0.91	5
11) Plastic clay Eocene, marine Paris Basin	79	1.82	0.65	0.81	15

edge the crystal structure is interrupted and free bonds produce potential fields which differ from those of the basal surfaces. The preferential adsorption of gold particles, phosphates and other matter (see MARSHALL 1964, p. 326), as do structural considerations, make it appear likely, that the clay mineral particle edges carry positive charges. As a result electrostatic attractive forces act between particle edges and basal planes. The card-house structure (fig. 5-47 a) of many newly sedimented clay sediments results from these attractive forces.

The resistance which a clay mud offers to compaction occurs on the one hand from the repulsion of diffuse double layers, and on the other hand from the strength of the structure built from the surface-to-edge bonding. Low pressures act primarily in opposition to repulsion by the diffuse double layers, whose thickness is decreased by expulsion of water. The volume reduction produced in this manner is reversible. Upon removal of pressure, the clay adsorbs water again, in order to restore as much as possible the original state. This process is called swelling. The observed swelling pressures in clays can be considerable, being a little more than 10 kg/cm².

Compaction at higher pressures, which corresponds to those of actual diagenesis, is in contrast an irreversible process. It causes a change in particle fabric formed originally by the bonding action at edge-face points of particle contact. In this structure each mineral particle is bound to its neighbor at these contact points. The lower the porosity the greater the mean particle coordination number, that is, the number of bonding sites between a particle and its neighbors. As long as the system is in equilibrium the external pressure is balanced by the bonding forces. The pore solution is shielded from the external pressure by the particle framework. If the pressure is increased a small amount, the bonding forces at a few sites are exceeded, individual particles make adjusting movements, and the framework is collapsed somewhat. As a result the pressure is transferred to the pore solution, which is expelled from the compacting clay. The rate of this process is determined by the flow rate of the pore solution, that is, by the permeability. The flow rate is maintained until the particles adjust to new positions of greater bonding force or higher coordination number. Then a new structure of lower porosity results that can withstand the higher pressure. Since, with increasing compaction, the number of more stable contact sites, as well as total sites generally increases, the compressibility of the clay decreases with increasing compaction. At the same time the permeability is reduced, since the mean pore size available for pore water flow decreases as the particles are brought into closer proximity. Therefore, with decreasing porosity, the rate with which equilibrium is established following the pressure increase also decreases. Studies of the influence of chemical composition of pore solutions on clay compaction have indicated that chemical factors influence compaction only in the low pressure range, that is, up to about 10 kg/cm^2 (see summaries by MEADE 1964, 1966 of previous data). The effect of salinity is quite different for different clays. Fine grain illite resulted in a smaller pore volume during compaction when the NaCl-concentration was increased (BOLT, after MEADE). Coarse grain illite behaved oppositely, and montmorillonite like fine grain illite (MITCHELL, after MEADE). These differences are the same as were determined for the primary sediment volumes. Normal, or relatively coarse grain clays form more voluminous sediments in saline water than in fresh water, while very fine grain montmorillonite behaves oppositely. In addition the kind of exchange ion produces different effects in individual clays. Na-montmorillonite maintains, during compaction, a higher pore volume than Ca- or Al-saturated montmorillonite. An Al-saturated kaolinite maintains a higher volume than either Ca- or Na-kaolinite (SAMUELS, after MEADE).

The electrolyte content of a pore solution has a significant effect on the amount of compaction apparently only as long as the work of compression mainly acts against the repulsive forces of the diffuse double layers surrounding the clay particles. This is because the potential and thickness of those double layers are determined by the ionic species and concentration. Electrolyte concentration apparently has little effect on the strength of the bonding sites holding clay particles together. Therefore, at higher pressures the salinity of pore solutions does not influence the compaction attained at a given pressure. At pressures between 30 and 3000 kg/cm^2 v. ENGELHARDT & GAIDA (1963) found the same compaction as with pure water for kaolinite and montmorillonite clays in NaCl (0.2—4.6 n)- and CaCl$_2$-solutions (0.6—1.2 n).

On the other hand the rate of compression as well as the structure of compacted clays at high pressures clearly are influenced by the electrolyte content. For pure kaolinite and montmorillonite clays, compressed under loads of several hundred kg/cm², v. ENGELHARDT & GAIDA (1963) determined by a X-ray method, that clay mineral platelets are oriented better perpendicular to the load, the higher the NaCl-content of the pore solution. Fig. 5-46 shows the progress of compaction with time of a montmorillonite clay at 126 kg/cm², saturated with water and NaCl-solutions of different concentrations. The compaction progresses more rapidly, that is, pore solution flows more rapidly through the compacted clay, the higher the salt concentration. Kaolinite clays show the same behavior. This phenomenon probably can be explained by assuming that the permeability of the compacted clay structure is greater, the higher the salinity of the pore solution.

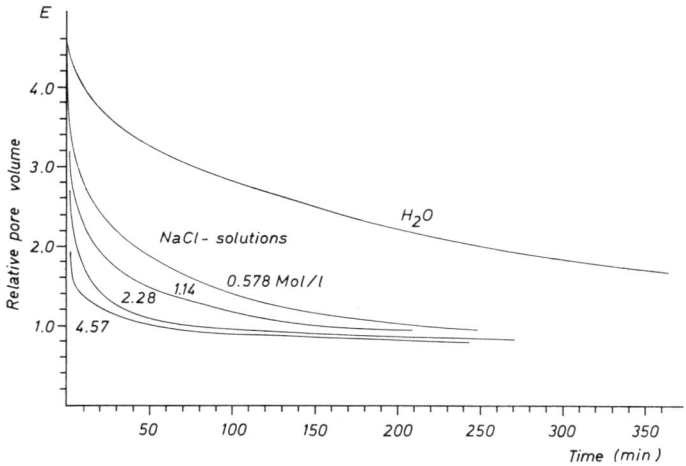

Fig. 5-46. Compaction of montmorillonite muds at a pressure of 126 kg/cm² and with different NaCl-concentrations in the pore solutions (v. ENGELHARDT & GAIDA 1963).

The fact that pore solution salinity does not influence the total porosity achieved at a given pressure, although permeability increases and the fabric collapses, can best be explained by an aggregate structure, which depends on pore solution salinity (fig. 5-47). Aggregates of individual particles, such as occur upon flocculation of a dilute clay suspension, form a fabric whose pore space consists of small pores within and large voids between aggregates. Electrolyte content of the pore solution promotes the formation and stability of the aggregate. Therefore, clays with pore solutions of low salinity contain small aggregates or free particles, which can compact relatively unhindered perpendicular to the directed pressure. Permeability, because of the small pore channels, is low. At higher salt concentrations the aggregates are large. Preferred orientation is rendered difficult and permeability is higher, since large channels occur between aggregates. With increasing compaction the aggregate structure is gradually destroyed, and a homogeneous fabric should form at the highest pressures, at which individual clay mineral platelets are arranged preferentially perpendicular to the

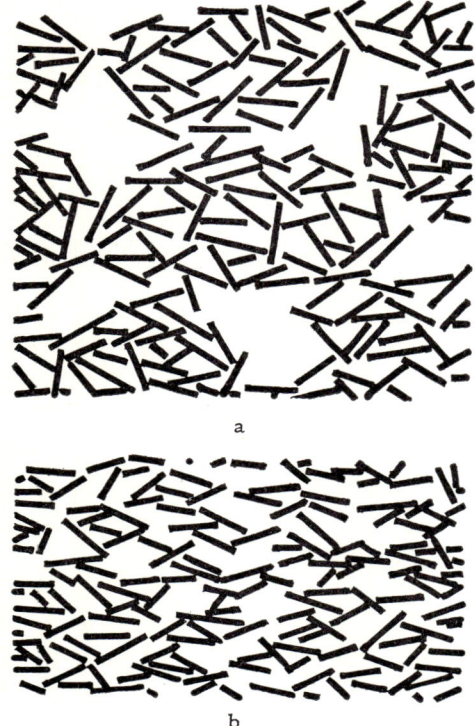

Fig. 5-47. Two-dimensional model of the fabric of compressed clays. a) Pore solution with high electrolyte content; b) Pore solution with low electrolyte content (v. ENGELHARDT & GAIDA 1963).

directed pressure. In compaction experiments with hydrostatically confined cylindral samples, such textural development with compaction has not yet been demonstrated clearly. Results of v. ENGELHARDT & GAIDA (1963), according to which the orientation of kaolinite and montmorillonite clays should have increased over the pressure range between 100 and 1000 kg/cm², have not been confirmed by accurate measurements (THIEM 1967). This indicates that the degree of orientation of platelets within a compressed sample varies so greatly that there is no definite indication of a regular increase in orientation with compaction. It is presumed that the lateral confinement of the clay sample affects the particle orientation, which probably requires shear motion perpendicular to the directed pressure. In addition it is likely that the pore solution expelled during the experiment at the bottom of the clay cylinder, was compressed from above by the piston. In doing so it flowed through the most strongly compacted zone and probably disturbed particle orientation. In contrast, during natural compaction the pore solution flows from the strongly compacted zone at the bottom up into progressively more porous layers.

In contrast to experiments with confined samples, compression of laterally unconfined clay samples produces clay particle orientation perpendicular to the direction of pressure even at very low pressures. With this sort of experimental arrangement lateral shear motion is not inhibited and orientation is not disturbed by the flow of expressed pore solution (MEADE 1964 has reported on different experiments).

From the significant compressibility of clay muds under slight loads, as was established in laboratory experiments, it follows that the porosity of argillaceous sediments clearly should decrease with burial depth. At the same time experiments indicate that a regular decrease in porosity is to be expected only when the grain size does not change significantly with depth and when other factors, such as mineral and pore solution compositions, remain constant. Compaction of clays is very dependent on these factors at low pressures.

Actually the porosity of unconsolidated argillaceous sediments in the first few meters below the surface decreases considerably with depth. Some examples are shown in fig. 5-48, including sediments from Lake Zürich (ZÜLLIG 1956), marine sediments from off the California coast (EMERY & RITTENBERG 1952) and Black Sea sediments (SAWELJEW, after RUCHIN 1958). Although these examples represent very different depositional environments and compositions, their decreases in porosity with depth are quite similar. In the uppermost 8 meters the porosity decreases from values between 0.8 and 0.9 at the surface to about 0.6. The clay sediment looses during this first stage of compaction almost 30 % of its original water content.

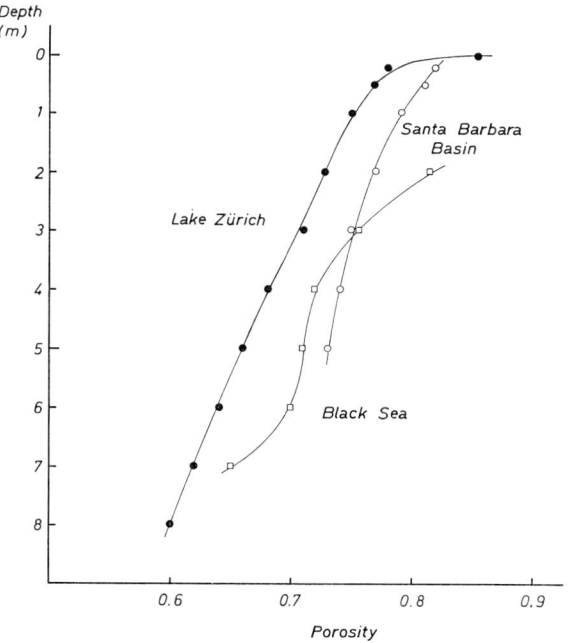

Fig. 5-48. Porosity variation in recent clay sediments as a function of depth. Lake Zürich (ZÜLLIG 1956); Santa Barbara Basin, Pacific, California (EMERY & RITTENBERG 1952); Black Sea (SAWELEJEW after RUCHIN 1958).

272 5.5 Diagenesis of clays

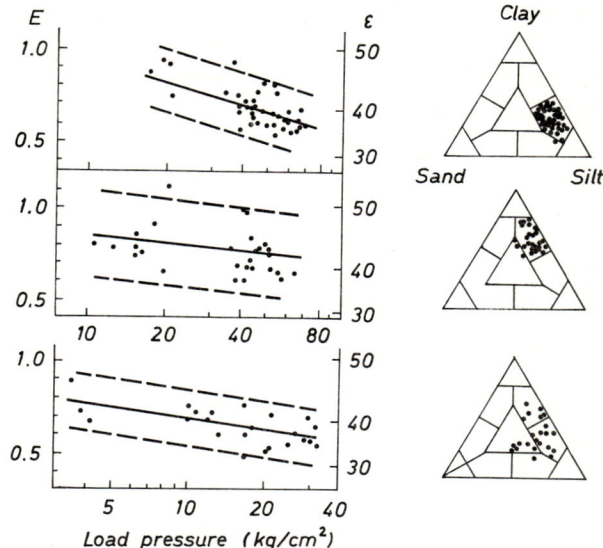

Fig. 5-49. Porosities of argillaceous, alluvial fresh water sediments from well drilling in the San Joaquin and Santa Clara Valleys, California (MEADE 1968).

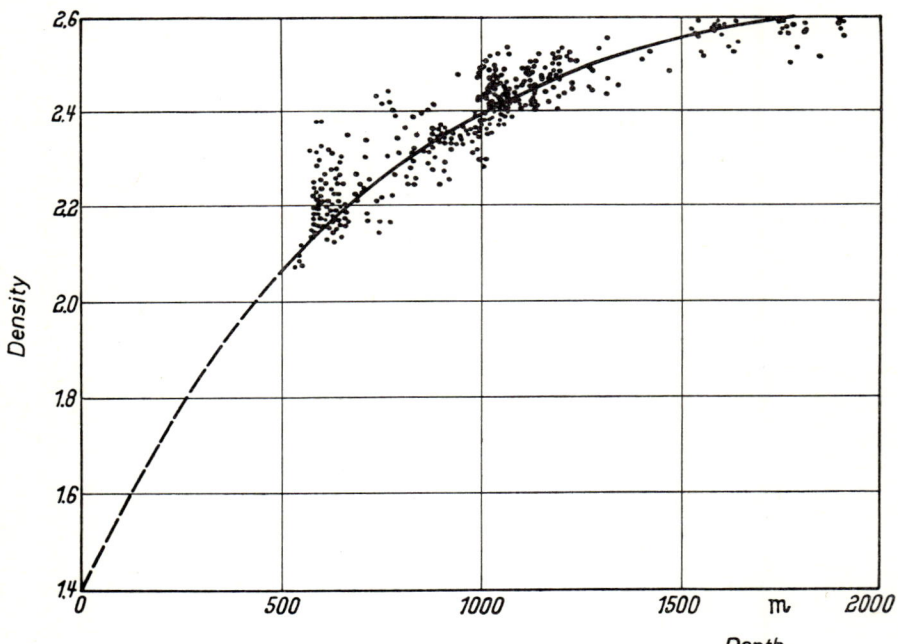

Fig. 5-50. Densities of shales from deep borings in Oklahoma, as a function of their maximum depth of burial (after ATHY 1930).

5.51 Mechanical diagenesis (compaction)

Not only the weight of the overlying strata determines early diagenetic compaction of argillaceous sediments, but also additional factors play a role. This is shown by the studies of MEADE (1968 a) of the porosity of young alluvial sediments of the San Joaquin and Santa Clara Valleys in California at depths down to 700 m. The variable grain size distributions and the equally variable diatom content of these sediments exert their influence here. If one compares sediments of about the same grain size, a decrease in pore volume is indicated, in spite of considerable scatter, with increasing depth of burial, as fig. 5-49 shows. Since, in these deposits, an excess artesian pressure prevails, the pressure actually prevailing in the sediments was calculated as the difference between the sediment load and the pore solution pressure measured in wells (hydrostatic + artesian pressure). On the average the relative pore volume in the argillaceous sediments of California, in the pressure range from 3 to 70 kg/cm², decreases about 0.3 units, corresponding to a reduction in total volume of about 15 %.

As a result of the advance of oil well drilling to great depths over the last few decades, it has been shown cleary that the porosity of shales in all sedimentary basins decreases systematically with depth. ATHY's (1930) published summary of 2200 density determinations on well core samples from Texas and Oklahoma (fig. 5-50) illustrates, in spite of the scatter, a regular decrease in shale density with depth. At

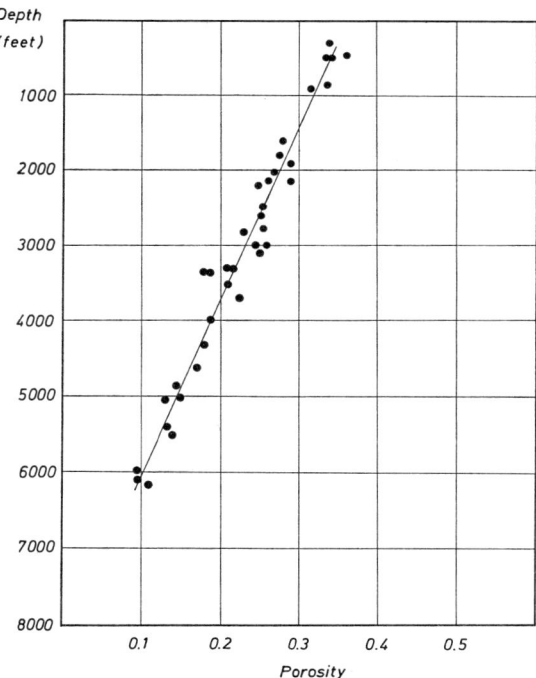

Fig. 5-51. Porosities of Tertiary claystones from a deep well in Venezuela, as a function of depth (after HEDBERG 1936).

a depth of about 2000 m a mean density of 2.6 is attained, which corresponds to a shale with vanishingly small pore volume. The data of ATHY include very dissimilar rocks, including not only those which were subjected to loading from overlying sediment, but also the effects of tectonic stress. In his classic work on the compaction of clays HEDBERG (1936) investigated the porosity of a uniform shale series from the Tertiary geosynclinal basin of Venezuela, from available well cores from depths to 2000 m. He selected this basin in order to isolate the effect of orogenic pressure as much as possible and to exclude other influences such as tectonic pressure and change in sediment composition. Values for sediments from essentially one well produced the relation between depth and porosity illustrated in fig. 5-51. If one plots instead of porosity the relative pore volume against the log of depth, one obtains for depths below 400 m an approximately linear relation, as fig. 5-52 shows.

Later investigations of shales from other formations and basins, illustrated in figs. 5-53 to 5-59, have, for burial depths between a few 100 and approximately

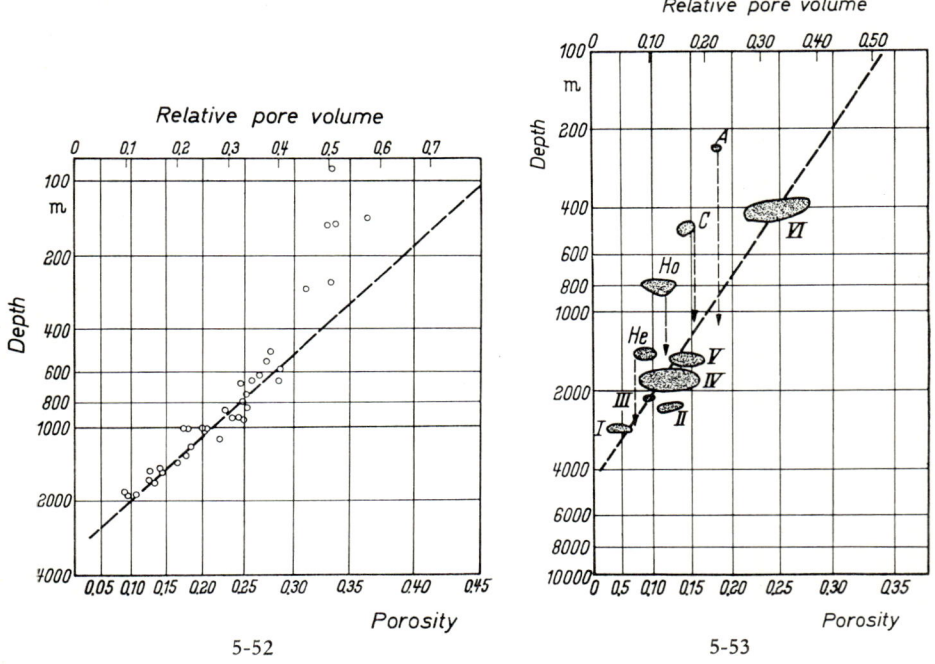

Fig. 5-52. Porosity (ε) and relative pore volume (E) of Tertiary claystones from Venezuela, as a function of the log depth (after HEDBERG 1936).

Fig. 5-53. Porosity (ε) and relative pore volume (E) of Lias α shales from deep wells in NW-Germany, as a function of present depth. The arrows indicate the original maximum depth of later uplifted strata, deduced from geologic considerations. I: Harsebruch; II: Wettenbostel; III. Bokel; IV: Hohne and Wesendorf; V: Eldingen; VI: Wesendorf; He: Gross Hehlen; Ho: Hohenassel; C: Calberlah; A: Abbensen (after measurements by FÜCHTBAUER & v. ENGELHARDT, v. ENGELHARDT 1960).

3000 m, confirmed this linear relationship. In the case of the Lias clays of northwest Germany (fig. 5-53) it can be shown that the porosity of shales can indicate the maximum burial depths to which they have been subjected at some previous time. Rocks which have been uplifted subsequent to deposition by tectonic movements, have preserved that porosity impressed on them at their greatest depth of burial.

The foregoing observations show that the dependence of relative pore volume of shales on maximum depth of burial, below about 400 m, may be represented quite well by the following empirical equation:

$$E_T = E_1 - b \lg T \tag{5.44}$$

E_T is the relative pore volume at T meters depth. b and E_1 are constants. E_1 is formally the pore volume at 1 m depth, and therefore, represents the primary pore volume for newly deposited sediment.

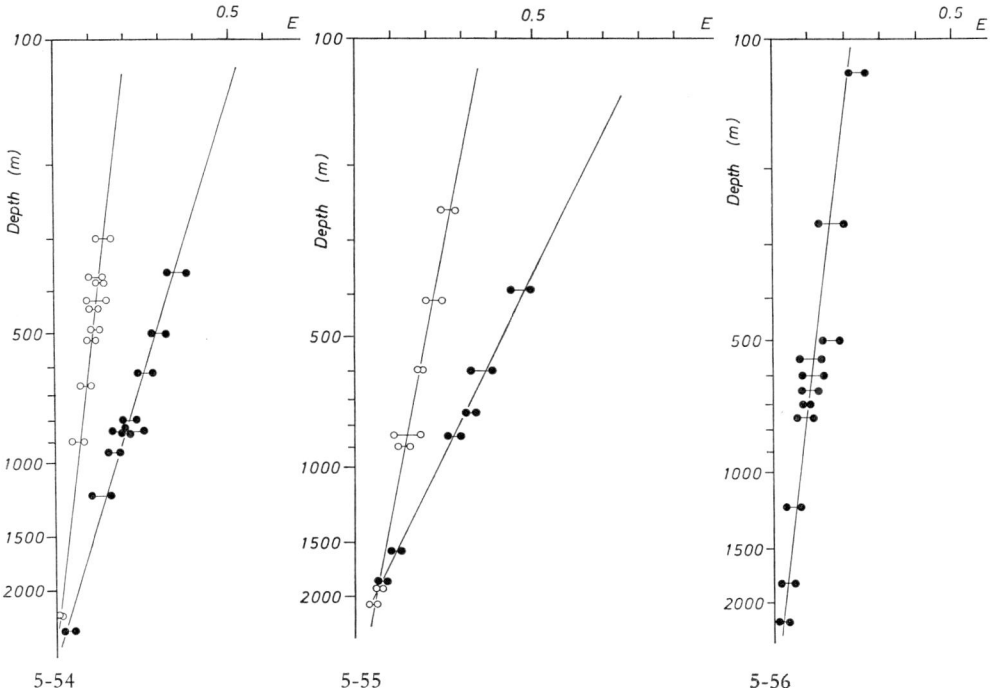

Fig. 5-54. Relative pore volumes of Keuper shales from SW-Germany, as a function of depth. ○ Lower Keuper (Lower Bunte Marl); ● Middle Keuper (Knollenmergel) (HELING 1967).

Fig. 5-55. Relative pore volumes of Lias shales from SW-Germany, as a function of maximum depth. ○ Lias α and β; ● Lias δ (HELING 1967).

Fig. 5-56. Relative pore volumes of Dogger (α, ζ) shales from SW-Germany, as a function of maximum depth (HELING 1967).

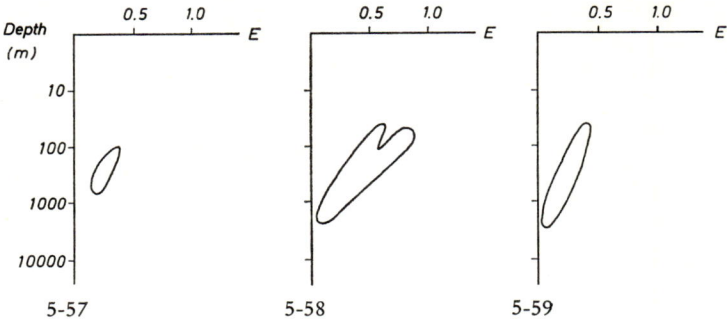

Figs. 5-57—5-59. Relative pore volumes of Tertiary clay sediments of the Rhine Valley Graben, as a function of depth. Fig. 5-57: Pliocene; Fig. 5-58: Miocene; Fig. 5-59: Oligocene (after HELING 1969).

For the formations studied, E_1- and b-values are summarized in table 5-15. As can be seen from fig. 5-60, b and E_1 are related closely. High E_1-values correspond to high b-values. There is, therefore, an analogy to the experimentally determined relation between compression index β and the initial pore volume E_1 (equation 5-43, fig. 5-43). Originally porous clays are compacted upon burial more than are less porous clays. The initial porosity of a clay sediment depends on grain size, mineral composition, and pore solution salinity. The different values of E_1 and b for individual clay formations may be explained in part from these relations. As an example, the difference in compaction behavior of lower and middle Keuper clays apparently is due to the high content of montmorillonite mixed-layers in middle

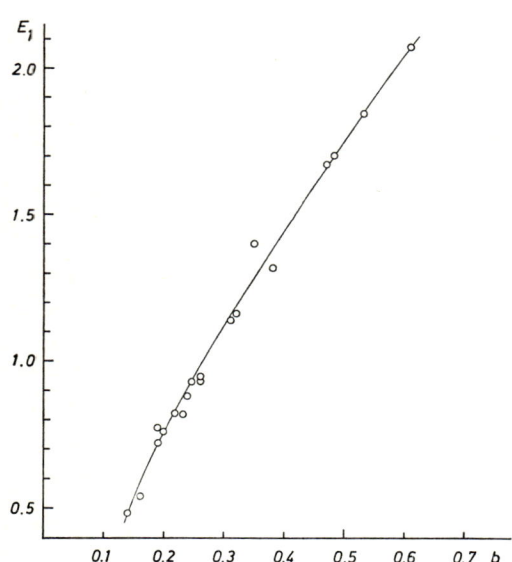

Fig. 5-60. Initial pore volumes (E_1) and compaction constants (b) for the clay sediments of table 5-15.

5.51 Mechanical diagenesis (compaction)

Table 5-15. Constants for equation (5.44) for different clay sediments.

Formation	Minerals	b	E_1	ε_1	MD (μ)
Lower Keuper, SW-Germany (Heling 1967)	I, c, k	0.16	0.54	0.35	3
Middle Keuper, SW-Germany (Heling 1967)	ML, i, c, k	0.38	1.32	0.57	2
Lias α, β NW-Germany (v. Engelhardt 1960; Füchtbauer & Goldschmidt 1963)	I, K, c	0.32	1.16	0.51	...
Lias α, SW-Germany (Heling 1967)	K, i, c	0.23	0.82	0.45	10
Lias δ, SW-Germany (Heling 1967)	K, i, c	0.61	2.07	0.67	3
Dogger, SW-Germany (Heling 1967)	I, c, k	0.14	0.49	0.33	12
Tertiary, Venezuela (Hedberg 1936)	...	0.53	1.84	0.65	...
Tertiary, Po Basin (after Storer 1959)	...	0.48	1.70	0.63	...
Tertiary, Rhine Valley Graben (Heling 1969)					
Young Tertiary (Fresh water)	i, ml, k, c	0.22	0.82	0.45	1.5
Upper Hydrobia beds (brackish)	i, ml, k, c	0.35	1.40	0.51	2.0
Lower Hydrobia beds (brackish)	i, ml, k, c	0.47	1.67	0.63	2.5
Cerithian beds (marine)	i, (ml), k, c	0.31	1.14	0.53	5.0
Bunte Niederröden beds (brackish)	i, (ml), k, c	0.19	0.72	0.42	6.0
Cerena marls (marine)	i, k, c	0.22	0.82	0.45	6.0
Meletta beds (marine)	i, k, c	0.20	0.76	0.43	7.0
Fischschiefer (marine)	i, k, c	0.19	0.77	0.43	9.0
Upper Pechelbronn beds (marine)	i, k, c	0.25	0.93	0.48	17
Middle Pechelbronn beds (marine)	i, k, c	0.88	0.24	0.19	13
Lower Pechelbronn beds (marine)	i, k, c	0.26	0.94	0.48	12
Lymnäen Marl (marine)	i, k, c	0.26	0.93	0.48	12

C = chlorite; I = illite; K = kaolinite; ML = mixed-layer illite/montmorillonite.
Upper case indicates large and lower case small amounts of the corresponding minerals.

Keuper shales, which suggests a high initial porosity. The difference between the Lias α and δ clays of southwest Germany is related to the fine grain size of the latter, which also produced a higher initial porosity. The E_1-values of the Tertiary sediments of the Rhine Valley graben, investigated by HELING (1969), as they relate to median diameter, are plotted in fig. 5-61. In spite of the scatter, there appears to be an effect of depositional environment, from which the marine clays of equivalent grain size exhibit a higher initial porosity than fresh or brackish water sediments. For both groups E_1 decreases with increasing grain size.

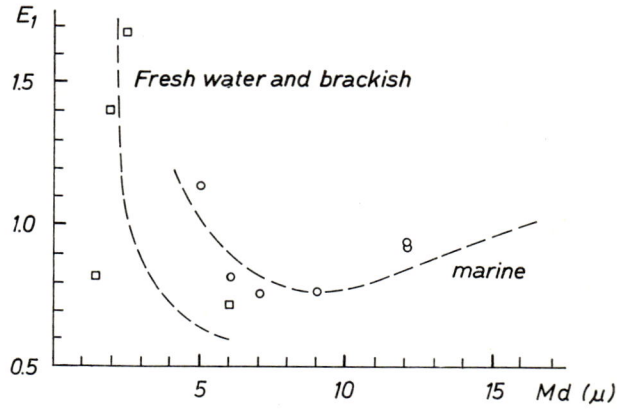

Fig. 5-61. Relation between initial pore volume (E_1) and median diameter (Md) of marine and non-marine sediments of the Rhine Valley Graben (after HELING 1969).

Equation (5-44) represents an empirical approximation, describing, in the range from several hundred to about 3000 m, the dependence of relative pore volume on depth of burial. Probably, there will be no generally valid function representing decrease in porosity at greater depths, since there mineral neoformation and recrystallization play a more significant role than mechanical compaction. Also, equation (5.44) does not apply at very shallow depths. Between the surface and a few hundred meters depth, porosity decreases more rapidly. Of course only a few data are available at shallow depths, but the marked porosity decrease at shallow depths can be ascertained by observations from recent sediments (fig. 5-48).

Fig. 5-62 represents an attempt to illustrate the depth decrease in relative pore volume for an average clay sediment over the entire depth range to 3000 m. For the near-surface region porosity measurements were used, taking the mean values at each depth from recent clay sediments from the Pacific, Black Sea, and Lake Zürich (fig. 5-48). From depths below about 200 m, the corresponding curve calculated from equation (5.44) is noted, based on the measured mean values for the Lias of northwest Germany and the Tertiary beds of Venezuela. This compaction curve will serve as a model in the following considerations.

5.51 Mechanical diagenesis (compaction) 279

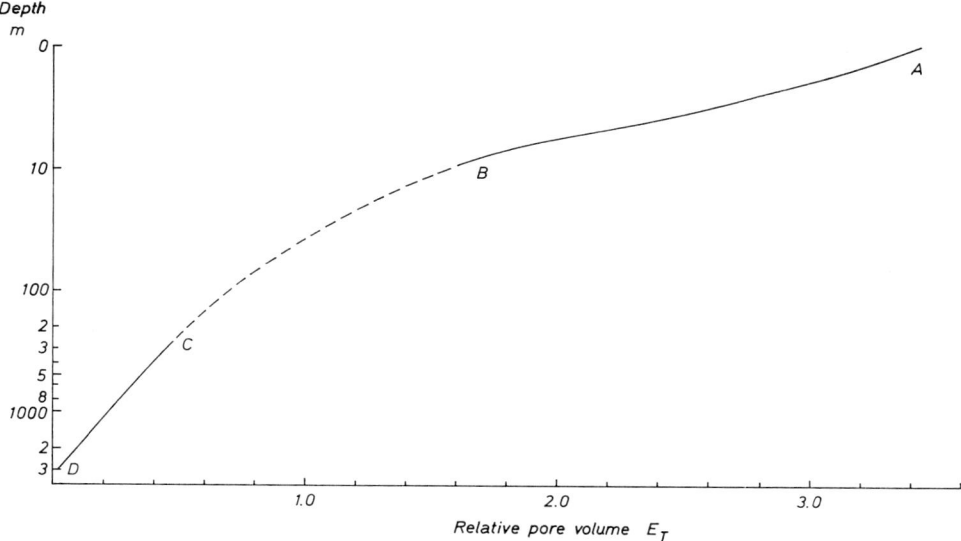

Fig. 5-62. Relative pore volume (E) of clay sediments, as a function of depth of burial. A—B: Mean values for Recent sediments, fig. 5-48; C—D: Mean for the Lias of NW-Germany (fig. 5-53) and Tertiary of Venezuela (fig. 5-52).

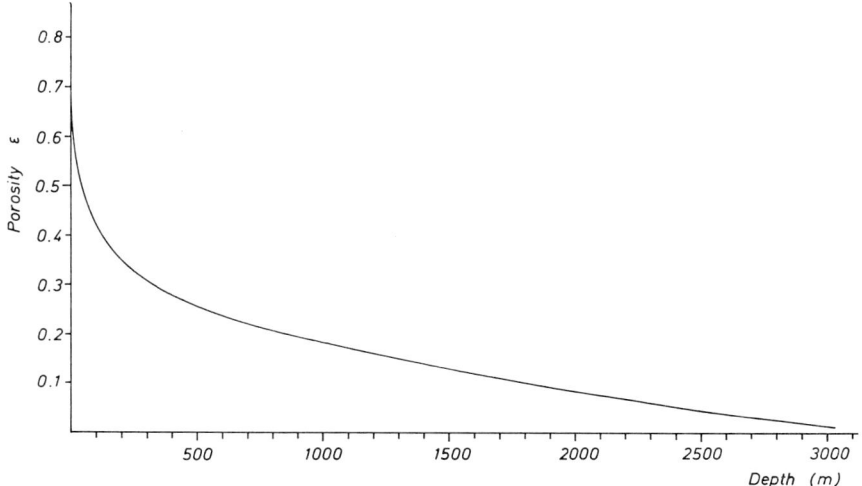

Fig. 5-63. Porosity of clay sediments as a function of depth of burial (from fig. 5-62).

From fig. 5-62, the dependence of porosity on depth is derived, and plotted in fig. 5-63. The volume V_{lT} (m³) (fig. 5-64) of pore solution contained per m² of surface to a depth T (m) is obtained by graphic integration of the porosity function as follows:

$$V_{lT} = \int_0^T \varepsilon \, dT \quad |m^3/m^2| \tag{5.45}$$

For the mean porosity $\bar{\varepsilon}_T$ of all strata above a depth T, the following expression applies (fig. 5-65):

$$\bar{\varepsilon}_T = \frac{1}{T} \int_0^T \varepsilon \, dT \tag{5.46}$$

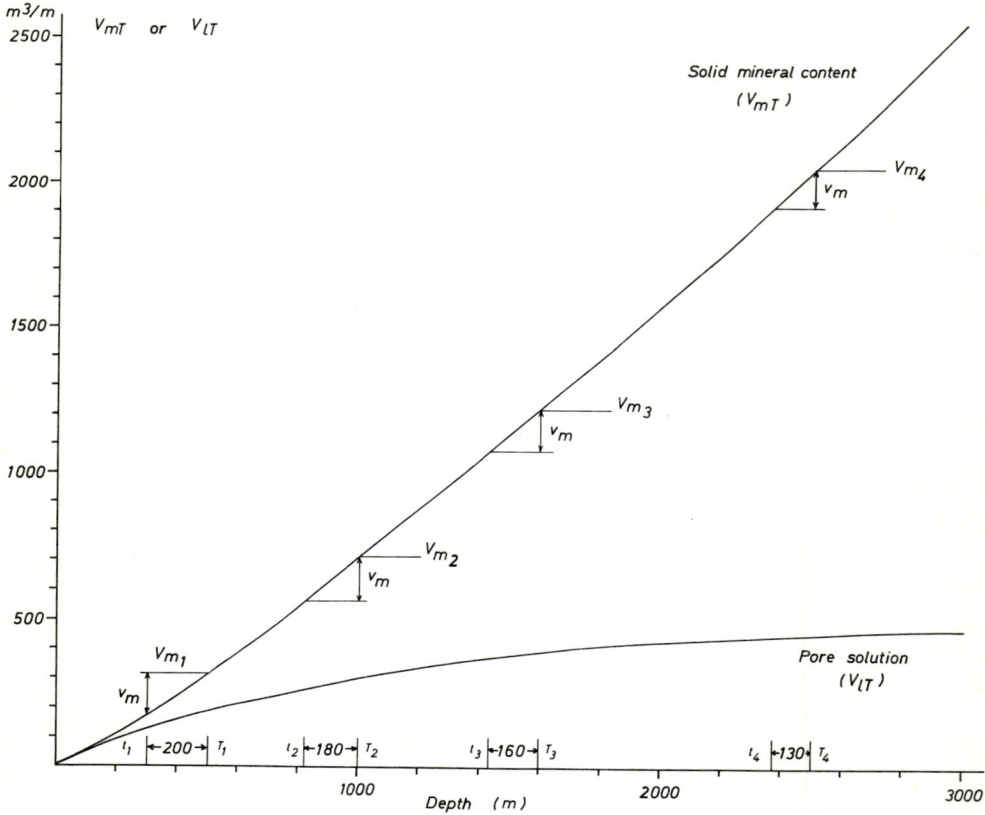

Fig. 5-64. Amount of mineral solids (V_{mT}) and pore solution (V_{lT}) in m³/m² in a T m thick sequence of homogeneous clays, after attaining compaction equilibrium (from fig. 5-62).

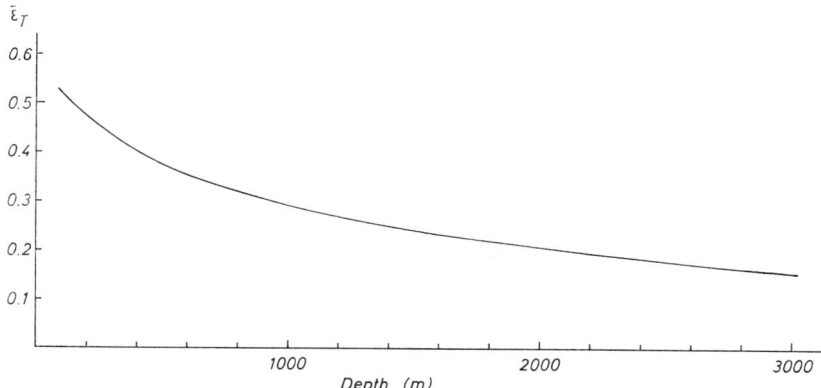

Fig. 5-65. Averaged porosity ($\bar{\varepsilon}_T$) for all strata above the depth T in a sequence of homogeneous clays, after attaining compaction equilibrium (from fig. 5-62).

The volume V_{mT} (m³) of solid mineral matter per m² of surface to depth T (m) in the sediment is:

$$V_{mT} = (1 - \bar{\varepsilon}_T)\, T = T - \int_0^T \varepsilon\, dT \quad |m^3/m^2| \tag{5.47}$$

In fig. 5-64 V_{mT} is plotted as a function of depth.

In order to compare the depth decrease in shale porosity with compaction experiments, depth must be translated into pressure. The pressure resulting from compaction cannot be measured. It can be calculated only by taking into account the rock porosity, density of the rock or its mineral constituents, and the density of the pore solution. The calculation can be based on the following different assumptions which lead to different results:

1. All pores of the rock are interconnected. In the pore solution hydrostatic pressure prevails. The compaction pressure results from the weight of the solid components of the rocks, reduced by the buoyancy in the pore solution. The pressure (kg/cm²) which a layer of thickness T (m) and mean porosity $\bar{\varepsilon}_T$ exerts on an underlying clay, when one considers the rock weight on the total cross section is:

1a) $\quad G = (\varrho_M - \varrho_L)(1 - \bar{\varepsilon}_T) \cdot 9.81 \cdot 10^{-2} \cdot T \quad |kg/cm^2| \tag{5.48}$

where ϱ_M = density of the mineral components of the layer, and ϱ_L = density of the pore solution.

If one considers the rock weight only on that portion of the loaded clay surface consisting of solid particles, and if the porosity of the clay is ε_T, the following expression gives the load pressure:

1b) $\quad G = \dfrac{(\varrho_M - \varrho_L)(1 - \bar{\varepsilon}_T) \cdot 9.81 \cdot 10^{-2} \cdot T}{(1 - \varepsilon_T)} \quad |kg/cm^2| \tag{5.49}$

2. The load pressure is calculated, ignoring buoyancy, as the weight of solid constituents and the pore solution filling the pore space. One considers this weight over the total cross section, obtaining for the load pressure exerted by a layer of thickness T:

2a) $G = [\varrho_M - (\varrho_M - \varrho_L) \bar{\varepsilon}_T] \cdot 9.81 \cdot 10^{-2} \cdot T$ $|kg/cm^2|$ (5.50)

If one bases the weight of rock with its pore solution only on that portion of the loaded clay surface consisting of solid particles (porosity $= \varepsilon_T$), one obtains the following:

2b) $G = \dfrac{[\varrho_M - (\varrho_M - \varrho_L) \bar{\varepsilon}_T] \cdot 9.81 \cdot 10^{-2} \cdot T}{(1-\varepsilon_T)}$ $|kg/cm^2|$ (5.51)

These methods of calculation have been used by various investigators. For example, HEDBERG (1936) calculated the pressure according to 2a. WELLER (1959) adopted expression 1a for calculating pressure to depths of 500 feet. For greater depths or for porosities below 0.37 he used the following formula, which takes into account a partial effect of buoyance:

3. $G = [(\varrho_M - \varrho_L)(1 - \bar{\varepsilon}_T) + (1 - \bar{\varepsilon}_T/0{,}37)\varrho_L] \cdot 9.81 \cdot 10^{-2} \cdot T$ (5.52)

MARSHAL & PHILIPP (1970) assumed expression 1b for shallower depths and 2b for greater ones.

Fig. 5-66 shows the calculated load pressures for an ideal sediment, based on the assumptions of 1a, 1b, and 2a. The values for ε_T were taken from fig. 5-65. As can be seen, quite different values are obtained, depending on whether the buoyancy of the mineral matter is ignored (2a) or not (1a, 1b). The load pressure is higher, if one does not base the weight on the total cross section (1a) but only on the solid framework (1b).

When one presumes that clay compaction results from overburden pressure, and that the pore solution is expelled by collapse of the mineral particle framework, then during the process of compaction a pressure must prevail in the pore solution which is greater than the hydrostatic pressure. When it is assumed further that as a result of the corresponding increase in load pressure the entire pore solution finally can be expelled, all pore space must be interconnected. The pore solution can be expelled, flowing against the hydrostatic pressure, only as long as it is subjected to an excess hydrostatic pressure. The rock pressure will then be borne by the particle framework. In a series of strata, in which all horizons have attained compaction equilibrium, one should be able to calculate the effective pressure at a given depth, and in equilibrium with the prevailing porosity from expression 1a. Because of the interconnecting pores, the buoyancy effect must be taken into account, and the rock pressure is borne by the particle framework and not by the pore solution.

In the deeper zones of a subsiding basin compaction equilibrium can be maintained, if the flow rate of the pore solution can keep up with the rate of subsidence. Because of the very low permeability of compacted clays, it is possible in young

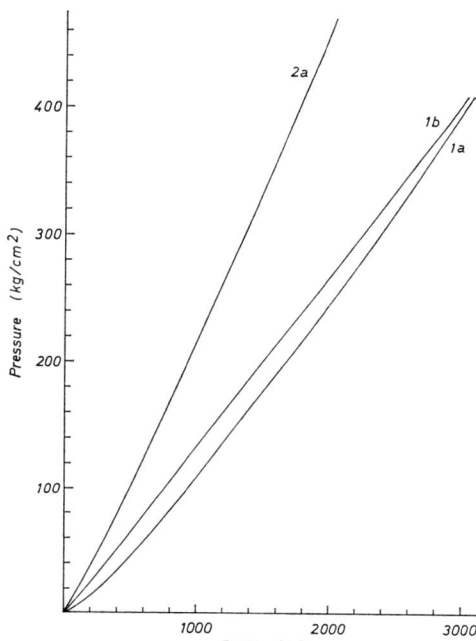

Fig. 5-66. Load pressure as a function of depth in a series of homogeneous clays, after attaining compaction equilibrium. Based on assumptions 1a, 1b and 2a (see text).

sedimentary basins that equilibrium is not yet attained, so that at deeper horizons the pore solution is under pressures higher than hydrostatic pressure. If the shale layers are interbedded with permeable sandstones, the elevated pressure is transmitted to the pore filling of the sand and can be measured in deep drill holes.

In fact, in different sedimentary basins, filled with thick series of predominantly young argillaceous deposits, pressures in excess of hydrostatic have been observed. In sands at 2700 m depth, in the Ventura Basin of California, which is filled with young Tertiary clays, pressures were established which correspond to a gradient of 0.23 atm/m. This is significantly higher than the hydrostatic gradient (approx. 0.11 atm/m, WATTS 1948). In the Tertiary basin of the Gulf Coast (Louisiana) abnormal pressures have been observed in numerous oil wells (up to 0.21 atm/m, DICKINSON 1953). The elevated pressures were measured in sands interstratified in a thick clay sequence. Also in the upper Italian Po Basin, filled with Miocene and Pliocene sediments, pressures in excess of hydrostatic have been established (FACCA 1951). In all these basins compaction is still not completed. The excess hydrostatic pressures show that pore solutions flow upward and compression has not yet been completed.

Fig. 5-66 shows that the load pressure in a uniform sequence of argillaceous rocks over a considerable depth range increases almost linearly with depth. The relative pore volume (E) of an argillaceous sediment decreases at depths over several hundred meters essentially linearly with the log of depth (fig. 5-62), and, therefore, also with the log of the load pressure (fig. 5-67).

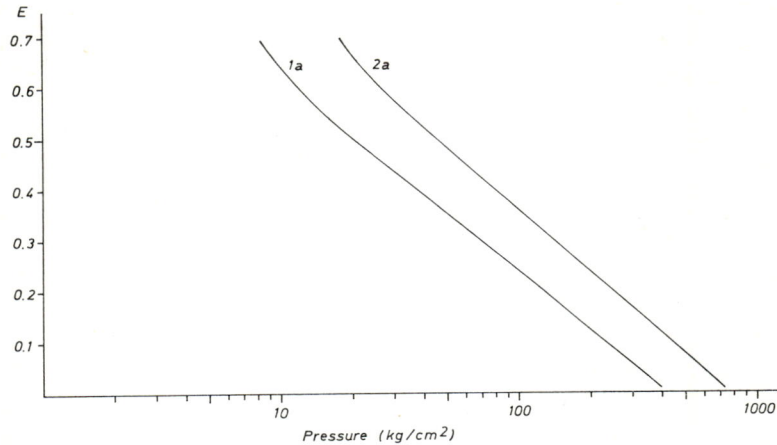

Fig. 5-67. Relation between relative pore volumes of clay sediments illustrated in fig. 5-62 and the load pressures calculated according to assumptions 1a and 2a.

Thus the formal correspondence of equations (5.43 and 5.44) is understandable. That is, the relative pore volume for many clays in compaction experiments is proportional to the log of the pressure and in nature, at depths below a few hundred meters, to the log of depth. Not only in compaction experiments, but also during natural compaction, clays behave, at least within a certain pressure range, so that the decrease in pore volume with increasing pressure is inversely proportional to the pressure.

$$\frac{dE}{dP} = -\text{const.} \ \frac{1}{P} \qquad (5.53)$$

Equations (5.43) and (5.44) follow from this differential relation by integration.

In spite of the formal correspondence, a difference between the mechanisms of experimental and diagenetic compaction may exist. A comparison of fig. 5-40, 5-44, 5-45, and 5-63 shows that for some clays in compaction experiments about the same pressures must be applied, as occur during diagenesis, to attain a given compaction. In other cases one has the impression that the same compaction by natural processes is accomplished at lower pressure than by experiment. Different factors may be responsible for such a difference. First of all diagenetic compaction takes place very slowly with a very gradual increase in load pressure, while by experiment the load is applied suddenly. Secondly, during diagenetic compaction the lateral boundary conditions always present experimentally are lacking. Diagenetic compaction probably in most cases will proceed with participation of shearing movements with one component perpendicular to the load pressure. Finally, and especially at considerable depths, diagenetic compaction is not simply a mechanical process as we have been assuming. Chemical processes of mineral neoformation and recrystallization play a

5.51 Mechanical diagenesis (compaction)

greater and greater role with increasing depth. These processes also gradually cause a reduction in pore volume.

Use of the previously depicted theoretical and empirical information about the compaction of argillaceous sediments is of interest with respect to mainly two matters:
1. Volume loss and water expulsion of an individual clay layer during subsidence.
2. Compaction history of a layered sequence of argillaceous sediments in a subsiding basin.

1. If h_0 is the thickness of a clay layer before burial, and h_T its thickness after subsiding to depth T (in m), according to equations (5.2) and (5.3), the following expression applies for the reduction in thickness of this layer:

$$\frac{h_T}{h_0} = \frac{1 + E_T}{1 + E_0} = \frac{1 - \varepsilon_0}{1 - \varepsilon_T} \tag{5.54}$$

E_0, ε_0, and E_T, ε_T are the relative pore volumes and porosities initially and at depth T. If the clay is compressed sufficiently so that it looses all its porosity, the layer thickness equals $(1 - \varepsilon_0)h_0$ or $h_0/(1 + E_0)$ (maximum compaction). If the initial porosity (E_0, ε_0) is known and if the dependence of porosity (E_T, ε_T) on depth of burial is given, one can calculate the volume reduction a clay bed experiences during subsidence. In this case it is assumed that a single bed in a thick series of homogeneous clays is involved. For this model clay, having the average properties of natural clay sediments, the reduction in thickness during burial is as illustrated in fig. 5-68. Assuming an initial porosity of $\varepsilon_0 = 0.775$ ($E_0 = 3.45$), the layer thickness after maximum compression ($\varepsilon = 0$) equals $0.225 \cdot h_0$. As can be seen from fig. 5-68, the greatest volume decrease takes place at relatively shallow depths. The major portion of the original entrapped pore solution is expelled at burial depths of only a few hundred meters.

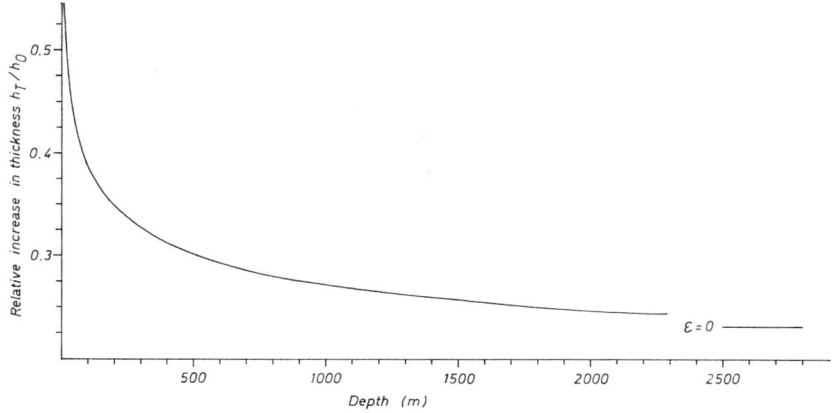

Fig. 5-68. Decrease in thickness of a clay bed (h_T/h_0) as a function of depth of subsidence T. h_0 = depositional thickness. h_T = thickness at depth T (from fig. 5-62).

2. If continuous sedimentation of clay with simultaneous subsidence takes place in a basin, then each layer, first deposited as a fluid mud at the surface, is continuously buried by new sediment. To the extent to which it is buried, each bed experiences a continuous reduction in porosity and thickness (fig. 5-68). A mathematical model for this process has been described by MARSAL & PHILIPP (1970). The pressure on all the older clay layers is continuously increased by deposition of new material. They are compressed and a corresponding amount of pore solution must leave the sediment packet. If no permeable strata or joints are present, by which the expelled solution can ascend upward, the compaction flow must permeate the entire clay sequence.

In a homogeneous clay sequence which was deposited originally with the porosity ε_0, and whose thickness after reaching compaction equilibrium is T meters, the porosity from the surface downward decreases from ε_0 to ε_T. For the model sediment under consideration, this reduction in porosity, the porosity averaged over the total thickness, and the volume of pore solution and solid clay phase contained in the layer sequence, are plotted in figs. 5-64 and 5-65. The compaction produced by additional clay deposition, the increase in thickness of the entire packet, and the upwardly directed pore water flow are best described, referring to fig. 5-64. When a total of $V_{m1} = 314$ m³/m² of water-free clay material was deposited, the sediment packet had a thickness of $T_1 = 500$ m. It contained, at an average porosity of $\bar{\varepsilon}_T = 0.372$ (fig. 5-65), a total $V_{l1} = 186$ m³/m² of pore solution. The clay was deposited originally as mud with a porosity of $\varepsilon_0 = 0.775$. Therefore, originally the following amount of water was contained in the sediment clay:

$$V_{l0} = \frac{(1 - \bar{\varepsilon}_T)\varepsilon_0}{(1 - \varepsilon_0)} T = 1082 \text{ m}^3/\text{m}^2$$

During the course of subsidence and sedimentation a volume of water expelled through the actual surface which is equal to the difference between V_{l0} and V_{l1}:

$$V_{l0} - V_{l1} = \frac{(\varepsilon_0 - \bar{\varepsilon}_T)}{(1 - \varepsilon_0)} T = 896 \text{ m}^3/\text{m}^2$$

So much clay sediment may be deposited accompanying subsidence that the sediment thickness increases from 500 to $T_2 = 1000$ m. According to fig. 5-64 this thickness corresponds to a volume of solid clay matter of $V_{m2} = 707$ m³/m². For an increase in thickness from 500 to 1000 m, 393 m³/m² of clay solids would be sedimented. The 1000 m thick layer sequence has a mean porosity of $\varepsilon_T = 0.293$ and contains 293 m³/m² of pore solution. During the increase in sediment thickness from 500 to 1000 m, 1244 m³/m² of pore solution would be expelled by compaction.

To illustrate the pore water flow in the sediment packet, let us keep track of a given marker horizon, which at the time the sediment thickness was 500 m, was at depth $t_1 = 300$ m beneath the surface. From fig. 5-64 it is seen that between the depth $T_1 = 500$ m and $t_1 = 300$ m, $v_m = 140$ m³/m² of water-free clay is present. After deposition of additional clay, increasing the sediment thickness to 100 m, the marker horizon also migrates downward to a depth t_2. Since only pore solution can disappear from the sediment, between the subsurface and the marker horizon one notes from fig. 5-64 that $t_2 = 820$ m, since between $T_2 = 1000$ m and $t_2 = 820$ m, 140 m³/m² of clay solids are obtained. By subsidence the sediment between the subsurface and the marker horizon has experienced a reduction in thickness of 20 m. As a result 20 m³/m² of pore solution have flowed upward through the marker horizon.

In a corresponding manner the further reduction in sediment thickness between the subsurface and the marker horizon and the pore water flow through the marker horizon with continuing sedimentation can be inferred from fig. 5-64. If the total sediment thickness reached a value $T_3 = 1600$ m, the marker horizon would lie at a depth $t_3 = 1430$ m. During this period of sedimentation an additional 20 m³/m² of solution would flow through the marker horizon. If the total thickness increases to $T_4 = 2500$ m, the marker would lie at $t_4 = 2370$ m and the sediment between it and the subsurface would be 130 m thick. While subsiding to this depth 30 m³/m² of pore solution would have permeated the marker horizon.

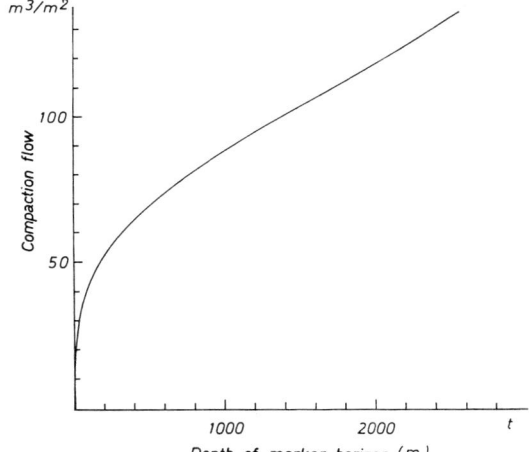

Fig. 5-69. Compaction flow in a subsiding clay sediment. Flow of solutions (in m³) per m² through a marker horizon as it subsides to the depth T (from fig. 5-62).

Finally one can ascertain also from fig. 5-64 the sediment thickness, or depth to the base of the sedimentary column at the time when the marker horizon was located at the surface. This depth $T_0 = 260$ m. The amount of pore solution which flowed through the marker horizon is plotted in fig. 5-69, as a function of the horizon depth t. This illustration can be perceived also as a picture of the compaction flow at different depths in a subsiding sediment packet. The amount of pore solution flowing through a unit area per meter of subsidence is given by the slope of the curve. Compaction flow is greatest at shallow depths and diminishes downward.

If the compaction flow is not diverted through sandy strata or joints, the pore solution originally entrapped between clay particles is not only reduced in amount during compaction but is replaced progressively by solutions ascending from below.

With increasing compaction the porosity is not only diminished, but the pore size becomes smaller and smaller, since individual particles move closer together and aggregates are homogenized. HELING (1970 b) investigated the pore size distributions in Tertiary argillaceous sediments of the Rhine Valley Graben of different age and depths of burial, determining capillary pressure curves with mercury as the displacing phase. For the same samples the grain size distributions and specific internal surface areas were determined, the latter by the BET method. The mean pore radius decreases proportionally with porosity with depth of burial. The pore radius distribution showed that to about 1000 m depth, both large and small pores diminished in size about equally. At the greater depths the smallest pores disappear. As would be expected for purely mechanical compaction, consisting only of particle rearrangement, to depths of 1000 m the specific internal surface area is almost independent of depth of burial. At depths below 1000 m a marked decrease in internal surface area is initiated, which indicates that primary individual particles coalesce to form larger and compact structures. HELING (1967) related this process to the formation of illite from primary montmorillonite observed at these depths. This study shows that the purely mechanical model for clay compaction is correct only for shallow depths. Below a certain level, which is at about 1000 m for the Rhine Valley Graben, the processes of neoformation of clay minerals begin to play an increasingly greater role.

One would expect that consolidation of clay would lead not only to compaction, but simultaneously to a progressively better orientation of clay mineral platelets in the plane perpendicular to the pressure direction. Quantitatively such orientation can be measured by X-ray textural analysis (see, for example, GEHLEN 1960). The invariable axial symmetrical clay fabric always assumes a simpler form than the general case (TAYLOR & NORRISH 1966; LIPPMANN 1970). Such analysis gives the X-ray intensities of basal reflections of clay minerals as a function of the angle which the normal to the clay platelet forms with respect to the normal to an S-plane of the rock. Macroscopically the S-plane is a bedding plane, a joint surface, or a plane of foliation. On the surface of a sphere, whose equatorial plane coincides with the S-plane, the normal directions of individual clay mineral particles appear as points. The position of each point is defined by two angles, the polar angle φ, which designates the angular distance with respect to the normal of the S-plane, and an azimuth angle ψ, measured on the equator. The intensity of a basal reflection measured over an infinitesimal angular range $d\varphi \cdot d\psi$ is proportional to the volume of mineral particles whose face normals have this position. If the intensity function $I(\varphi, \psi)$ is known for all values of φ and ψ, the mean intensity per unit surface of the hemisphere can be calculated. In the case of a clay of equal size particles, this intensity is proportional to the total number of particles per unit volume:

$$\bar{I} = \frac{\int_0^{2\pi} \int_0^{\pi/2} I(\varphi, \psi) \sin\varphi \, d\psi \, d\varphi}{2\pi} \tag{5.55}$$

Because the orientation in shales in axially symmetrical, the intensity is dependent on the pole angle, not on the azimuth. Therefore, one can write:

$$\bar{I} = \int_0^{\pi/2} I(\varphi) \sin\varphi \, d\psi \tag{5.56}$$

If the clay platelets lie predominantly parallel to the S-plane, the function $I(\varphi)$ for $\varphi = 0$ has a maximum:

$$I_{max}(\varphi) = I(\varphi = 0) = I(0) \tag{5.57}$$

As a measure of texture one defines the ratio of this maximum intensity to the averaged intensity:

$$\text{Texture index } t = \frac{I(0)}{\bar{I}} \tag{5.58}$$

The texture index is equal to the ratio of the intensity measured at $\varphi = 0$ in an oriented clay to the intensity at $\varphi = 0$ for the same clay with random orientation. This is the same as the ratio of the number of particles with preferred orientation parallel to the S-plane to the number parallel to S in randomly oriented clay.

To determine the texture index one must measure the function $I(\varphi)$ for as great an angular range as possible and calculate $I(\varphi) \sin\varphi$ for different φ values. If one plots $I(\varphi) \sin\varphi$ against φ, the averaged intensity $\bar{I}$ is obtained by graphical integration.

The intensity I decreases with increasing pole angle φ more rapidly, the higher the texture index t. The intensity is a function, therefore, of φ and t. LIPPMANN (1970) showed that $I(\varphi, t)$ can be represented in most cases by the function

$$I(\varphi, t) = t \cdot \bar{I} (\cos\varphi)^{t-1} \tag{5.59}$$

From this one obtains:

$$\log I(\varphi, t) = \log(t \cdot \bar{I}) + (t-1)\log\cos\varphi \tag{5.60}$$

This formulation facilitates the determination of texture index t. If one plots the log of the measured intensity $I(\varphi)$ against $\log \cos\varphi$, one obtains a straight line, whose slope is $(t-1)$.

Fig. 5-70 shows as an example the texture functions of different artificially produced montmorillonite clay specimens as measured by THIEM (1967). The continuous curves correspond to the function specified by LIPPMANN (1970).

Systematic measurements of texture index of shales of different degrees of compaction still have not been carried out by this quantitative method. Instead more qualitative methods have been used, which are based on intensity measurements from sections cut parallel and perpendicular to the S-plane. KAARSBERG (1959) used the intensity ratio of the (002) and (110) reflections of illite in sections parallel to the

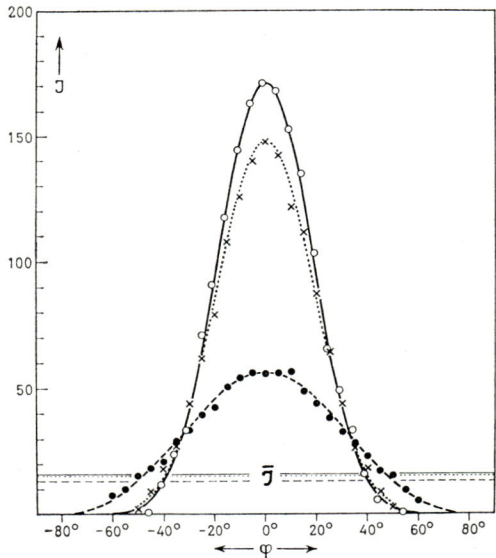

Fig. 5-70. Textural measurements of artificially compressed montmorillonite specimens (THIEM 1967) and the texture function according to LIPPMANN (1970).

S-plane as a measure of preferred orientation. MEADE (1964) used as orientation ratio the ratio of (001) and (020) intensities of montmorillonite:

$$\text{Orientation ratio (MEADE)} = \frac{I\,(001):I\,(020) \text{ parallel S}}{I\,(001):I\,(020) \text{ perp. S}}$$

The orientation ratio for randomly orientated fabric is 1.

GIPSON (1966) used the following orientation ratio:

Orientation ratio (GIPSON) = $(A_{002}:A_{020}):([B_{002}+C_{002}]:[B_{020}+C_{020}])$. ($A_{002}$ and A_{020} = intensities of the (002) and (020) reflections of illite in the S-plane; B and C represent the same intensities, measured in two mutually perpendicular sections also perpendicular to the S-plane). A randomly oriented fabric gives a value of 1.

ODOM (1967) defined a fabric index as the ratio of intensities of basal reflections of clay minerals (illite, kaolinite, chlorite) measured from sections cut parallel and perpendicular to the S-plane:

$$\text{Fabric index (ODOM)} = \frac{I\,(\text{basal}) \text{ perp. S}}{I\,(\text{basal}) \text{ perp. S} + I\,(\text{basal}) \text{ parallel S}}$$

For a randomly oriented fabric the fabric index equals 0.5. In Pennsylvanian shales ODOM (1967) found a correlation between fabric index and macroscopic fabric. Rocks with shaly or slaty structure had fabric indices between 0 and 0.3. Massive rocks (claystones) had indices between 0.35 and 0.5. Rocks with fabric indices between 0 and 0.15 show especially fissle structures.

In addition to macroscopic criteria (schistosity) and X-ray methods, electron optical methods have been used also to ascertain orientation in shales (replica methods: GIPSON 1965; scanning electron microscopy: MATTIAT 1969).

KAARSBERG (1959) found in a series of Cretaceous shales from different deep well cores and outcrops in the United States, an increase in preferred orientation of illite parallel to the S-plane, with increasing rock density. He concluded that the preferred orientation in these rocks increased with increasing depth of burial. Many observations of other investigators (see MEADE 1964, 1968a; GIPSON 1965, 1966; ODOM 1967) show, however, that this is not generally true. In cores from deep wells, shaly and massive claystones are interbedded, without exhibiting systematic increase in schistosity with depth (see RUBEY 1931; WHITE 1961). In addition to orogenic pressure, still other influences must affect the orientation of clay mineral particles. Available data are not yet sufficient to survey completely the role of these factors. It is only possible to consider a few of the probably most important factors (see MEADE 1964).

One significant influence appears to be the conditions which prevailed during deposition. In shales of the coal-bearing Pennsylvanian sequence in Illinois, consisting of variable amounts of illite, kaolinite, chlorite, and mixed-layer minerals, ODOM (1967) found an alternation of thin beds of well-oriented, shaly and massive, poorly oriented rocks. There is no relation between mineral composition and fabric. However, a dependence of form on organic content was established. The most strongly oriented rocks are especially rich in organic matter. Claystones (underclays) lying under coal beds and low in organic matter are massive and show random orientation. This was noted also by O'BRIEN (1964) for Pennsylvanian underclays in the Illinois Basin. GIPSON (1965) observed the same relationship by electron optical studies of claystones in the Pennsylvania of Iowa. INGRAM (1953) established generally that claystones with higher organic contents are the most platy. Apparently compaction produces an oriented fabric preferentially in clays which were deposited originally with a high degree of preferred orientation. These would be clays sedimented as individual dispersed particles, and should result when no coagulation takes place. This suggests deposition under non-saline solutions, low sedimentation rate, or also hindrence of flocculation by protective colloids. Probably the formation of carbonaceous shales is promoted by organic colloidal matter which promotes the sedimentation of individual clay mineral particles and thereby promotes the formation of primary oriented fabrics in the fresh sediment. Deposition from saline water without protective colloids leads in contrast to randomly oriented sediments formed of larger aggregates, which, during compaction become massive, poorly oriented claystones. The interbedding of massive and shaly claystones observed in Pennsylvanian strata, which are not all marine, probably indicates alternations of depositional conditions. Thus the Pennsylvanian underclays form during marine regression under brackish water conditions (O'BRIEN 1964). Flocculation of the clay material supply leads to random fabrics, while in a low-saline, organic-rich environment, deposition of unflocculated clays occurred after the formation of coal.

In addition to depositional conditions, the grain size of the clay mineral particles is important. Coarse grain kaolinitic and illitic clays experience better orientation by compaction than fine grain ones, such as montmorillonitic clays. MEADE (1968a) found in Pliocene to Recent fresh water clays from California, consisting mainly of montmorillonite, no increase in orientation with depth of burial to depths of 600 m. All of these sediments are only slightly oriented (orientation ratios between 1 and 3). The highest orientation ratio (1.5—3.0) was found in lake sediments, the lowest (1) in alluvial fan sediments.

It has been presumed until now that compaction of clay results exclusively from the weight of overlying sediments, under the influence of a vertically directed pressure. However, there are indications that for compaction, laterally directed pressure components also play a role. LANGHEINRICH (1966) observed and measured the deformation of ammonite shells *(Psiloceras)* in clays of the lower Lias near Goettingen, which is not tectonically stressed. The shells are compressed laterally about 16%, whereby the long axes of the deformed shells are uniformly aligned. From observations of fossilization, it can be concluded that the deformation took place in an early stage of diagenesis. Apparently the still plastic sediment undergoing reduction in pore volume is caused to slide on the slightly inclined sea floor.

From seismic velocities in the Tertiary strata of the subalpine Molasse basin, LOHR (1969) could deduce that the 4000 m thick, unfolded basin fill has been compacted to a greater extent from the laterally directed mountain-forming forces of alpine folding than from the normal load pressure. The seismic velocity of uniform basin sediments increases as a rule almost exactly linearly with depth. In the Molasse basin one observes first a more marked increase in velocity with depth than in the corresponding strata of the Rhine Valley Graben, which was subjected to no lateral pressure. Secondly the velocity increase with depth clearly increases approaching the zone of folded Molasse at the border of the Alps. Fig. 5-71 shows the increase in mean seismic velocity with proximity to the Alps.

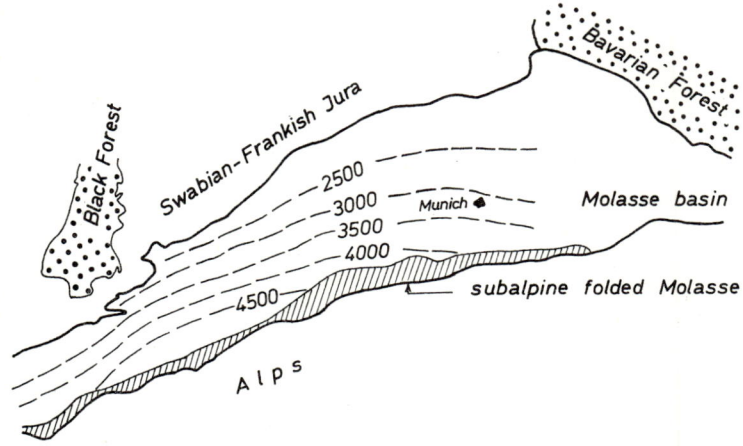

Fig. 5-71. Mean seismic velocities (m/sec) in the Tertiary subalpine Molasse basin (LOHR 1969).

5.52 Chemical diagenesis

In the foregoing section the compaction of clays was considered as a mechanical process, taking place under the influence of orogenic pressure and consisting merely of rearrangement, deformation, and fracture of primary particles. In reality mechanical and chemical processes always act together, even though in the diagenesis of clays, in contrast to sands, mechanical processes alter the fabric much more markedly than chemical processes. Therefore it was appropriate first to deal with these mechanical processes without consideration of chemical reactions.

Factors which give rise to the chemical diagenesis of clays are the change in chemistry of pore solutions which penetrate the sediments during compaction, as well as the increase in temperature and pressure. Different investigators have differentiated an early stage of diagenesis during shallow sediment burial and a later stage during deeper burial (see DAPPLES 1959, 1967 [three stages]; MÜLLER 1967, 1970).

Although it is difficult to designate exactly the boundary between these two stages, still it is important to distinguish the conditions prevailing during shallow sediment burial from those which prevail at greater depths.

Following shallow burial the sediment still has a high porosity and correspondingly high permeability. In a subsiding sedimentary sequence of sufficient thickness the compaction flow is relatively strong, minerals come in contact with relatively large amounts of solution, and dissolved material is removal as other matter is introduced in solution. If organic matter, which survived subaquatic weathering, with its usual oxidizing conditions, becomes part of the sediment, reducing conditions (negative Eh) prevail on account of the exclusion of atmospheric oxygen in the pore solution. The salinity of the pore solution is still relatively low, corresponding more or less to the composition of the solution entrapped during deposition. Marine clays contain solutions similar to sea water; continental clays contain solutions corresponding to continental waters. The changes in chemistry, discussed in the section on diagenesis of formation waters, take place to a large extent during this early stage of diagenesis. Examples are the fixation of magnesium and potassium and the reduction of sulfate. Temperature and pressure are still low. If one sets as the approximate limit for early diagenesis a porosity of 0.3, this stage corresponds to a burial of about 300 m (see fig. 5-63). At this depth, as fig. 5-69 shows, a significant decrease in the intensity of compaction flow results.

In late stages of diagenesis porosity and sediment permeability are quite low. Compaction flow is reduced so that minerals come into contact with smaller amounts of solution than before. Pore solution salinity is increased due to the increasing action of ion filtration, and its composition in most sediments approaches more and more a pure NaCl-solution. The nature and rates of chemical reactions are determined by the increase in temperature and pressure. Diffusion processes and reactions in the solid state gain importance. With final loss of the pore volume at a depth of about 6000 m the diagenetically characteristic reactions between solid phases and pore solutions cease; the region of metamorphism is reached, in which reactions proceed essentially by means of grain boundary diffusion and diffusion in the solid state.

Processes of chemical diagenesis affect mainly the clay mineral constituents. Certain clay minerals disappear and others take their place, as a result of alteration or neoformation. Newly formed or altered clay minerals are, however, difficult to recognize as such. It is usually difficult to determine whether a change in mineral composition with depth is diagenetically related, occurs as a result of subaquatic weathering, or indicates a change in composition of the primary source material.

Concretions are typical formations of early diagenesis. They are particularly widespread in marine shales of all formations. Concretions are spherical, ellipsoidal, or irregularly formed, often flattened accumulations of diagenetic deposited minerals. They either displace or enclose the detrital components of the clay. Most common are concretions of calcite, siderite, dolomite, and gypsum. In addition barite, calcium phosphate, pyrite, and marcasite form concretions. SiO_2-concretions (flint, chert) occur mainly in limestones.

According to their relation to the surrounding rock, different kinds of concretions can be differentiated (ILLIES 1949).

1. Concretions which originate by the filling up of pore space with a diagenetic mineral.

2. Concretions which result from the mechanical displacement of sediment as a result of crystal growth.

3. Concretions which originate from the metasomatic replacement of the surrounding rock.

Concretions of the third type are widespread in limestones and do not occur in shales. Among the most common concretions in clays are carbonate concretions of type 1 and pyrite and marcasite which are essentially type 2.

Concretions can form only when the sediment is not homogeneous with respect to physical-chemical conditions. At certain particular points conditions must prevail which cause precipitation of a solid phase. This can happen, because at these points localized chemical reactions occur which lead to formation of a given phase, whose components are dissolved in the pore solution. As long as such a localized reaction continues, the components of the new phase must diffuse to and precipitate at the site of the concretion. Another possibility for concretion formation arises when the pore solution is supersaturated weakly with respect to a given phase, so that the probability of spontaneous nucleation is low. Then precipitation can be concentrated at a certain site where appropriate nuclei occur.

In any case, the substance forming concretions must diffuse from the surrounding rock to the concretionary center. The often observed concentric internal structure of concretions indicates the sometimes non-uniform growth process, which progresses outward from within. The rate at which concretions grow depends on the rate of diffusion, that is, on the diffusion coefficients and concentration gradient.

In the first of the two cases mentioned, some distance away from the concretion saturation prevails, and at the site of the concretion a lower concentration, which is determined by the reaction initiated there.

In the second case supersaturation prevails at some distance from the concretion. The solution is just saturated at the surface of the concretion. BERNER (1968 a) calculated the growth rate for a spherical concretion for such a diffusion model. He considered two cases:

1. Concretionary growth in a stationary pore solution,
2. Concretionary growth in flowing pore water.

The following expressions apply:

Growth in a stationary solution:
$$t = \frac{R^2}{2\,v\,D\,(c_\infty - c_R)} \quad (5.61)$$

Growth in flowing solution:

$$t = \frac{(R - D/0.715\,U)\,(1 + RU/D)\,0.175 + D/0.175\,U}{1.175\,U\,v\,(c_\infty - c_R)} \quad (5.62)$$

t is the time a sphere of diameter R has grown, v is the mole volume of precipitating matter, D is the diffusion coefficient, c and c_r are the concentrations of dissolved constituents at a great distance from and on the surface of the concretion, U is the rate of flow of the pore water. Fig. 5-72 illustrates both equations assuming the growth of calcite concretions. The formulae are valid only for the case where individual concretions are sufficiently far from one another that their diffusion halos do not overlap, and where the same supersaturation always prevails at the surface of the concretion. Because of the simplifying assumptions and the likely variable conditions in individual cases, such calculation can yield only the order of magnitude of growth rates. Nevertheless, it is certain that concretions can grow in geologically short times. In addition it is shown that the flow of pore solution increases

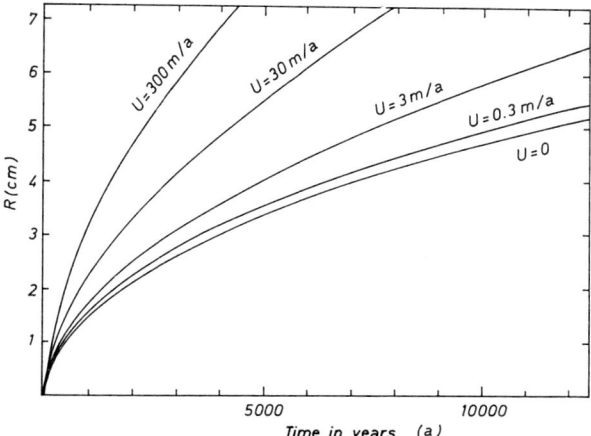

Fig. 5-72. Growth of concretions after the formula by BERNER (1968) as a function of pore solution flow rate (U in m/year).

the growth rate less than about one order of magnitude. For concretion formation it is not necessary that compaction flow brings in dissolved material.

That the very abundant carbonate concretions do not form at the sediment surface, but rather within the sediment is recognized by the lack of induced precipitation by benthonic organisms. Within these concretions the remains of organisms are frequently found. In contrast to the deformed fossils of the surrounding clay they remain intact and undeformed. Therefore, the concretions formed quite early and the fossils they contain before deformation could be pretected, due to compaction of the sediment. The early formation of concretions is indicated also from the behavior of the fine bedding, which continues through the clay and the concretions. The bedding is horizontal within the concretions and draws closer together in passing into the clay, where the bedding accommodates itself to the curved upper and lower surfaces of the concretion. From the separation of such fine bedding planes within the concretion and the clay, the amount of compaction which the clay underwent after formation of the concretion can be deduced. For calcareous concretions in the Lias ε of southwest Germany (Swabian Alb) EINSELE & MOSEBACH (1955) inferred such a volume decrease of $1/3$ to $1/4$. A further indication for the early formation of concretions comes from a comparison of the detrital material entrapped in concretions with the mineral composition of the surrounding clay. This shows that the mineral components of the clay are more strongly diagenetically altered.

From these observations it follows that carbonate concretions form by filling the pore space of unconsolidated sediment with carbonate. From the carbonate content one can estimate that the sediment porosity at the time of concretion formation was about 0.7—0.8. This indicates that the concretions were formed a few meters below the sediment surface or at least began to grow there (LIPPMANN 1955; SEIBOLD 1962). Since the pore volume of the clay diminished increasingly during growth, the carbonate content decreases from the center outward (SEIBOLD 1962; KNOKE 1966).

The formation of carbonate concretions shows that calcium and iron carbonates are very mobile in many argillaceous sediments during a very early stage of diagenesis. This observation may be related to the fact that a higher hydrogen ion concentration has been found in the pore solutions of recent sediments, compared to sea water. Such solutions are supersaturated no longer in calcite, as in sea water. Rather they are weakly undersaturated or just saturated. In some marine shales calcareous fossil remains show signs of early diagenetic partial or complete dissolution. For example in the concretionary Posidonian shales (Lias ε) of southwest Germany, ammonite shells, consisting of relatively soluble aragonite, are completely dissolved. The shells are preserved as very thin, completely flat films of organic matter. *Aptychus,* pelecypods, and the hard parts of belemnites and crinoids, consisting of calcite, are more or less wholly preserved, although they show some signs of solution (EINSELE & MOSEBACH 1955).

Concretions are formed from carbonates at certain places in a sediment, where locally carbonate is precipitated over a period of time. Since carbonate solubility is lowered by an increase in pH, CORRENS (1950) and others after him (WEEKS 1953;

LIPPMANN 1955; BERNER 1968 b) have suggested that concretions are formed where organisms are decomposing with the formation of amines and ammonia. For a period of time these compounds act to increase alkalinity. This explanation is supported by the fact that concretions often contain gross remains of animals, and sometimes also plants. Also the occurrence of phosphate and pyrite in them indicates the contribution of organic matter. Concretions with no present visible signs of organic remains formerly may have enclosed the bodies of soft-bodied invertebrates or plants which had no hard parts.

Concretions contain carbon only in the form of carbonate. Therefore, with the decomposition of organic matter, CO_2 as well as NH_3 will have been produced. CO_2 increases the solubility of carbonate. The formation of carbonate concretions, therefore, must have been associated with special conditions, by which ammonia and amine production were sufficiently high to precipitate carbonate. BERNER (1968 b) showed experimentally that during the anaerobic decomposition of fish remains in sea water, the pH rises significantly with the formation of ammonia and amines, and that simultaneously the calcium and magnesium contents of the solution decrease, while carbonate increases. Calcium and magnesium were not precipitated in these experiments as carbonates, but rather as salts of long-chain fatty acids. Such conditions are known from the natural decomposition products (cadaver wax) of recent cadavers in water (BERGMANN 1963). Later chemical cleavage could form calcium and magnesium carbonate as well as hydrocarbons from these substances. BERNER (1968 b) suggested that at least some carbonate concretions form in this way.

The low nitrogen content of organic matter makes it questionable whether concretion formation can be explained alone by the decomposition of organic matter. If carbonate deposition takes place according to the reaction

$$NH_3 + Ca^{++} + HCO_3^- \rightarrow CaCO_3 + NH_4^+$$

each mole of precipitated calcium carbonate corresponds to one mole of ammonia or nitrogen. The nitrogen content of plants, because of their lower protein content, is less than that of animals. It is, on the average, 3 weight% of dried plant material (VINOGRADOV, after BOWEN 1966). The average content of animals is about 10 weight% of dried material (various authors, after BOWEN 1966). Therefore, the decomposition of 100 g dried plant material can lead to a maximum of 20 g $CaCO_3$, while 100 g dried animal matter can give a maximum of 70 g $CaCO_3$. A spherical concretion of 10 cm diameter contains an average of 80% or 1.13 kg $CaCO_3$. To precipitate this amount at least 1.6 kg of dried animal or 5.7 kg plant material would be necessary. When it is noted that the dried material makes up only about 1 weight% of an organism, unreasonably large amounts of plant or animal remains would be necessary to form a single concretion. It must be assumed that additional processes have played a role in concretion formation. Possibly only the innermost core of the carbonate concretion was precipitated as a result of organic decomposition and then the surface of the initial concretion acted as a nucleus for further precipitation during later stages of diagenesis. This would require at least a slight supersaturation in the pore solution. Supersaturation can be caused by aragonitic hard parts of organisms embedded in the clay, since a solution saturated with respect to aragonite is supersaturated in calcite. In such a case calcareous concretions grow at the expense of dissolving aragonitic shells.

The formation of calcareous concretions produces an accumulation and concentration of carbonate, which had been distributed originally as large and small shells more or less uniformly throughout the sediment. Often concretions are concentrated in certain horizons. Probably such horizons represent especially carbonate-rich primary sediments. Adjacent concretions in a given horizon can merge together, forming a transition into continuous limestone beds.

The formation of calcareous concretions results in a separation of calcium and magnesium, as a result of reduced solid solution of magnesium in calcite at low temperatures. The concretions studied by SEIBOLD (1962) from the southwestern German Lias, as well as those from the Devonian of the Harz, studied by KNOKE (1966), contain less magnesium than the carbonates of the surrounding clay.

Studies of the stable carbon isotope distribution in carbonate concretions up to now have not been able to further clarify the mechanism of formation. HOEFS (1960, and his literature citations) found an enrichment of C^{12} in lower Cretaceous concretions, which may originate from organic carbon. In contrast C^{13} is enriched in Devonian concretions which he studied. This may have resulted from isotopic equilibrium with methane.

In addition to calcite, siderite also frequently forms concretions (clay ironstones). In the lower Cretaceous clay studied by LIPPMANN (1955), siderite concretions were formed later than calcite ones. The iron originated from the clay. It may have been derived by the reduction of primary trivalent iron deposited in the sediment, with organic matter serving as the reducing agent. In any case the formation of siderite concretions has taken place in a reducing environment (negative pH) under neutral or weakly acid conditions.

Carbonate concretions frequently are penetrated by radial cracks and open crevices, which become narrower or even closed from the outside inward. The volume of concretions with such shrinkage cracks (septarian) has decreased during diagenesis. They formed first as a soft, highly porous calcareous mud and experienced compaction during burial by the expulsion of water. This is expressed in part by deformation of the still plastic body (flattened form) and partly by the formation of open fissures in the already lithified structure. Since these fissures as a rule taper inward, the volume loss was greater in the interior than in the outer zone of the concretion. This is understandable, since the calcareous mud precipitated in the first phase of formation was more porous than in the later, formed under greater sediment cover. In the fractures of septarian concretions crystals of different minerals are found, which formed during even later stages of diagenesis. From concretions in lower Cretaceous clays LIPPMANN (1955) found calcite, pyrite, barite, whewellite (calcium oxalate) and sphalerite. These minerals have precipitated from the pore solution of the shale. This was probably an effect of the difference between the orogenic pressure and the hydrostatic pressure in the pore solution, and related to the pressure dependence of solubility.

In argillaceous rocks which contain carbon compounds, sulfate in the pore solution and protein sulfur are reduced to sulfide by bacteria. The analysis of pore solutions extracted from recent marine sediments shows that sulfate reduction takes place in the earliest stages of diagenesis. The sulfide so-produced is fixed mainly as iron sulfide in the sediment. This occurs either finely dispersed or is concentrated in FeS_2-concretions as pyrite or marcasite. Iron sulfide concretions, related to their origin from protein sulfur, are frequently associated with organic remains. In contrast to carbonate concretions, the often radially formed iron sulfide concretions enclose little detrital material. Therefore, they were not precipitated in the pore space of sediments, but grew by pushing aside the still plastic enclosing clay. In a similar fashion clear gypsum crystals grew in many clays. Gypsum, pyrite, and marcasite, therefore, are able, in contrast to carbonates, to exert a so-called growth pressure and grow in opposition to the orogenic pressure. As CORRENS (1939, 1949) showed, this requires not only a certain supersaturation, which corresponds to a solubility increase as a result of pressure. In addition the condition must be fulfilled that the supersaturated solution can penetrate into the capillary space between the growing crystal and the adjacent mineral particles (see section 5.42). This depends on the properties of the boundary surface, or on the boundary surface tension. Thus, for example, a growing alum crystal is able to grow at 1.75 times saturation against a pressure of 40 kg/cm², when the growing octahedral face borders a glass surface. In contrast, no growth is observed under the same conditions when the octahedral face is in contact with a mica cleavage plane or when a cube face of alum grows against glass (CORRENS & STEINBORN 1939). In a similar way the properties of the contributing boundary surfaces, not yet understood in detail, may result in some minerals, like gypsum and iron sulfides, being able to exert a growth pressure in clays, whereas others, such as carbonates, generally cannot. Cone-in-cone structures may have originated by the growth pressure of carbonates.

The alteration of detrital mineral constituents proceeds during the early diagenesis of clay essentially along the lines of subaquatic weathering. Thus feldspars are dissolved and kaolinite and chlorite form anew. These processes may be ascertained by comparing the mineral compositions of concretions and their enclosing clays. The mineral particles entrapped early within the concretions were shielded from diagenetic changes and represent the detrital mineral composition. In Devonian shales, KNOKE (1966) found that plagioclase and quartz were more strongly dissolved and corroded than when they were contained within the calcareous concretions. In the interior of the concretions chlorite was found, which increased in amount from the center outward and into the shale. In the same direction the iron content of the chlorite increased. These observations show that diagenetic formation of chlorite began during concretionary growth, during which, at first magnesium-rich and later increasingly iron-rich chlorites were formed. A portion of the magnesium is tied up in the early formed chlorite, causing it to disappear early from the pore solution. FÜCHTBAUER & GOLDSCHMIDT (1963) could trace the early diagenetic formation of kaolinite and chlorite in calcareous concretions in Lias clays of southern and northern Germany. Within the concretions appreciable illite and little kaolinite was found; in the surrounding clay, much more kaolinite and chlorite. In all cases a gradual change in clay mineral ratios from the center of the concretions outward was established. This transition indicates progressive diagenesis during the concretionary growth process.

In which direction the original mineral composition of clay sediments is changed as a result of pressure and higher temperature during later diagenesis, can be read in general outline from a comparison of the average compositions of shales of different geologic ages. In young clays minerals of the kaolinite, montmorillonite and chlorite groups, along with illite and mixed-layer minerals occur in variable ratios. In contrast, the geologically older and presumably more strongly diagenetically altered shales as a rule are much more nearly monomineralic. Illite (muscovite) and chlorite predominate. Kaolinite, montmorillonite, and mixed-layer minerals, especially those with expandable layers, are absent or uncommon (GRIM 1953, ECKHARDT 1958; WEAVER 1959; KELLER 1963; MILLOT 1964). Compatible with these findings are the changes in chemical compositions of shales with age, which have been established by VINOGRADOV & RONOV (1957) from a large number of samples from the Russian platform. In these rocks the K-content increases from Tertiary to Cambrian from about 2% to about 4%, and the SiO_2/Al_2O_3-ratio decreases from 5 to about 3.

These observations indicate that montmorillonite, kaolinite, and mixed-layer minerals are decomposed and illite (muscovite) as well as chlorite, naturally synthesized during the course of later diagenesis. In different sedimentary series, composed of uniform materials, these diagenetic processes can be followed exactly.

The diagenetic transformation of montmorillonite and montmorillonite-illite mixed-layers into illite can be traced in the over 4000 m thick section through the Tertiary sediments of the Gulf Coast (Texas). In the upper 1000 m montmorillonite, with accessory illite and chlorite, is the predominant clay mineral. Below 1000 m montmorillonite begins to diminish in amount and the expandable layers of the mixed-layer minerals decrease with depth, while the illite content increases. At depths between 3000 and 4000 m the loss of montmorillonite and expandable phases is complete. Below 5000 m no expandable minerals occur. The clays at these depths consist of illite with a small amount of chlorite. The chlorite content in this section does not change with depth, although the chlorite crystallinity increases (BURST 1959, 1969; POWERS 1959, 1967). Quite analogous diagenetic transformations were observed by VAN MOORT (1971) in Mesozoic shales from New Guinea which were revealed in wells drilled to depths of 4500 m. The clays, which to depths of 1000 m consist of mixed-layer minerals with 60% montmorillonite and 40% illite, contain at 3000 m depth, mixed-layers with only 20% montmorillonite and 80% illite.

The transformation of montmorillonite into a mica is a complex process. Potassium, which can be fixed between the three layer silicate sheets, must be supplied, and simultaneously a corresponding amount of tetrahedral silicon is replaced by aluminum. Interlayer water is liberated during the potassium fixation. To remove these water layers solely by physical means, as VAN OLPHEN (1963) showed, pressures are necessary which are not available at depths of 3000 to 4000 m. From desorption experiments VAN OLPHEN concluded that the expulsion of the last of the four water layers from the montmorillonite structure requires a pressure of 5400 at. Values of similar order of magnitude were found experimentally by STEINFINK & GEBHART (1962) for the dehydration of vermiculite by pressure.

Therefore, pressure alone does not set the transformation of montmorillonite to mica into motion. A chemical reaction is involved. The necessary activation energy is supplied by the increased temperature which prevails in the transformation zone. According to well log measurements, the temperature at the depth at which montmorillonite layers disappear in the Tertiary of Texas, is about 100 °C (BURST 1969). It is still not certain where the potassium and aluminum, which must be supplied for mica formation, originate. The potassium content of the pore solution, according to rough estimates by POWERS (1959) is too low, as it is also for aluminum. In the case of Gulf Coast pelitic sediments, PERRY & HOWER (1970) have shown that potassium and aluminum are supplied by the decomposition of detrital K-feldspars and micas contained in the sediments.

Since most clays generally contain very little feldspar, it is possible that potassium and aluminum migrate from neighboring sand layers in which feldspar is being dissolved. The material balance for sandstones, in which feldspar is dissolved and kaolinite and quartz deposited, show that potassium is released in excess of that fixed in the sandstone itself. When little or no kaolinite is formed, aluminum could be made available also. For each mole of potassium, entering the mica structure, one mole of aluminum must be supplied and one mole of silicon removed to maintain charge equivalency. The diagenetic transformation of expanding clay minerals, therefore, supplies free SiO_2, resulting either in silicification of the clay or deposition in sandstones (TOWE 1962).

Transformation of the montmorillonite to the mica structure leads to chemical dehydration of the sediment, which takes place at depths, at which mechanical compaction essentially has ceased. In clays rich in montmorillonite and expandable mixed-layer minerals, this water released at great depths may be expelled, in turn resulting in the expulsion from the clay of dispersed liquid hydrocarbons and their accumulation as oil deposits in permeable strata (POWERS 1967; BURST 1969).

The formation of illite from illite-montmorillonite mixed-layer minerals and the recrystallization of poorly ordered detrital illite leads to an increase in the crystallinity of illite. This is expressed in an X-ray powder diagram by increasing sharpness of the 10 Å interference (WEAVER 1960). As KUBLER (1966) showed, one can use as a measure of illite crystallinity, and with it the degree of diagenesis, the width of the 10 Å reflection at half peak height. In a section through the upper Cretaceous of the Douala Basin in Cameroon DUNOYER DE SEGONZAC, FERRERO & KUBLER (1968) found, between 1000 and 4000 m depths, a linear decrease in the 10 Å reflection width to negligible values in the deepest samples. Simultaneously the amount of illite-montmorillonite mixed-layer clay decreases, until at 4000 m depth it is gone.

Kaolinite, derived from continental weathering and formed in the earliest stages of diagenesis, is unstable at greater depths of burial. This observation, based on kaolinite distribution in sandstones, is substantiated also in shales. Deeply buried shales contain generally little or no kaolinite. In the above mentioned upper Cretaceous section studied by Dunoyer de Segonzac and coworkers (1968), kaolinite disappears completely at a depth of about 3000 m, while illite and chlorite take its place. In Carboniferous shales from wells of Munsterland in northwest Germany (Stadler 1963) kaolinite is the predominant mineral to a depth of 3300 m. Below 3300 m the clays consist of chlorite and sericite. Of course, the change in mineral composition in this case may have resulted from a change in detrital source material, but it is also possible that it reflects the lowered stability of kaolinite at great depths. Füchtbauer & Goldschmidt (1963) studied samples from deep wells in northwest Germany and showed that in the Alb, Dogger and Lias shales the kaolinite to chlorite ratio decreases continuously with depth between 400 and 3500 m.

In addition to illite, chlorites are the most important diagenetic minerals in some shales (Eckhardt 1958). Chlorite formation begins even during submarine weathering, is observed in some clays during early diagenesis, and continues to the greatest degree at greater depths, where montmorillonite, expandable minerals, and kaolinite become unstable. The earlier formed chlorites are magnesium-rich, the later formed ones often rich in iron. In most cases trioctahedral chlorites are involved, although, dioctahedral, aluminum-rich varieties have been identified. Weaver (1959) found dioctahedral chlorite as a diagenetic transformation product in middle Ordovician bentonites from Pennsylvania, Virginia, Alabama, and Kentucky. In this case it can be demonstrated that the chorite was formed by interlayering of magnesium and/or aluminum octahedral layers in montmorillonite.

In middle Keuper clays and marls of southwest Germany Kromer (1963) found as the main mineral constituent a dioctahedral chlorite (sudoite), which seemingly was formed diagenetically. This mineral contains about 1 Al to 3 Si in tetrahedral sites. In the octahedral layer some magnesium occurs along with aluminum. In this case sudoite has formed at the expense of kaolinite, and partly also from illite-montmorillonite-chlorite mixed-layer minerals.

Diagenetically formed dioctahedral chlorites may be more abundant in clays than as yet realized, since their identity in mixtures containing other dioctahedral minerals is difficult. The chemical constituents of dioctahedral chlorites are similar to those of kaolinite. Still, either SiO_2 must be removed, or aluminum added for dioctahedral chlorite to form from kaolinite. This is because a part of the tetrahedral sites are occupied by aluminum in the chlorite. If magnesium and iron occur in addition to aluminum in the octahedral interlayers, these components must be supplied also.

The greater amount of diagenetic chlorites are trioctahedral, that is, octahedral sites are occupied by magnesium and iron. Since the chemical compositions of these diagenetic chlorites are not known exactly, the origin of their constituents is not known in detail. The iron and magnesium contents of pore solutions are certainly not sufficient for chloritization, especially in the later stages of diagenesis. Iron and magnesium can originate only from primary iron-magnesium minerals, such as glauconite, detrital biotite, and vermiculite. The chemical milieu for chloritization is characterized by weakly alkaline conditions (pH 7 to 9), and low potassium content.

In this respect the pore solutions of those shales in which chlorite forms, such as those of the Alb, Dogger, and Lias of northwest Germany, may differ from those, like the Gulf Coast Tertiary, which form illite from montmorillonite. Perhaps also the Gulf Coast, as well as the northwest German pore solutions, may have been slightly alkaline, but the former had a higher potassium content and/or lower iron and magnesium content than the pore solutions in the northwest German clays.

In clays containing abundant pyroclastic material, diagenesis leads to the formation of zeolites, already noted for the diagenesis of sandstones.

While in normal clays the potassium concentration in pore solution is low, so that only dioctahedral micas form or are reconstituted, in clays with volcanic constituents, sufficiently high potassium concentrations can occur such that diagenetic formation of alkali feldspars occurs. An example is the lower Permian shale of the Saar-Nahe region (HEIM 1960).

In clays in saline sedimentary sequences, whose pore space was filled with highly concentrated sea water at the time of deposition, a more intensive diagenesis of magnesium-rich minerals occurs. For example, in the salt clays of the Zechstein formation of northwest Germany, the formation of trioctahedral chlorites is initiated very early, and takes place apparently at the expense of detrital illite (FÜCHTBAUER & GOLDSCHMIDT 1959; NIEMANN 1960). From chemical analyses, the chlorites of the 30 m thick Gray Salt Clays of Reyhershausen near Goettingen were shown to be magnesium chlorites, whose aluminum contents decrease continuously from top (penninite) to bottom (sheridanite) (NIEMANN 1960). In some clays from saline formations corrensite occurs along with magnesium chlorite or in place of it. Corrensite is a regular mixed-layering of chlorite and expandable saponite. A high content of very early formed corrensite is characteristic of saline clays of the middle Keuper (LIPPMANN 1954, 1960 a; SCHLEMMER 1971). Argillaceous strata in Zechstein anhydrite in northwest Germany contain diagenetically formed talc as the main mineral.

Occasionally pyrophyllite has been reported as a late diagenetic mineral in shales. Examples include the Pennsylvanian of Utah (EHLMAN & SAND 1959), the Ordovician of the Rhenish Schiefergebirge (SCHERP & STADLER 1968), the Devonian of the Armoricanian Massif (DUNOYER & MILLOT 1962; MELOV & PLUSQUELLEC 1967) and the Silurian and Devonian of the Sahara (CHENNAUX & DUNOYER DE SEGONZAC 1967). In these occurrences pyrophyllite probably has been formed at the expense of kaolinite, with introduction of SiO_2 and removal of water. According to the experimental studies of ALTHAUS (1966 a) pyrophyllite forms from kaolinite and quartz in a neutral environment at 390°. The temperature of formation is only slightly pressure-dependent. WINKLER (1967) considered the occurrence of pyrophyllite as critical for the beginning of the greenschist facies of regional metamorphism. Corresponding to natural observations, the investigations of ALTHAUS (1966 b) suggest that pyrophyllite can form also at lower temperatures. ALTHAUS established that the formation temperature is lowered considerably in the presence of acidic solutions. Pyrophyllite forms at 260° in 10 n HCl or H_2SO_4. The basis for this effect is the lowering of the activity of water by the high acid concentration. It would be expected that a similar effect is produced by high salt concentrations. Also the formation of pyrophyllite is facilitated, if SiO_2 occurs dissolved in the pore solution rather than as quartz. The rarity of diagenetically formed pyrophyllite apparently is related to the fact that kaolinite usually is no longer present at great depths of burial. The aluminum introduced into chlorites or micas during diagenesis apparently is no longer available for pyrophyllite formation.

In summary the following principles apply for the course of chemical diagenesis of normal clays (see DUNOYER DE SEGONZAC 1969). Aluminum passes increasingly from octahedral positions (kaolinite) into the tetrahedral positions of chlorites and micas, as SiO_2 is liberated. Magnesium from the pore solution, as well as adsorbed and dissolved potassium, are tied up in minerals of the chlorite and mica groups. Intercrystalline water from expandable layers is expelled. When the diagenetic processes take place within the clay, essentially isochemically, chemical exchange with the pore solutions of permeable beds may take place to a certain degree. From the decomposition of feldspars, released potassium migrates from sandstones into the clay, while the clays give up dissolved SiO_2. As a result of these processes, diagenesis converts the initially variably composed clays finally into the simpler mineral association of dioctahedral mica and predominantly trioctahedral chlorite.

5.6 Diagenesis of carbonates

5.61 Mechanical diagenesis

Calcilutites and calcarenites after deposition possess approximately the same porosities as clays, silts, and sands. GINSBURG (1957) found that aragonitic muds from Florida Bay have at the surface a porosity of 0.87 and at a depth of 3.4 m porosity of 0.75. Calcareous muds around the Bahama Islands have porosities between 0.60 and 0.70 at the surface (BATHURST 1971; MILLER & RICHARDS 1969). KÖGLER (after FÜCHTBAUER 1970) found that for calcareous muds in the Gulf of Oman $\varepsilon = 0.70$ at the surface and $\varepsilon = 0.60$ at 1—3 m depth. Calcarenites have initial porosities, which like normal sands lie between 0.4 and 0.5 (EBHARDT 1968).

Compaction experiments, with calcareous muds, carried out in cylinders with a moving piston (TERZAGHI 1940; PARASNIS 1960; EBHARDT 1968) result in the same approximately linear relation between relative pore volume and pressure, exhibited by clays:

$$E = E_1 - \beta \lg p/p_1 \qquad (5.63)$$

E is the relative pore volume at pressure p. E_1 (relative pore volume at $p_1 = 1$) and β are constants. Some experimental data are plotted in fig. 5-73. Table 5-16 gives some experimentally determined values for the constants E_1 and β.

These constants are of similar magnitude to those measured for clays (see section 5.51). Therefore, mathematically the mechanical compaction of calcareous muds is analogous to that of clays. That the processes are different in detail from those

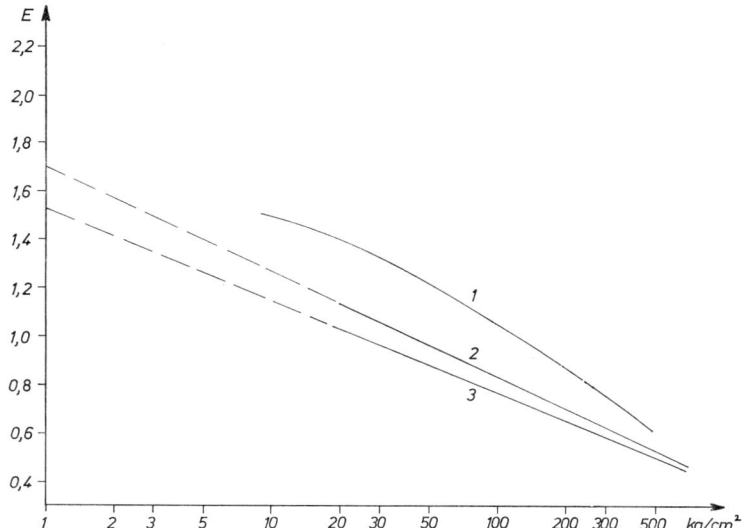

Fig. 5-73. Experimental compaction of carbonate muds 1: Globigerina mud (PARASNIS 1960); 2: Carbonate silt, Red Sea; 3: Carbonate silt, Florida Bay (EBHARDT 1968).

involving clays and sands, is shown by the experiments of FRUTH, ORME & DONATH (1966) on compaction of lime muds of different grain size. They used an apparatus with three pistons which produced uniform hydrostatic pressures up to 1000 atm. Sands, which consist mainly of quartz, can be compacted mechanically only slightly, since the sand grains can withstand high load pressures. In contrast FRUTH et al. (1966) could show in thin sections of a compacted oolitic calcarenite, that the ooides break and penetrate one another under the effect of pressure. The calcareous shells of organisms behave likewise. Although the effect of pressure solution probably is not ruled out entirely, the absence of styolitic contacts indicates that the particles flow plastically under the influence of pressure. From these observations it is concluded that mechanical compaction of carbonate sediments, because of the lower strength of calcite, takes place with considerable involvement of fracturing and plastic flow, while the mechanical compaction of quartz and silicate clastic sediments results predominantly from a reorientation of particles.

In experiments on mechanical compaction of clay and carbonate muds, compaction is produced by differential pressure within a chamber that permits the exit of pore water, but still confines the solid sediment. The type of deformation might, in the terminology of SANDER (1948, p. 41) best be described by a conjugate shear model (after gliding on two shear planes). In this case the special condition applies, that the total volume decreases as a result of the expulsion of water. A spherical volume of a homogeneous mass before deformation experiences distortion to a rotation ellipsoid, flattened in the direction of applied pressure (SANDER 1950, p. 112). The diameter of the circular section of this ellipsoid, oriented perpendicular to the pressure direction, maintains the diameter of the original sphere during deformation. In nature the compaction of clays by a load pressure occurs largely according to this model, as verified by experiment. Fossil shells deposited in the clay undergo brittle or plastic deformation, which corresponds to the deformation of a sphere into a compressed ellipsoid of rotation.

Table 5-16. Compaction of carbonate muds.

No.	% $CaCO_3$	MD $|\mu m|$	E_1	β
1	93	—	2.20	0.58
2	75	20	1.70	0.43
3	86	7	1.54	0.38
4	—	—	1.45	0.35
5	—	—	1.03	0.23

1: Globigerina mud, Pacific Ocean (PARASNIS 1960); 2: Globigerina mud, Red Sea (EBHARDT 1968); 3: Aragonite mud, Florida Bay (EBHARDT 1968); 4: Aragonite mud, Bahama Banks (TERZAGHI 1940); 5: Aragonite mud, Bahama Banks (TERZAGHI 1940).

That the compaction of carbonate sediments takes place naturally in a different manner than for clays, although carbonate and clay muds behave similarly in experimental compaction, is indicated by the state of preservation of fossils in limestones. As mentioned by many authors (WELLER 1959; PRAY 1960), but unfortunately still not systematically studied, shells and fragile skeletal remains of organisms embedded in limestones show as a rule no or only slight indications of deformation. They frequently remain completely or almost completely intact and almost never are as flatly compressed or broken as they are in clays. This usual slight deformation of fossils shows that carbonate sediments are distorted only during a short initial phase of diagenesis. Relatively early this sort of compaction ceases as a result of processes which essentially eliminate the original pore volume. As a result unconsolidated carbonate sediments with porosities between 0.40 (calcarenites) to about 0.70 (calcilutites) form limestones with porosities below 0.05. This can involve only chemical processes, which, because of their slowness do not occur during experiment.

5.62 Isochemical diagenesis

5.621 General

The most important processes of chemical diagenesis of carbonate sediments occur isochemically, that is, without change in the main chemical components. A characteristic of isochemical diagenesis of limestone is, as the state of preservation of fossils indicates, that it is initiated very early. As a result the rock fabric is so stabilized that internal shear motions which deform fossils cease. Exactly when this stage of mechanical compaction is brought to an end by chemical lithification, depends on different factors which as yet are not understood sufficiently. Some of these can be demonstrated by specific examples.

In the Wetterstein limestone (middle Triassic) of the Alps GERMAN (1966) found no indication of compaction. *Pelecypod* shells remained unbroken and former cavities in micrites are not distorted by compression. In the Senonian chalk of Northern Ireland WOLFE (1968) noted different behavior of different stratigraphic members. In the fine-grain White Chalk member (mean grain size 0.5—1.5 µm) one can infer a mean compaction of 22 % from the distortion of macrofossils. Similar compaction was deduced for the *Inoceramus* chalk, which contains small prisms of disintegrated *Inoceramus* shells in a groundmass with mean grain size between 0.5 and 2.5 µm. The bioturbation chalk, with prolific borrow structures (mean grain size 0.5—1.5 m) shows in contrast only slight indications of compaction, since the macrofossils are compressed only about 1—7 %.

According to a study by ZANKL (1969) the clay content of a limestone is important in the interplay of mechanical compaction and chemical infilling of its pore space. ZANKL determined the amount of hydrochloric acid-insoluble residue ("clay content") in limestones in which the amount of mechanical compaction was known from the deformation of certain fossils. As a measure of compaction the length/height ratios of shells and casts of the pelecypods *Pinna* and *Rolliera* were used. The results, based on Triassic and Jurassic limestones, are illustrated in fig. 5-74. Apparently

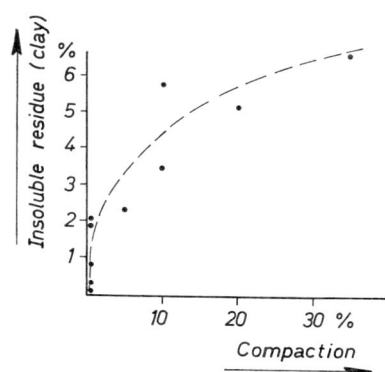

Fig. 5-74. Relation between compaction and the clay content of limestones (ZANKL 1969).

there is a systematic relation between mechanical compaction and clay content, the two increasing together. Compaction proceeds further and is interrupted later by chemical infilling of the pore space, the greater the clay content of the sediment. The cause of this interrelationship, which should be confirmed by further investigation, is still not clear. It is possible that even a very low clay content is sufficient to inhibit carbonate deposition in the pores of the rock.

There are two possible means of filling up the pore space in carbonate sediments by chemical processes:

1. Exogenic cementation. Carbonate is introduced from without and precipitated in the pore space. In this case the original volume of a porous sedimentary

layer is not changed. The carbonate is brought in either by diffusion in solution or by actual fluid flow. This then raises the question, as to where the carbonate comes from and what are the conditions which bring about precipitation from solution.

2. Endogenic cementation. The pore space is filled as a result of redistribution of carbonate components making up the sediment. In this case the sediment experiences a reduction of its original volume equivalent to the amount of pore space lost.

5.622 Exogenic isochemical diagenesis

The exogenic cementation of carbonate sediments falls, as a rule, into an early stage of diagenesis, since it is initiated before the sediment is significantly mechanically compacted. Early exogenic cementation can result either from the action of sea water on the sea floor and in coastal regions, or from rainwater on the continents. Both kinds of early exogenic cementation are transitional between submarine or subaerial weathering and diagenesis. Under certain conditions exogenic cementation can take place during later stages of diagenesis, namely, when lithified, but still porous, limestones reach greater depths. In some sedimentary basins the penetration of artesian surface waters into deep horizons is indicated. As a result of a temperature increase, such waters can precipitate $CaCO_3$ at depth. Analogous to the later diagenetic precipitation of carbonates observed in sandstones, porous limestones in this way can be supplied with late carbonate cement.

The lithification of loose carbonate sediments has been investigated most extensively in the Persian Gulf (SHINN 1969; DE GROOT 1969; TAYLOR & ILLING 1969). In water to depths of 60 m calcarenites are lithified at the surface over wide expanses of the Persian Gulf. By means of radiocarbon age determinations, and by identification of clay vessels found in these limestones, it has been determined that the lithification occurred in the last millennium and is going on even today. The hard layers are covered here and there by nonlithified calcarenites. In coastal lagoons several 5—10 m thick lithified layers are found, interlayered with nonlithified calcarenites. Apparently conditions for surface lithification always did not prevail. That the process proceeds from the surface down into the sediment is recognized by the decrease in lithification from the surface downward. The thickness of solid layers clearly is related to the sediment grain size. In fine-grain and relatively impermeable layers the lithification can penetrate to lesser depth than in coarse grain sediment. Therefore, the lithified layers are thicker in coarse grain sediments than in fine-grain ones. The lithifying cement consists predominantly of acicular and fibrous aragonite. In addition, newly formed Mg-rich calcite of very fine grain size occurs.

From the fabric and the mode of occurrence of these lithified layers it is clear that precipitation of calcium carbonate from the warm, slightly supersaturated sea water was involved. This percolated from the surface into the sediment as far as permeability or the closing of porous channels by lithification permitted. The process is only possible when simultaneous sedimentation is slight. The hard layers indicate periods of minor sediment supply; unlithified layers indicate more rapid sedimentation.

That aragonite is precipitated from sea water corresponds to expectations. More difficult to explain is the formation of Mg-calcite. Also, why is the precipitation of calcium carbonate induced at the surface and within the sediment? Since aragonite is never found as an overgrowth on quartz grains, the presence of calcium carbonate substrates of high surface area may have induced nucleation. In addition the effect of the decomposition of organic matter in the sediment, especially the activity of sulfate reducing bacteria, should be taken into consideration. In the presence of organic matter these bacteria reduce sea water sulfate, simultaneously producing carbonate ions according to the following gross reaction:

$$SO_4^{-2} + 2\,C + H_2O \rightarrow CO_3^{-2} + CO_2 + H_2S$$

In the upper parts of sedimentary layers the concentration of carbonate ions, and along with it the supersaturation for calcium carbonate, is increased by sulfate reducing bacteria. To what extent the activity of algae plays a role in precipitating carbonate, especially Mg-calcite, still is not clarified.

Also in the deep sea different examples have been noted in recent years of lithification of carbonate sediments by the action of sea water. In all cases pelagic coccolithic muds, with globigerina and some pteropods, are involved. In the Mediterranean at 2055 m depths, 1–8 cm thick crusts have been observed, which are underlain by unconsolidated carbonate mud, and whose hardness decreases from the surface downward (FISCHER & GARRISON 1967; GARRISON & FISCHER 1969). The lithified carbonate is a very fine grain Mg-rich calcite. Aragonite shells are noted to be partially dissolved. Lithification by the precipitation of Mg-calcite was observed also in carbonate sediments from Atlantic seamounts (FRIEDMAN 1964, MILLIMAN 1966; CIFELLI, BOWEN & SIEVER 1966). According to FISCHER & GARRISON, not only precipitation of Mg-calcite cement, but also partial replacement of aragonite and calcite skeletal remains by Mg-rich calcite has taken place. Precipitation of iron and manganese oxides on these crusts indicates that a very low sedimentation rate is a prerequisite for reaction of the loose sediment with sea water. The factors which permitted the precipitation of calcium carbonate and especially the formation of Mg-calcite on the deep sea bottom are not known still.

That the temperature of the bottom water plays a determining role is illustrated by the findings of BARTLETT & GREGGS (1969) from drill cores from the North Atlantic. The cores consist of alternating, up to 6 cm thick, layers of lithified and unlithified carbonate sediments. The lithified layers consist of hardened globigerina-pteropod mud and contain a warm water fauna. The unlithified layers contain a cold water fauna, consisting of corals, gastropods, bivalves, and pteropod shells in a fine grain matrix. Age determinations show that the upper boundary of the lithified layers corresponds to periods of about 30,000 years without sedimentation. Precipitation of calcium carbonate from sea water was possible, therefore, only at the higher temperatures and during periods of low sedimentation.

Hard layers in calcilutites, which are rich in planktonic foraminifera and pteropods, are observed at the bottom of the Red Sea at depths between 384 to 1737 m They owe their origin perhaps to special local conditions (GEVIRTZ & FRIEDMAN 1966). In drill cores a repeated sequence of 1 to 5 mm thick hard layers were observed, which are lithified by aragonite. GEVIRTZ & FRIEDMAN (1966) relate the aragonite precipitation to the hotter and highly concentrated solutions which have been found in the deepest parts of the Red Sea. The alternating sequence of hard and unlithified layers reflects apparently variations in temperature and bottom water salinity.

From various formations hard carbonate layers are known, which probably formed under conditions similar to those of recent formations in the Persian Gulf, where sea water reacts with unlithified carbonate sediment. The surfaces of these so-called hardgrounds are bored into, corroded, and populated by bottom organisms. Therefore, they represent periods of very slow sedimentation. From the Ordovician of southern and western Sweden, LINDSTRÖM (1963) described 2 to 20 cm thick, calcilutite-hardgrounds of this kind, interbedded with marl and clay beds. From the Triassic limestones of the Bavarian Alps PICHLER (1963) and GERMAN (1966) describe indications for submarine and synsedimentary lithification. Limestone beds in the Dogger of the Paris Basin with bored surfaces suggested to PURSER (1949) lithification of calcarenites at the sea bottom, during periods of reduced sedimentation. In the Turonian chalk of England, studied by BROHMLEY (1965, after BATHURST 1971), hardgrounds of coccolith-micrites alternate with layers of only slightly lihtified chalk. Their surfaces are bored, corroded and populated by various sessile organisms and covered by glauconite and calcium phosphate. Lithification gradually decreases with depth, grading into irregularly distributed concretions in the transition to unlithified chalk.

Of greater importance than exogenic cementation of carbonate sediments on the sea bottom, is the exogenic cementation experienced by carbonate sediments when they are uplifted above sea level and become part of the land surface. This gives rise to two kinds of diagenetic lithification. In the tidal region so-called beachrocks form, as well as other limestones resulting from the cementing effects of rainwater.

Recent beachrocks are known from numerous coastal regions and have been described by many authors. They form when unconsolidated carbonate sediments are saturated from time to time by sea water, which evaporates in the pores. A cement forms, consisting in part of aragonite and in part of Mg-rich calcite, both of which are compatible with the marine environment (FRIEDMAN 1968).

If unlithified marine carbonate sediments come under the influence of CO_2-rich rainwater, transformations are initiated which are the result of increased solubility by CO_2 and the instability of Mg-rich calcite and aragonite. According to studies by various authors of recent to Pleistocene carbonate sediments, especially from Bermuda, Florida, the Bahamas, and the Mediterranean (STEHLI & HOWER 1961; FRIEDMAN 1964; GINSBURG 1957; LAND 1966, 1967; GAVISH & FRIEDMAN 1969) the following picture summarizes lithification of marine carbonate sediments by meteoric waters.

5.62 Isochemical diagenesis

The initial stage is provided by the unlithified sediment, whose carbonate components consist of remains of organisms of aragonite (molluscs, corals) and Mg-rich calcite (foraminifera, algae, echinoderms), with minor low Mg-calcite (fragments of older limestone). In the first stage a low Mg-calcite is deposited in the pore space, mainly at grain contacts. This must have been introduced from without by rainwater, since no dissolution of aragonite has taken place yet. Simultaneously Mg-calcite transforms into low-Mg-calcite, preserving all of the details of the original organic structures. The addition of low-strontium calcite cement results in a decrease in strontium content of the rock. The origin of the newly precipitated carbonate from atmospheric CO_2 brings about a noticeable increase in the light isotopes of oxygen and carbon (δC^{13} falls from $+2 - +3$ to about 0; δO^{18} falls from -1 to -2, GAVISH & FRIEDMAN 1969). According to GAVISH & FRIEDMAN (1969) rocks in this stage in Israel are 7000 to 10,000 years old.

In the second stage the pore space becomes filled more or less completely with low-Mg-calcite cement. Aragonite is dissolved and in the resulting cavities drussy calcite precipitates. δC^{13} decreases further to -5.5 to -6 and δO^{18} to about -4. Strontium experiences a further decrease because of the solution of Sr-rich aragonite. On the Mediterranean coast of Israel rocks of this stage are 80,000 to 100,000 years old (GAVISH & FRIEDMAN 1969).

The addition of calcium carbonate during this diagenetic stage must balance a dissolution of carbonate. The lithified carbonates, whose pore space is filled more or less, are carbonate acceptors. Donors are apparently in most cases carbonates which were dissolved at the surface by CO_2-rich surface water. On Bermuda red soils and limestones, from which aragonite is dissolved without formation of calcite cement, are the sources for the calcite deposited in deeper horizons (LAND 1967). On Bermuda Pleistocene limestones of the last stage have porosities still of 0.20 (LAND 1966). In other cases limestones of greater or less density form. For complete cementation of a limestone bed, somewhere a primary carbonate sediment of corresponding thickness must be dissolved.

It is understandable that calcium carbonate near the surface is dissolved in meteoric solutions and that aragonite and Mg-rich calcite mainly are affected by dissolution. It is still not clear why solutions saturated in calcium carbonate precipitate calcite in deeper horizons. That the supersaturation of inflowing solutions cannot have been high is indicated by the side by side dissolution of aragonite and Mg-rich calcite and precipitation of low-Mg-calcite. GAVISH & FRIEDMAN (1969) observed in the first and second stages of diagenesis on the Israeli coast the replacement of quartz by calcite, indicating slightly alkaline conditions.

Deep drilling on the Eniwetok and Bikini Atolls have shown impressively how aragonitic carbonate sediments can be transformed into calcite limestone by the influence of meteoric waters (SCHLANGER 1963). The up to 1500 m deep drill cores contain up to three zones of solid limestone which consist of calcite. In rocks the original aragonitic shells are dissolved. The resulting pore spaces are coated with fibrous calcite or filled with coarse sparry calcite. Also the fine grain groundmass is recrystallized completely to calcite. As a result of transformation to calcite, the original aragonitic corals have lost their strontium, from which it can be concluded that calcitization took place through solution. The solid limestone ledges grade downward gradually into unlithified carbonate sediments which consist of a mixture of various calcitic and aragonitic fossil fragments. The well section covers a time span reaching back from the present into the Eocene. The hard limestone layers represent periods when the surface of the island lay high above sea level. Unlithified carbonate sediments came into range of rainwater activity, which dissolved carbonate at the surface and caused cementation by calcite in deeper horizons and replacement of aragonite by calcite. In those horizons which could not be reached by fresh water, the sediments remained unaltered and preserved their aragonite content over a considerable period of time.

5.623 Endogenic isochemical diagenesis

It remains controversial as to whether exogenic cementation, especially by continental surface waters, has played a role in the early diagenetic formation of dense limestones from porous carbonate sediments. There can be no doubt that unlithified carbonates are cemented and lithified by the influence of rainwater. Still it is certainly not valid to conclude from such observations, that, with the exception of some submarine-formed hardgrounds, all dense limestones form in this way. Almost all early diagenetically lithified limestones, while still unlithified and before burial to greater depths, must have been exposed to the influence of continental surface waters. At least for limestone beds which continually comprise part of the basin fill this is very unlikely. There is also difficulty in accounting for the origin of the large amounts of carbonate which must be supplied by surface waters in order to cement a porous carbonate sediment.

Therefore, it can be assumed likely that many carbonate sediments were diagenetically lithified early without introduction of pore cement from surface waters. If no carbonate is introduced from without, the cementation must occur endogenically, that is, from the carbonate supply within the bed itself. Redistribution of carbonate by endogenic cementation takes place by dissolution of primary particles and redeposition within the pore space. As a result of porosity of the sediment is lost through microscopic processes and pore solutions are expelled without causing shear movements, such as are able to deform fossils during compaction of clays.

The redistribution of carbonate from primary grains into the pore solution can result only from pressure solution. Pressure solution in sandstones was described in section 5.42. It occurs in sandstones only at depths of burial below 1000 m. Carbonate minerals, however, are much more susceptible to pressure solution than quartz. So-called "indented" limestone pebbles, which are formed in Pleistocene conglomerates at burial depths of only a few meters, are impressive indicators of this phenomenon (MORAVIETZ 1958). Therefore, it can be assumed with certainty that pressure solution also will change very early the fabric of fine grain carbonate sediments. In fact sutured contacts and other clear indications of pressure solution are observed in many limestones with coarser constituents, for example, in oolithic limestones (USDOWSKI 1962 and others, see BATHURST 1971). Electron micrographs of the groundmass of micritic limestones show interfingering contacts also of the finest carbonate particles (see, for example observations by WOLFE 1968 of chalks).

5.624 Stylolites

Macroscopic indications of pressure solution are widely distributed in limestones as stylolites. An extensive literature exists on the morphology, occurrence, and formation of stylolites in limestones (WAGNER 1913; STOCKDALE 1922, 1926; SHAUB 1939, 1949; BUSHINSKIY 1961; DUNNINGTON 1967; TRURNIT 1967 a, b; PARK & SCHOT 1968). Especially informative is the summary by TRURNIT (1967 a), which unfortunately is not yet published.

Stylolite surfaces are usually argillaceous, irregularly shaped boundary surfaces between two rock units, which interfinger by virtue of small conical or columnar projections. In sections perpendicular to this surface they appear as irregular jagged bands (sutures). Stylolites and stylolite surfaces form in certain planes where pressure solution acts according to the RIECKE principle, whose mechanism was depicted in section 5.42. The irregularity of the solution surface arises from slight localized differences in pressure solubility. The partner with least solubility penetrates as a stylolite into the more soluble one. Observations of pressure solution contacts between individual mineral grains lead to the conclusion that pressure solution surfaces are less irregular the greater the pressure solubility difference between the two partners (TRURNIT 1967).

As in the case of pressure solution at contacts between individual grains, appreciable dissolution along a stylolite surface can occur only if the removal of dissolved material is facilitated. The solution must be able to flow through the rock and a conduit of flow must be available along the stylolite surface. The clay seam which occupies the stylolite surface represents the insoluble residue which accumulates there, as well as the conduit for flow.

The most common and extensive stylolite surfaces run parallel to the bedding. They form from the effect of load pressure and may develop from thin sedimentary layers with a higher clay or coarser clastic content. As MORAVIETZ (1958) and TRURNIT (1967) observed, they start to form at isolated points of pressure solution between individual coarse grains. These grow gradually into a continuous surface, continuing through the cement which is precipitated in pore space next to the pressure contacts. Alternately grain contacts, initially involving point contacts, enlarge to include the entire cross section of the grain, until the pressure sutures of adjacent grains meld together. Because of the high susceptibility of carbonate minerals to pressure solution, this process begins at shallow depths of burial, in still unlithified sediment. Compaction and lithification of the sediment, as well as formation and development of extensive, continuous stylolite surfaces, are processes proceeding simultaneously. The dissolution of carbonate along stylolite surfaces can continue in the finally compacted and lithified rock, only as long as conduits remain accessible through which fluids may flow. As long as the sediment is subsiding and is being exposed to increasing load pressure, this flow is supported by the compaction flow, which in limestones uses the stylolite surface as a conduit for its ascension. In the carbonate sediment it is augmented by the solution which is expelled from the pores as they are filled with cement. In lithified limestones, later uplifted to the surface, circulating surface water can be responsible for continuing dissolution along stylolite surfaces.

Stylolite surfaces orientated at considerable inclination to the bedding do not extend very far and are not so common. They are dissolved by tectonic pressure. Since they develop along ruptures, which serve as fluid conduits, they can form only in rocks that have developed a certain strength and brittleness.

The carbonate dissolved at the stylolite surface is precipitated first in the vicinity of the solution surface as pore cement. Therefore, porosities and permeabilities in limestones rich in stylolites are especially low. After the surroundings are lithified in this way, the dissolved carbonate can be introduced along stylolites to more distant parts of the rock where pore space is still available. The amount of carbonate mobilized and redeposited in this way is difficult to estimate. From the height of stylolites STOCKDALE (1926) estimated that lower Mississippian limestones have lost 13 to 34% of their original volume as a result of pressure solution. These values and similar estimates by other investigators represent only minimum values, because the stylolites themselves were formed by pressure solution. A reliable estimate of dissolved carbonate probably can be deduced from the amount of insoluble residue concentrated in the stylolite surface.

In any case it is certain that appreciable amounts of carbonate can be mobilized by pressure solution at stylolite surfaces. An essential part of the carbonate necessary for endogenic cementation may stem from this source. A limestone penetrated by stylolites parallel to the bedding is quite different from the original carbonate sediment. The latter may have contained calcareous and non-calcareous (clastic) components homogeneously mixed together. By pressure solution along stylolite surfaces a redistribution of carbonate and clastic components has taken place. The carbonate from the dissolved part of the sediment has precipitated in the pore space of the rock layers between stylolite surfaces, while the clastic components are concentrated as a seam along the surface.

If a stylolite front encounters a less soluble bed, the stylolitic cones are dissolved and a planar pressure solution surface is formed which is very difficult to recognize (TRURNIT 1967). Dissolution of carbonate and accumulation of the non-calcareous components can take place on such planar surfaces. Layers, marked by thin clay partings, so common in all limestone formations, and the regularly alternating sequence of calcareous and marly layers, in many cases may have been produced by later redistribution of carbonate and clay components by diagenetic pressure solution, rather than by a rhythmic sequence of argillaceous and carbonate sedimentation.

5.625 Transformation of Mg-calcite and aragonite into calcite

Recent marine carbonates consist of three $CaCO_3$-minerals, aragonite, Mg-rich calcite (Mg-calcite) and low-Mg calcite (calcite). The relative amounts of the three phases depend on the physical and chemical conditions of the invironment and on the kinds of organism which were involved in carbonate deposition. Aragonite and Mg-calcite are the main constituents of carbonate sediments in shallow seas. Lagoonal sediments are richest in aragonite. Sediments in the detrital areas of reefs have the highest Mg-calcite content. In contrast deep sea sediments with high proportions of planktonic foraminifera and coccolithophorids are rich in calcite (FRIEDMAN 1964). Depending on the evolution of plant and animal varieties, in the geologic past the calcite:aragonite ratios of carbonate sediments were subject to considerable variation (LOWENSTAM 1963). In the Paleozoic there were a great many organisms with calcite shells; in the Mesozoic aragonite played a greater role and in the Cenozoic the center of gravity swung again in favor of calcite, as a result of development of foraminifera and coccolithophorids.

In contrast to marine carbonate sediments, those formed in fresh water lakes are composed of calcite (for examples see FÜCHTBAUER in the second part of this series).

In contrast to the variability in mineral composition of unlithified carbonate sediments, limestones of all geological epochs, both marine and non-marine, consist of calcite. Aragonite occurs in diagenetically altered limestones only where special circumstances have promoted its preservation. Examples are carbonate sediments which were saturated at any early stage with oil (FÜCHTBAUER & GOLDSCHMIDT 1963) and asphalt (STEHLI 1956) or embedded in impermeable shales. Mg-calcite is even less stable, and is found only in young sediments.

As was indicated in section 4.2, Mg-calcite and aragonite are metastable under conditions prevailing at the surface and during diagenesis. The only stable phase of $CaCO_3$ is calcite. In sea water, however, the growth of calcite nuclei is inhibited so strongly that Mg-calcite and aragonite precipitate, with and without the assistance of organisms, in place of calcite. The inhibiting factor is the presence of Mg-ions, which are absorbed preferentially on the surfaces of calcite nuclei. At 20 °C aragonite precipitates instead of calcite, if the Mg:Ca-ratio in solution is greater than 3 to 4. At higher temperatures the critical ratio is smaller (LIPPMANN 1973). The prerequisite for transformation of metastable phases into calcite is a reduction in the Mg:Ca-ratio in the pore solution.

Where sea water remains as the pore solution, diagenetic calcitization can not take place. This is apparently the situation which prevails in the subsurface of atolls in the Pacific, where at depths of over 1000 m unlithified aragonitic calcilutites and calcarenites were encountered (SCHLANGER 1963). On the other hand Mg-calcite and aragonite transform rapidly into calcite when marine carbonate sediments come into the sphere of influence of Mg-free rainwater. This rapid transformation is substantiated experimentally. Fine aragonitic mud is transformed in the presence of pure water into calcite, since a solution, saturated with respect to aragonite, is undersaturated in calcite. The reaction rate depends on temperature. Complete transformation is attained at 100° in one day, at 70° in two days, and at 23° in 100 days. At 3° between 10 and 20% aragonite is transformed after 100 days (FYFE & BISCHOFF 1965; TAFT 1967). The reaction product consists of calcite rhombohedra of equal size, indicating that the rate of crystal growth decreases with crystal size. On the surface of growing calcite crystals apparently growth-inhibiting impurities are accumulated.

Since entrapped sea water inhibits the transformation of aragonite, and since unlithified aragonite sediments occur in Marshall Island drill cores, and since calcitization of marine carbonates by surface waters has been observed in many places, some authors are led to believe that this is a general phenomenon for diagenetic calcitization. It was mentioned previously that it is very unlikely that all marine limestones consisting now of calcite were exposed in the early diagenetic stage to the action of surface waters. Neglecting this unlikelihood, to conclude that diagenetic calcitization takes place by surface waters exclusively is only warranted if surface waters are the only source of pore solutions of low Mg-content. This is certainly not the case.

When discussing pore solutions, it was shown that one of the first changes experienced by sea water entrapped in sediments is a decrease in Mg-content. Waters encountered in all kinds of sediments are deficient in magnesium relative to sea water, although the content of dissolved solids is increased. Magnesium during the course of diagenesis is removed from pore solutions, primarily by chloritization and dolomitization. The compaction flow which permeates the sediments in a subsiding basin, consists of solutions with low Mg-content, which probably are able to cause calcitization of aragonitic carbonates. Just how much solution is supplied from the compaction flow and has permeated a given limestone layer, depends on the geologic situation. As was discussed in early sections, it can be considerable, especially when the carbonate is interlayered with very much thicker clay layers. The Marshall Island example is not typical of most ancient limestones, which were deposited in shelf regions or in geosynclinal troughs. In the subsurface of the Pacific atolls there are no clays. The carbonate sediments lie on basalt. Here no compaction flow from the subsurface can displace sea water entrapped in the pores of calcareous sediments. Primary aragonite is preserved over a time period extending back to the Eocene.

Lithification of limestones goes hand an hand with diagenetic dissolution of Mg-calcite and aragonite and precipitation of calcite. In primary calcitic sediments these reactions do not take place. They can exist also in an essentially unlithified state at greater depths of burial. This applies, for example, to chalks, which consist essentially of calcitic coccolithophorids and foraminifera. In northwest Germany, slightly lithified chalks with porosities between 0.4 and 0.3 occur even to depths of 1500 m.

Mg-calcite is affected first during the calcification of carbonate sediments containing metastable $CaCO_3$-minerals. Pseudomorphs of calcite after Mg-calcite are formed which preserve the fabric and organogenic structures. For example, Bermuda

Island limestones, calcitized by surface waters, have preserved the finest structure of algae, which originally consisted of Mg-calcite (FRIEDMAN 1964).

Whether aragonite dissolution and calcite precipitation take place in microscopically adjacent or distant places, depends, it seems, primarily on the flow rate of pore solution, that is, on the permeability of the rock (FÜCHTBAUER in second part of this series). In highly permeable calcified fossiliferous limestones aragonite shells frequently are found to be selectively dissolved. The dissolved carbonate has been precipitated in the resulting void space or at grain contacts, often in a quite coarsely crystalline form (FRIEDMAN 1964). The large crystals indicate slow crystallization from weakly supersaturated solutions. In only slightly permeable matrix-rich limestones pseudomorphs of calcite after aragonite are found, in which the fine structures of aragonitic shells are preserved (FÜCHTBAUER in second part of this series). Also transformation of very fine grain aragonitic material into calcite may result from dissolution of aragonite and deposition of calcite at the same site (pseudomorphic replacement).

Because of variable solubility of foreign ions in the aragonite and calcite structures, the transition of aragonite to calcite does not proceed isochemically with repect to trace elements. The Sr-content of limestones is affected strongly by the aragonite to calcite transformation. While recent aragonitic sediments contain between 8000 and 10,000 ppm Sr, the Sr-content of most ancient limestones is about 500 ppm. This corresponds to the reduced ability of the calcite structure to accept Sr (KINSMAN 1969). Strontium must have been carried away in the solutions which brought about the aragonite to calcite transformation. According to analyses by HAGLUND, FRIEDMAN & MILLER (1969) uranium appears to have similar behavior. The aragonite to calcite transformation results in a loss of about 50% of the original uranium.

As a result of the greater density of aragonite ($d = 2.93$) pseudomorphic replacement of aragonite by calcite ($d = 2.71$) leads to a volume increase of about 7.5%. The pore volume of an aragonitic sediment decreases about this amount upon calcitization.

5.626 Recrystallization of calcite

In addition to the transformation of metastable phases into calcite, different processes of calcite recrystallization take place in limestones. By dissolution of calcite crystals and growth of new ones, new fabrics form which differ from the primary ones in orientation and size. Recognition of diagenetic recrystallization of calcite is difficult, since the original character of the sediment usually is not known and because often it is not possible to distinguish which fabric formed by the transformation of metastable phases. Usually it is hardly possible to distinguish cement precipitated in the pore space from recrystallized material.

Apart from very early diagenetic micritization, that is, the transformation of coarsely crystalline aragonite and calcite particles into a structureless and very fine grain crystalline aggregate, a process which apparently originates from the activity of algae (descriptions by FÜCHTBAUER in second part of this series), diagenetic recrystallization involves enlargement of original crystallites. Such recrystallization is observed frequently in fossil fragments (echinoderm shells, crinoids, foraminifera), whereby the new crystallites protrude from the fossil shells into the host rock. Some limestones with crystallite dimensions over 50 μm may have been produced from originally very much finer grain calcilutites. On the other hand a great many fine grain limestones with grain sizes of a few μm have been preserved without experiencing recrystallization. To such limestones belong, for example, the Solnhofen lithographic limestones of the south German Jurassic. Which factors promote the diagenetic recrystallization of limestones and which inhibit it, still is not known. BAUSCH (1968) and also FÜCHTBAUER (in second part of this series) presume that the clay content of the sediment plays a role. BAUSCH found in the Frankish Jurassic that limestones with more than 2% insoluble residue were finely crystalline (2—10 μm); those with less than 2% insolubles were coarsely crystalline (50—250 μm). Fine clay films between crystallites may inhibit their growth.

By X-ray methods SIEMES (1966) confirmed a slight orientation of the 5 μm crystallites (c-axis in the bedding plane) in the Solnhofen lithographic limestone of upper Jurassic age. To what extent this is a depositional fabric, forming during sedimentation in the sense of SANDERS (1950) or results from diagenetic recrystallization still is not known.

5.63 Allochemical diagenesis

5.631 Dolomitization and dedolomitization

As already indicated in section 4.3 dolomite does not form directly from sea water under normal conditions, although it is strongly supersaturated with respect to dolomite. The reason is the strong hydration of Mg-ions, by which the nucleation and growth rates of dolomite crystals are reduced considerably.

Dolomite rocks nevertheless form an important fraction of the carbonate rocks of the earth's crust. Fig. 5-75 shows that the abundance of dolomite rocks increases

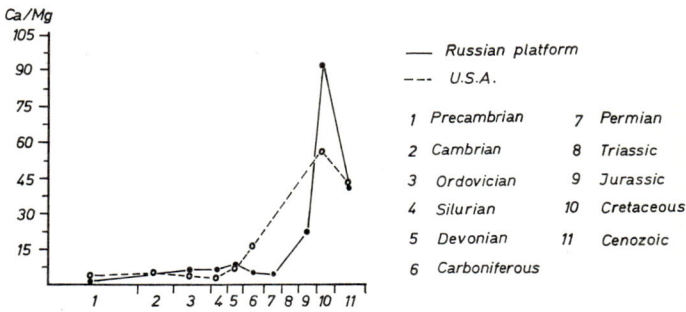

Fig. 5-75. Abundance of dolomitic rocks of different ages.

on the whole with age. Only the Cretaceous formations deviate from this rule by a significantly lower dolomite content. All dolomitic rocks must have formed diagenetically by the action of Mg-bearing solutions on primary precipitated limestones. Fabric, fossil content and geologic association indicate that dolomitization can take place both during very early stages of diagenesis as well as later (FÜCHTBAUER in second part of this series). Since it has not been possible yet experimentally, under the conditions of diagenesis, to convert limestone to dolomite, the diagenetic process of dolomitization must be deduced theoretically and from observations on the fabrics and mineral contents of dolomite rocks. A prolific literature has dealt with the question of dolomitization; which cannot be dealt with in detail here. Recent summary publications have been provided by FAIRBRIDGE (1957), STRAKOV (1956, 1958), USDOWSKI (1967), FRIEDMAN & SANDERS (1967), BATHURST (1971), LIPPMANN (1973).

Of special importance to early diagenetic dolomitization are the observations of recent and subrecent formation of dolomite from primary carbonate sediments in shallow sea and coastal areas of various seas, as well as the Pacific atolls.

During the last decade recent dolomitization has been described from different regions of the earth. Summaries of this early diagenetic dolomitization in the Carribbean, on the Florida Keys, on the Bahama Islands, in the Persian Gulf, in South Australia, in the Karabugas Bay of the Caspian Sea, in the Balkan Lakes and elsewhere, have been given, for example, by FÜCHTBAUER in the second part of this series, by USDOWSKI (1967), by FRIEDMAN & SANDERS (1967) and by BATHURST (1971). In all these cases dolomite forms by the action of evaporated sea water on unlithified carbonate sediments which consist mainly of aragonite. We are dealing here with a reaction occurring in the boundary region between synsedimentary processes (submarine weathering) and diagenesis. Since dolomitization must take place mainly in fresh sediment beneath the surface, we are dealing with early diagenesis. This sort of early diagenetic dolomitization in a saline environment apparently depends not only on an increase in salinity, but especially on a marked reduction in the Ca:Mg-ratio, which was brought about by $CaCO_3$-precipitation and the formation of anhydrite or gypsum. In many occurrences the latter minerals are encountered along with dolomite.

The findings from deep well cores obtained from some Pacific atolls have played a major role in discussions of dolomitization. Here in a thick Tertiary to Recent carbonate sequence individual dolomitized horizons have been encountered. In addition to the well-known section from Funafuti, now drill sections are available from the islands of Kita-Daito-Jima, Bikini and Eniwetok (SCHLANGER 1963). From all of these drillings a more complex picture emerges than had been derived earlier from the much cited Funafuti section. On Funafuti carbonate below a depth of 212 m is dolomitized completely. In contrast on Kita-Daito-Jima the uppermost portion of the section to a depth of 120 m consists of dolomite. On Bikini there are no pure dolomite horizons. On Eniwetok dolomite is found only in the oldest layers drilled (Eocene), about 1500 m below the surface. On the basis of the findings at Funafuti REULING (1934) had advanced the idea, soon taken up by others, that dolomitization came about by increased hydrostatic pressure. The newer data show this is not tenable.

The dolomitized horizons in the atoll sections are distributed not only at different depths, but also their fabrics exhibit considerable variability (SCHLANGER 1963). Especially on Funafuti and Kita-Daito-Jima the pseudomorphic replacement of primary carbonates by dolomite takes place. In doing so the organogenic structures are preserved in detail. On Eniwetok carbonate dissolution and dolomite crystallization were spacially and/or temporally separate processes. In samples from Eniwetok dolomite occurs as relatively large rhombohedra, which either are dispersed in the limestone or, with complete dolomitization, form a sugary texture. In other samples from Eniwetok red algae or mollusc shells are replaced selectively by large dolomite crystals or the framework of corals are dissolved and septal cavities filled with dolomite. Since the red algae consisted primarily of Mg-calcite on Eniwetok the observed preferential dolomitization of algal structures may have been induced by localized Mg-enrichment. One general observation can be applied to the samples from all of the atolls: Dolomite occurs together with Mg-rich and low-Mg calcite, but never with aragonite. Therefore, the dolomitizing solutions have attacked preferentially the metastable aragonite in these sediments.

The early diagenetic dolomitization of primary carbonate sediments took place in these atolls as they did under more saline conditions in coastal and shallow sea areas by the action of solutions. Where, as in the Pacific atolls, primary Mg-calcite was available, a part of the magnesium necessary for dolomitization originated from this material. This primary Mg-supply, however, cannot account for complete dolomitization of very thick strata.

The main amount of dolomite must have been formed by the interaction of $CaCO_3$-minerals and the pore solutions. The reaction can be summarized by the following gross equation:

$$2\ CaCO_3\ (solid) + Mg^{+2} \rightarrow CaMg(CO_3)_2\ (solid) + Ca^{+2}$$

According to this equation dolomitization at the expense of calcite or aragonite is promoted by a low Ca:Mg-ratio. As has been indicated earlier, even in sea water the Ca:Mg-ratio is so low that dolomite should form. Reduction of this ratio below that of sea water will promote diagenetic dolomitization. This can happen either by increasing the Mg-concentration or by lowering the Ca-concentration. There is apparently no way to increase magnesium relative to calcium either in the Pacific atoll sediments or in the shallow sea and coastal regions. In contrast Ca-ion concentration can be reduced.

Where dolomite is formed under high salinity conditions in the shallow sea and coastal regions, paragenesis of anhydrite and dolomite is observed. Here apparently evaporation of sea water proceeds so far that $CaSO_4$ begins to precipitate. As a result the Ca:Mg-ratio is decreased so much that dolomite can form from calcium carbonate.

Another possible means of reducing the Ca:Mg-ratio is provided by increasing the concentration of CO_3^{-2}-ions. Because of the low value of the solubility product $[Ca^{+2}]\ [CO_3^{-2}]$, an increase in CO_3^{-2}-ion concentration must lower the Ca^{+2}-ion concentration. In considering this mechanism, according to LIPPMANN (1973) the following gross equation for dolomitization should be written:

$$CaCO_3 \text{ (solid)} + Mg^{+2} + CO_3^{-2} \rightarrow CaMg(CO_3)_2 \text{ (solid)}$$

Since magnesium is available in appreciable amounts in diagenetically unaltered pore solutions, diagenetic processes which produce CO_3^{-2}-ions, bring about dolomitization. LIPPMANN (1968, 1973) has indicated that oxidation of organic matter by sulfate-reducing bacteria supplies CO_3^{-2}-ions in the following ways:

$$SO_4^{2-} + 2C + H_2O \rightarrow CO_3^{2-} + CO_2 + H_2S$$
$$SO_4^{2-} + 4H_2 + CO_2 \rightarrow CO_3^{2-} + 3H_2O + H_2S$$

Based on these equations (LIPPMANN 1973) the oxidation of 13 g of organic carbon or 32 g of a carbohydrate would give rise to 100 g dolomite.

That the organic matter in carbonate sediments on the average is oxidized considerably more than in argillaceous sediments is apparent from a comparison of organic contents of fresh and diagenetically altered clays and carbonates. According to the summary by HUNT (1961) (table 5-17) recent carbonate sediments contain about the same amount of organic matter as recent clays. In contrast carbonate rocks contain significantly less organic matter than shales.

Table 5-17. Organic content of recent an ancient sediments (HUNT 1961).

	Wt. % Organic matter
Recent clays[2]	1,5
Shales[2]	2.0
Recent carbonates[1]	1.7
Limestones[2]	0.2

[1] Gulf of Batabano, Cuba.
[2] Average for samples from different regions.

The amount of dolomite which can form as a result of the activity of sulfate-reducing bacteria is limited on the one hand by the magnesium content of the pore solution and on the other hand by the organic content of the sediment. Preferentially dolomitized layers in the Pacific atoll drill cores originally may have contained especially large amounts of organic matter. It is questionable whether the original organic content in a bed could have been sufficient to account for its complete dolomitization. In no case can the Mg-content of the pore solution in a sediment layer be sufficient for complete dolomitization. Therefore, magnesium must have been introduced into the dolomitized layer by pore solution flow or by diffussion.

Introduction of magnesium by flowing pore solutions must be assumed especially for the late diagenetic dolomitization of limestones. Numerous examples of dolomitization, clearly extending out from joints, which are connected to anticlines or are inclined to the primary bedding (see FÜCHTBAUER in second part of this series), show that late diagenetic dolomitization in lithified sediments is common. In these cases the sediments must have been permeated by pore solutions which introduced magnesium. To a considerable extent compaction flow from clay sediments is involved. Nothing is known about the chemistry of these solutions. One can observe only that solutions contained now within the pore space of consolidated sediments are considerably impoverished in magnesium relative to previously entrapped sea water (section 5.32). They represent the residual solutions resulting from entrapped sea water after migration through a thick sedimentary sequence and after dolomitization.

How these transformations take place kinetically, that is, what factors may have facilitated the inhibited transformation of limestone to dolomite, are not known. CO_3^{-2}-production by sulfate-reducing bacteria or the reduction of Ca-concentration by anhydrite precipitation, which are important during early diagenetic dolomitization, certainly are not involved in later diagenetic dolomitization. Inasmuch as late diagenetic dolomitization takes place at greater depths, then temperature dependence of the Ca:Mg-ratios of solutions in equilibrium saturation with calcite and dolomite (fig. 4-7) may be important. Rising temperature displaces the Ca:Mg-ratio of equilibrium solutions to smaller values. Furthermore, the time factor is important. In geologically long periods of time thermodynamic equilibrium probably is maintained, in spite of the low reaction rate which produces dolomite by interaction of limestone with Mg-bearing solutions. The slow reaction rate makes it impossible to observe this reaction experimentally or when normal sea water reacts with recent carbonate sediments.

The mineralogical composition of the limestone plays a role in diagenetic dolomitization. It was observed in the Pacific atoll sections that aragonitic layers are dolomitized, while calcitic horizons were not altered. Also during late diagenesis aragonitic limestones are susceptible preferentially to dolomitization. The rarity of dolomite rocks in Cretaceous formations (fig. 5-75) may be related to the fact that Cretaceous limestones are formed primarily of calcitic foraminifera and coccolithophorids.

The isotopic composition of carbon and oxygen in dolomites can provide information on their origin under certain conditions. BERNER (1965 a) ascertained an enrichment of O^{18} in Funafuti, Kita-Daite-Jima, and Eniwetok dolomites, and concluded that dolomitization results from saline solutions derived from the evaporation of sea water. Dolomites forming by carbonate production by sulfate-reducing bacteria should be enriched in light C^{12}, corresponding to the isotopic composition of organic matter. Actually an enrichment in light carbon has been observed in various dolomites (SPOTTS & SILVERMAN 1966; MURATA, FRIEDMAN & MADSEN 1969; DEUSER 1970; TAN & HUDSON 1971).

Dolomite rocks are frequently more porous than surrounding limestones (see FÜCHTBAUER in second part of this series). Since the limestone to dolomite transformation with its addition of magnesium and removal of calcium is coupled with a volume decrease, secondary porosity can develop from late diagenetic dolomitization of a previous lithified limestone. Dolomitization by this reaction results in a volume loss of 12.9 % for calcite and 5.8 % for aragonite. In some dolomites appreciably higher porosities are found (up to 0.30, see Füchtbauer in second part of this series). This can indicate either dolomitization of still highly porous sediment or dissolution of carbonate in the course of dolomitization. If dolomitization on the other hand, takes place by addition of carbonate and magnesium, for example, from the activity of sulfate-reducing bacteria, a considerable volume increase occurs (74.4 % from calcite, 88.4 % from aragonite, as related to the original volume). Porous dolomites formed in this manner can occur only when dolomitization takes place in still highly porous sediment (Lippmann 1973).

Some dolomites contain more calcium (up to 60 mol%) than corresponds to the ideal composition. The Ca-excess is noted by an increase in lattice parameters and can be determined by X-ray methods from interplanar spacings (Goldsmith & Graf 1958). Ca-rich dolomites exhibit broadened basal reflections which indicate irregularities in the ordering of successive Ca- and Mg-planes in the structure. According to studies by Goldsmith & Graf (1958), as well as by Füchtbauer & Goldsmith (1965), dolomites from evaporite rock series (for example, Zechstein dolomites of northwest Germany) as a rule have ideal stoichiometric compositions and are well-ordered. Dolomites in normal limestones (for example, the upper Jurassic of northwest Germany) usually have excess calcium and are disordered. Recent dolomites formed in highly saline environments have stoichiometric compositions. Ordered, stoichiometric dolomites appear to be typical of early diagenetic formation under saline conditions; disordered dolomite with exess calcium characterize later diagenetic formation.

The diagenetic dissolution of dolomite is referred to as dedolomitization. Usually it proceeds simultaneously with calcite precipitation (recalcitization). It always involves late diagenetic processes, which occur as a reversal of dolomitization, by solutions with high Ca:Mg-ratios. Different examples of dedolomitization and recalcitization have been described by Friedman & Sanders (1967) and by Füchtbauer (in second part of this series). Rocks which have been dedolomitized without a corresponding deposition of calcite contain void spaces formed from completely or partially dissolved dolomite rhomohedra. The upper Jurassic limestones of Frankenalp in southern Germany described by Bausch (1965) are examples. In most cases the dissolved dolomite is replaced by calcite, which forms either finely crystalline pseudomorphs after dolomite crystals or larger crystals which enclose dolomite relics. In the first case the solution may have been more supersaturated with respect to calcite than in the second case.

From observations of Lang (1964) and Bausch (1965) in the upper Jurassic of southern Germany, dedolomitization and recalcitization can involve large volumes of rock. From dolomitic limestones, pure coarse grain limestones form which occur especially frequently in the vicinity of larger dolomite bodies. Lang and Bausch presume that here, during a late diagenetic stage, material transfer has taken place, whereby from a limestone with more or less uniformly distributed dolomite content, a rock body is generated which consists both of pure calcitic and completely dolomitized regions.

Dedolomitizing and recalcitizing solutions can have various origins. In saline rocks the occurrence of sulfate minerals attests to the action of sulfate-bearing solutions which react with dolomite to form calcite and dissolved $MgSO_4$ (FRIEDMAN & SANDERS 1967). In most occurrences dedolomitization and recalcitization must have been caused by descending surface waters. This applies for the upper Jurassic of southern Germany, where FRITZ (1966) confirmed a reduction in dedolomitized limestones of O^{18} and C^{13} relative to the dolomites and primary limestones.

5.632 Formation of SiO_2-concretions (chert)

Chert concretions are common constituents of carbonate rocks. They consist of chalcedony or quartz and sometimes quartzine. The very variable and irregular forms of these formations exhibit a relation to the bedding inasmuch as they are flattened or elongate in the bedding plane and are concentrated in narrow layers parallel to the bedding. Concretionary layers are interrupted by denser, concretion-free zones. In many limestones, such as the upper Senonian chalks of central Europe, the concretionary layers lie at regular distances from one another. In some limestones the chert concretions in a layer can merge into a more or less continuous layer. Less often chert bodies run obliquely to the bedding or joints are filled with chert.

Concentric structures, which indicate gradual growth, or layered structures, like those of calcareous concretions in clays, are rare in chert concretions. Most cherts are structureless internally. Usually they contain various calcareous or siliceous fossil remains, polycrystalline carbonate particles and single dolomite and calcite crystals.

Earlier a number of authors advocated that chert concretions formed from SiO_2-gels which precipitated from sea water onto the surface of sediments (TARR 1917, 1926). This is not tenable, because of the very low SiO_2-content of sea water. All observations of the structures of chert concretions indicate that they form diagenetically under a certain amount of sediment burial.

In the literature various opinions have been offered as to when diagenetic formation of chert concretions takes place. This is particularly true for the flints in the chalks of the Baltic region. Especially GRIPP (1933, 1954) advocated a very late origin for flints through the influence of surface waters. An early diagenetic origin for most flints was advanced by W. WETZEL (1957), ILLIES (1949) and MÜLLER (1956). Observations cited in the recent literature suggest that the formation of most chert concretions, like the carbonate concretions in clays, was initiated very early and essentially concluded during an early diagenetic stage. Later on, even in finally lithified limestones, mobilization of SiO_2 occurred, leading to redistribution and late enrichment.

Very early formation of SiO_2-concretions is demonstrated by the preservation of skeletonless microfossils, for example, flagellates with fine cilia, as decribed by O. WETZEL (1933) from flints of the north German chalks. In addition, all observations indicate that most chert concretions were formed when the mechanical compaction of sediments was still in effect and before final lithification by chemical processes was completed. Indications are found both of mechanical compaction before concretion formation and of a continuation of compaction after their formation. To the former belong compressed fossil remains in chalk flints, which indicate formation at a certain depth below the surface of the soft sediment (GRIPP 1933, 1954; MÜLLER 1956). Other observations show that compaction still had not progressed far when some concretions formed. The SiO_2-mass is found often as a filling in echinoderm or other invertebrate shells in cavities that could have existed only at slight depths of burial. That compaction continued after concretion formation is indicated by the occurrence of unlithified or slightly lithified chalk fragments within flint concretions, surrounded by chalk that is lithified by pressure solution and reorientation of sediment grains (ILLIES 1954). Other indications of continuing compaction after concretion formation are the flattening and distortion of concretions, as described by MÜLLER (1956) from the north German upper Cretaceous. Also slickenside-like streaks on flint concretions were related by MÜLLER (1956) to the continuing compaction of the enclosing sediments.

Observations indicating early diagenetic formation of chert concretions are known from other limestones also. WILSON (1966) could distinguish the following phases of concretion formation in upper Jurassic limestones of southern England. The earliest formations are pseudomorphs of chalcedony after aragonitic shells, which preserve the finest organogenic structures, and therefore, were formed before aragonite dissolution. Later stages of SiO_2-deposition are indicated by the silification of ooids, which interpentrate each other as a result of earlier pressure solution, and by SiO_2-precipitation in pores which existed before formation of a fine grain cement, as well as the filling of cavities formed by the dissolution of aragonitic shells. Still later SiO_2-deposition replaced diagenetically formed calcite cement.

Of certain late diagenetic origin are the cracks and joints filled with chert in some limestones, which are oblique to the bedding (so-called fissure chert of the upper Senonian chalk of the Baltic region). They show that SiO_2 is mobile even in lithified sediment. Generally these cherts are much less common than the bedded chert nodules.

Since sea water entrapped as pore solution contains very little SiO_2 and a supply from other sources is concluded, at least for early diagenetic stages, the SiO_2 must originate within the sediment itself. SiO_2-suppliers are the shells and skeletal remains of marine organisms which induce precipitation of amorphous SiO_2 from sea water. Involved are diatoms, certain groups of flagellates, radiolaria, and siliceous sponges (see MÜLLER in the second part of this series). Calcitized and partially dissolved shells and skeletal parts of these organisms in the limestone surrounding the chert concretions indicate that the SiO_2, originally dispersed more or less uniformly in the sediment, first is dissolved and then accumulated at certain centers. The amorphous SiO_2 in shells or skeletal remains was isolated from interaction with sea water and pore solutions as long as they were sheathed by protecting organic membranes. As soon as these were destroyed, the SiO_2 could dissolve and be transported. Solution, transport, and deposition are determined by the dependence of SiO_2-solubility on physical-chemical factors, especially temperature and pH. As has been shown in an earlier section, SiO_2- solubility below pH 8 is practically pH-independent and rises markedly with pH above 8. Solubility increases with increasing temperature. A pore solution with the composition of sea water is undersaturated with respect to amorphous SiO_2. The dissolution of siliceous shells and skeletal remains in such a solution is expected, but it is necessary still to explain the concretionary precipitation of dissolved SiO_2.

According to our classification of concretions (section 5.52), chert concretions are in part those which form by filling up existing cavities, and in part those which occur by metasomatic replacement of enclosing limestone. In contrast to the carbonate concretions of clays, chert concretions on the first type are formed not by filling the pores in the sediment, but by filling up larger cavities, in the shells of echinoderms or other organisms. The predominant number of chert concretions form metasomatically by replacement of calcium carbonate, during which solutions introduce and precipitate SiO_2 as carbonate is dissolved and carried away (silicification).

Cherts consist of chalcedony or coarser crystalline quartz, less commonly quartzine. In many cases, for example, where finely crystalline quartz preserves the replaced shell and skeletal structures, the SiO_2 may have been deposited originally as crystalline quartz. Gel-like surface structures of many chert concretions, which seem to have been influenced by the effects of surface tension or produced by deformation or originally soft masses under the influence of load pressure, have been interpreted as the result of initial deposition in the form of a gel. Then the transformation into finely crystalline quartz must have taken place later. However, indications of significant shrinkage, which must accompany transformation of a hydrous gel into a crystalline quartz mass, are never found.

Conspicuous is the association of chert concretions with organic remains observed in many limestones, such as the upper Senonian chalk of the Baltic region (see, for example, MÜLLER 1956). The forms of concretions are determined in many cases by remains of organisms which serve as concretion centers. This is especially so for sponges, whose diversified forms are preserved in concretions. Various authors have connected the precipitation of chert concretions with the chemical conditions which prevail where organic matter is decomposed.

As was explained in the case of concretion formation in clays, the formation of chert concretions can take place only in a sediment that is physical-chemically inhomogeneous. Conditions must prevail at the center of concretionary formation which leads to dissolution of carbonate and precipitation of silica, while in the surrounding sediment the opposite is happening. Acting here apparently are the often mentioned opposing effects of pH on the solubility of calcium carbonate and SiO_2. According to this view, at the site of SiO_2-precipitation a lower pH should have prevailed than in the mass of the sediment. If the often observed association of chert concretions and organic remains indicates a causative relation, one could assume, that decomposition of organic matter in the concretionary limestone with the formation of CO_2 takes place, through whose influence SiO_2 is deposited and carbonate dissolved (see, for example, ILLIES 1949). If this interpretation is correct, the decomposition of organic matter in these limestones must have proceeded differently than in argillaceous sediments. In the latter case it was assumed that organic matter decomposed forming ammonia and amine, giving an alkaline reaction.

ILLIES (1949) held the compaction flow responsible for the formation of the bedded flint concretions in the Baltic chalks. At greater depths the slightly alkaline pore solutions in equilibrium with the carbonate should dissolve siliceous organic remains distributed in the sediment. The ascending solution at a given depth beneath the sediment surface, where decomposing organic remains release CO_2, should deposit dissolved SiO_2, simultaneously dissolving carbonate. This happened mainly in the vicinity of large accumulations of organic matter, where the CO_2-concentration is highest.

Near a pH of 8, slight variations in pH can alter the pH in favor of either SiO_2 or carbonate. Repeated alternations between replacement of carbonate by SiO_2 and of SiO_2 by carbonate, can be explained by slight changes in pH of the slightly alkaline pore solution flowing through the sediment. These phenomena have been described by WILSON (1966) in upper Jurassic limestones of southern England and by WALKER (1962) from Ordovician limestones in Wisconsin.

As in the case of carbonate concretions in clays, it is questionable whether precipitation of chert concretions can be related to the precipitating effect of organic matter decomposition alone. The organic matter originally contained in a concretionary layer will not have supplied enough CO_2 to precipitate all of the concretionary SiO_2. Still other mechanisms must be assumed to have affected a redistribution of previously widely distributed SiO_2 into concretions and layers. In any case a mixture of phases in which amorphous SiO_2 occurs in finely distributed form is thermodynamically less stable than one in which the SiO_2 forms large masses with a lower specific surface area. The diagenetic process, in ways still not understood in detail, has produced a thermodynamically more stable system from the original mixture.

5.633 Neoformation of silicates

Diagenetically formed silicates are rarer in carbonate rocks than in clastic ones. Since the pore space in carbonate rocks generally disappears early, usually early diagenetic formation must be involved. Alkali feldspars, as they also occur in argillaceous and sandy rocks (see section 5.42) are the most important neoformed silicates in limestones (concerning feldspars in limestones see FÜCHTBAUER in the second part of this series; also CROWLEY 1939; FÜCHTBAUER 1948, 1950, 1972; TOPKAYA 1950; CAROZZI 1953; BASKIN 1956; H. MÜLLER 1958; SCHÖNER 1960; KASTNER & WALDBAUM 1968).

That feldspars form diagenetically, is recognized by the usually idiomorphic development of small (< 0.2 mm) crystals and by carbonate inclusions. Albites are always very pure soda feldspar (< 3 mol% Or, < 1 mol% An, BASKIN 1956), tabular after (010) or (001) and frequently twinned as quadruplicates after the Roc

Tourné law. Low temperature albite is involved (incomplete Al-Si ordering), which is demonstrated by a large optic angle ($2\,V_z = 90—85°$, Füchtbauer 1950) and a γ^*-angle of the reciprocal lattice between $89—91°$ (Baskin 1956). They are usually smaller than the albites and correspondingly, when they are idiomorphic, are usually morphologically adularia. The optic angle is quite constant ($2\,V_x = 43°$, Füchtbauer 1950; Müller 1958; Schöner 1960).

The alkali content of diagenetic feldspars may come from the pore solutions. Therefore, it is understandable that Na-feldspar is more common in normal marine limestones than K-feldspar, although both feldspars occur together in some limestones. Observations by Füchtbauer (1972) on carbonate and sulfate sediments of the Zechstein formation of northwest Germany, indicate that feldspars form during an early stage of diagenesis. The chemistry of the depositional environment is important, therefore. In the evaporite cycles of this formation, the following sequence of diagenetic neoformations is related to increasing sea water salinity: sodium feldspar-quartz-potassium feldspar. Anhydrite, precipitated from strongly concentrated sea water, contains relatively appreciable K-feldspar.

Of the other silicates tourmaline should be mentioned. It occurs sometimes as long prismatic crystals in limestone. Tourmaline appears to be especially common in evaporite series, as in the dolomites of the Zechstein formation (Füchtbauer in the second part of this series).

References and Authors' Index

The numbers on the right within brackets indicate the page on which each reference was cited.

ALEKIN, A. A. (1962): Grundlagen der Wasserchemie (translated from Russ.). — Leipzig. [115]

ALEXANDER, G. B., HESTON, W. M. & ILER, H. K. (1954): The solubility of amorphous silica in water. — J. Phys. Chem., 58, 453—455. [37]

ALTHAUS, E. (1966 a): Die Bildung von Pyrophyllit und Andalusit zwischen 2000 und 7000 bar H_2O-Druck. — Naturwiss. 53, 105—106. [303]

— (1966 b): Der Stabilitätsbereich des Pyrophyllits unter dem Einfluß von Säuren. — Contr. Miner. Petrol., 13, 31—51. [303]

AMBERG, C. H. & MCINTOSH, R. (1952): A study of adsorption hysteresis by means of length changes of a rod of porous glass. — Canad. J. chem., 30, 1012—1032. [11]

AMBRASEYS, N. N. (1960): The seismic sea wave of July 9, 1956, in the Greek Archipelago. — J. Geophys. Res., 65, 1257—1265. [96]

ARRHENIUS, G. (1963): Pelagic Sediments. — In: M. N. HILL (General ed.), The Sea, Vol. 3: The Earth beneath the Sea. — Interscience Publ., New York, London, 655—727. [82, 123 f, 145]

ASCHENBRENNER, B. C. (1956): A new method of expressing particle sphericity. — J. sediment. Petrol., 26, 15—31. [102, 112]

ATHY, L. F. (1930): Density, porosity and compaction of sedimentary rocks. — Bull. amer. Assoc. Petrol. Geol., 14, 1—24. [272 ff]

AWASTHI, N. (1961): Authigenic tourmaline and zircon in the Vindhyan formations of Sone valley, Mirzapur district, Uttar Pradesh, India. — J. sediment. Petrol., 31, 482—484. [259]

BAAS BECKING, L. G. M., KAPLAN, I. R. & MOORE, D. (1960): Limits of the natural environment in terms of pH and oxidation-reduction potentials. — J. Geol., 68, 243—284. [19 f]

BAGNOLD, R. A. (1954): Experiments on a gravity-free dispersion of large solid spheres in a Newtonian fluid under shear. — Proc. roy. Soc. London, A 255, 49—63. [70, 74]

— (1957): The flow of cohesion-less grains in fluids. — Phil. Trans. roy. Soc. London, A 249, 235—297. [62 f, 70 ff, 74 f]

— (1962): Auto-suspension of transported sediment; turbidity currents. — Proc. roy. Soc. London, 265, 315—319. [97]

— (1963): Mechanics of marine sedimentation. — In: M. N. HILL (ed.), The Sea, Vol. 3. — Interscience Publ., New York, London, 507—528. [82, 88 f]

— (1965): The physics of blown sand and desert dunes. — Methuen & Co., London. [146 f, 150 ff, 154 ff, 158, 161]

— (1966): An approach to the sediment transport problem from general physics. — U. S. geol. Surv. prof. Pap., 422 — I. [71 f, 77 ff]

BANGHAM, D. H. (1934): The swelling of charcoal IV. — Proc. roy. Soc. London, A, 147, 175—188. [11]

BANGHAM, D. H., FAKHOURY, N. & MOHAMED, A. F. (1932): The swelling of charcoal II. — Proc. roy. Soc. London, A, 138, 162—168. [11]

— — — (1935): The swelling of charcoal III. — Proc. roy. Soc. London, A, 147, 152—175. [11]

BARREL, J. (1925): Marine and terrestrial conglomerates. — Bull. geol. Soc. Amer., 36, 279—342. [109]

BARTH, T. F. W. (1949): Frequency distribution of the minerals in two petrographic provinces. — J. Geol., 57, 55—61. [1]

BARTLETT, G. A. & GREGGS, R. G. (1969): Carbonate sediments: oriented lithified samples from the North Atlantic. — Science, 166, 740—741. [309]

BASKIN, Y. (1965): A study of authigenic feldspars. — J. Geol., **64**, 132—155. [251, 327 f]
BASSET, W. A. (1963): The geology of vermiculite occurrences. — Clays and Clay Minerals. Proc. 10th nat. Conf. on Clays and Clay Minerals. Pergamon Press, Oxford, 60—69. [42]
BATES, TH. F. (1962): Halloysite and gibbsite formation in Hawaii. — Clays and Clay Minerals. Proc. 9th nat. Conf. on Clays and Clay Minerals. Pergamon Press, Oxford, 315—328. [44 f, 51 f]
— (1964): Geology and mineralogy of the sedimentary Kaolins of the SE United States. A review. — Clays and Clay Minerals. Proc. 12th nat. Conf. on Clays and Clay Minerals. Pergamon Press, Oxford, 177—194. [45]
BATHURST, R. G. C. (1971): Carbonate sediments and their diagenesis. — Elsevier, Amsterdam, London, New York, 620 pp. [304, 310, 313, 319]
BAUSCH, W. M. (1965): Dedolomitisierung und Recalcitisierung in fränkischen Malmkalken. — N. Jb. Miner. Mh., 75—82. [323]
— (1968): Clay content and calcite crystal size of limestones. — Sedimentology, **10**, 71—75. [318]
BAYLISS, P. & LOUGHNAN, F. C. (1964): Mineralogical transformations accompanying the chemical weathering of clay-slates from New South Wales. — Clay Minerals Bull., **5**, 353—363. [45]
BELL, D. C. & GOODELL, H. G. (1967): A comparative study of glauconite and the associated clay fraction in modern marine sediments. — Sedimentology, **9**, 169—202. [120 f]
BERGMANN, W. (1963): Geochemistry of lipids. — In: E. INGERSON (ed.): Organic geochemistry. — Pergamon Press, Oxford, 503—542. [297]
BERNER, R. A. (1964): An idealized model of dissolved sulfate distribution in recent sediments. — Geochim. Cosmochim. Acta, **28**, 1497—1503. [238]
— (1965 a): Dolomitization of the mid-Pacific atolls. — Science, **147**, 1297—1299. [322]
— (1965 b): Activity coefficients of bicarbonate, carbonate and calcium ions in sea water. — Geochim. Cosmochim. Acta, **29**, 947—965. [177 ff, 180 f]
— (1967): Comparative dissolution characteristics of carbonate minerals in the presence and absence of aqueous magnesium ion. — Amer. J. Sci., **265**, 45—70. [182]
— (1968 a): Rate of concretion growth. — Geochim. Cosmochim. Acta, **32**, 477—483. [293]
— (1968 b): Calcium carbonate concretions formed by the decomposition of organic matter. — Science, **159**, 195—197. [297]
BERTHOIS, L. & PORTIER, J. (1956): Recherches expérimentales sur la mode d'usure des graviers. — C. r. Acad. Sci. Paris, **243**, 1778—1781. [104]
— — (1957 a): Recherches expérimentales sur le façonnement des graviers de quartz. — C. r. Acad. Sci. Paris, **244**, 362—364. [104]
— — (1957 b): Recherches expérimentales sur le façonnement des grains de sable quartzeux. — C. r. Acad. Sci. Paris, **245**, 1152—1154. [108]
BERTHOUD, A. (1912): Théorie de la formation des faces d'un cristal. — J. Chim. Phys., **10**, 624—635. [22]
BIEGER, T. (1952): Verwitterung von Albit. — Diss. Göttingen, 37 pp., cit. in CORRENS, Clay Min. Bull., **4**, 249—265. [28]
BILLINGS, G. K., HITCHON, B. & SHAW, D. R. (1969): Geochemistry and origin of formation waters in the Western Canada Sedimentary Basin, 2: Alkali metals. — Chem. Geol., **4**, 211—223. [211, 214]
BISCAYE, P. E. (1965): Mineralogy and sedimentation of recent deep-sea clay in the Atlantic Ocean and adjacent seas and oceans. — Bull. geol. Soc. Amer., **76**, 803—832. [122]
BISCHOFF, J. L. & FYFE, W. S. (1968): Catalysis, inhibition, and the calcite-aragonite problem. I. The aragonite-calcite transformation. — Amer. J. Sci., **266**, 65—79. [179]
BLACKWELDER, E. (1933): The insolation hypothesis of rock weathering. — Amer. J. Sci., **26**, 97—113. [13]
BLATT, H. (1959): Effect of size and genetic quartz type on sphericity and form of beach sediments, Northern New Jersey. — J. sediment. Petrol., **29**, 197—206. [110]
BLATT, H. & CHRISTIE, J. M. (1963): Undulatory extinction in quartz of igneous and metamorphic rocks and its significance in provenance studies of sedimentary rocks. — J. sediment. Petrol., **33**, 559—579. [4 f]
BLISSENBACH, E. (1952): Relation of surface angle distribution to particle size distribution on alluvial fans. — J. sediment. Petrol., **22**, 25—28. [109]
BLUM, R. (1843): Die Pseudomorphosen des Mineralreichs. — Schweizerbart, Stuttgart. [241]

BÖHMEKE, A. (1952): Verwitterung von Muskovit. — Diss. Göttingen 1952, 28 pp., cit. in CORRENS, Clay Min. Bull., 4, 249—265. [31]
BONATTI, E. & ARRHENIUS, G. (1965): Eolian sedimentation in the Pacific off northern Mexico. — Mar. Geol., 3, 337—348. [122]
BOSTRÖM, K. (1967): The problem of excess manganese in pelagic sediments. — PH. A. ABELSON (ed.): Researches in geochemistry, Vol. 2, 421—452. [124]
BOSWELL, P. G. H. (1933): On the mineralogy of sedimentary rocks. — Murby, London. [259]
BOUMA, A. H. & BROUWER, A. (1964): Turbidites. — Elsevier, Amsterdam, London, New York. [93]
BOWEN, H. J. M. (1966): Trace elements in geochemistry. — Academic Press, New York, 241 pp. [297]
BRAITSCH, O. (1962): Entstehung und Stoffbestand der Salzlagerstätten. — Mineralogie und Petrographie in Einzeldarstellungen, Springer, Berlin, Göttingen, Heidelberg. [167]
BREDEHOEFT, J. D., BLYTH, C. R., WHITE, W. A. & MAXEY, G. B. (1963): Possible mechanism for concentration of brines in subsurface formations. — Bull. amer. Assoc. Petrol. Geol., 47, 257—269. [217, 224]
BREHLER, B. (1951): Über das Verhalten gepreßter Kristalle in ihrer Lösung. — N. Jb. Miner. Mh., 110—131. [242 ff]
BRIGGS, L. J., MCCULLOCH, D. S. & MOSER, F. (1962): The hydraulic shape of sand particles. — J. sediment. Petrol., 32, 645—656. [63]
BRINDLEY, G. W. & DE KIMPE, C. (1961): Identification of clay minerals by single crystal electron diffraction. — Amer. Miner., 46, 1005—1016. [45]
BROWN, G. (1953): The dioctahedral analog of vermiculite. — Clay Minerals Bull., 2, 64—69. [43]
— (ed.) (1961): The X-ray identification and crystal structures of clay minerals. — Miner. Soc. London. [39]
BRYANT, J. P. & DIXON, J. P. (1964): Clay mineralogy and weathering of a red-yellow podzolic soil from quartz mica schist in the Alabama Piedmont. — Clays and Clay Minerals. Proc. 12th nat. Conf. on Clays and Clay Minerals. Pergamon Press, Oxford, 509—521. [45]
BRYDON, J. E., CLARKE, J. S. & OSBORNE, V. (1961): Dioctahedral chlorite. — Canad. Miner., 6, 595—609. [43]
BRYDON, J. E. & KODOMA, H. (1966): The nature of aluminum hydroxide-montmorillonite complexes. — Amer. Miner., 51, 875—889. [43]
BÜLOW, K. v. (1951): Schwermineralseifen an der mecklenburgischen Ostseeküste. — Arch. Lagerstättenforsch., H. 81, 1—63. [140]
BURRI, C. (1929): Sedimentpetrographische Untersuchungen an alpinen Flußsanden. — Schweiz. Miner. Petrogr. Mitt., 9, 205—240. [111]
BURST, J. F. (1958): "Glauconite" pellets: their mineral nature and applications to stratigraphic interpretation. — Bull. amer. Assoc. Petrol. Geol., 42, 310—327. [120]
— (1958): Mineral heterogeneity in "glauconite" pellets. — Amer. Miner., 43, 481—497. [120]
— (1959): Postdiagenetic clay mineral environmental relationships in the Gulf Coast eocene. — Clays and Clay Minerals. Proc. 6th nat. Conf., 1951, 327—341. [300]
— (1969): Diagenesis of Gulf Coast clayey sediments and its possible relation to petroleum migration. — Amer. Assoc. Petrol. Geol. Bull., 53, 73—93. [300 f]
BUSER, W. & GRÜTTER, A. (1956): Untersuchungen an Mangansedimenten. — Schweiz. Miner. Petrogr. Mitt., 36, 49—62. [123]
BUSHINSKY, G. I. (1967): Stylolites. — Izvest. Akad. Nauk. SSSR, Ser. Geol. 1961 (after TRURNIT 1967). [313]

CAILLEUX, A. (1942): Les actions éoliennes périglaciaires en Europe. — Mém. Soc. géol. France, n. s. 21, No. 46, 1—176. [157]
— (1952): Morphoskopische Analyse der Geschiebe und Sandkörner und ihre Bedeutung für die Paläoklimatologie. — Geol. Rdsch., 40, 11—19. [102 f, 111]
CAILLEUX, A. & TRICART, J. (1959): Initiation à l'étude des sables et des galets. — Paris, T. I, 376 pp. [157]
CARMAN, P. C. (1937): Fluid flow through granular beds. — Trans. Inst. Chem. Eng., 15, 150. [200]
— (1948): Some physical aspects of water flow in porous media. — Disc. Faraday Soc., 3, 72—77. [200]

— (1956): The flow of gases through porous media. — Academic Press, New York. [195, 200]

CAROZZI, A. (1953): Données micrographiques sur le crétacé supérieur helvétique. — Bull. Inst. Nat. Genevois, **76**, 1—76. [327]

CAROZZI, A. V. (1960): Microscopic sedimentary petrography. — John Wiley & Sons, New York, London, 485 pp. [247, 250]

CARRIGY, M. A. & MELLON, G. B. (1964): Authigenic clay mineral cements in cretaceous and tertiary sandstones of Alberta. — J. sediment. Petrol., **34**, 461—472. [252f]

CARROL, D & STARKEY, H. C. (1959): Effect of sea water on clay minerals. — Clays and clay minerals, Proc. 7th nat. Conf., Washington. Pergamon Press, Oxford, 80—101. [118]

CHAVE, K. E. (1954): Aspects of biogeochemistry of magnesium. I. Calcareous marine organisms; II. Calcareous sediments and rocks. — J. Geol., **62**, 266—283; 587—599. [179]

— (1960): Evidence on history of sea water from chemistry of deeper subsurface waters of ancient basins. — Bull. amer. Assoc. Petrol. Geol., **44**, 357—370. [208]

CHEBOTAREV, I. I. (1955): Metamorphism of natural waters in the crust of weathering. — Geochim. Cosmochim. Acta, **8**, 22—48; 137—170; 198—212. [208]

CHENNAUX, G. & DUNOYER DE SEGONZAC, G. (1967): Etude pétrographique de la pyrophyllite du Silurien et du Dévonien au Sahara. Répartition et origine. — Bull. Serv. Carte géol. Als. Lorr., Strasbourg, **20**, 159—210. [303]

CHILINGAR, G. V. & KNIGHT, L. (1960): Relationship between pressure and moisture content of kaolinite, illite and montmorillonite clays. — Bull. amer. Assoc. Petrol. Geol., **44**, 100—106. [265]

CIFELLI, R., BOWEN, V. T. & SIEVER, R. (1966): Cemented foraminiferal oozes from the Mid-atlantic ridge. — Nature, **209**, 32—34. [309]

CLARKE, F. W. & WASHINGTON, H. S. (1924): The composition of the earth's crust. — U.S. geol. Surv. prof. Pap. 127. [1]

CLAYTON, R. N., FRIEDMAN, I., GRAF, D. L., MAYEDA, T. K., MEENTS, W. F. & SHIMP, N. F. (1966): The origin of saline formation waters. I: Isotopic composition. — J. geophys. Res., **71**, 3869—3882. [223, 224]

COOMBS, D. S., ELLIS, A. J., FYFE, W. S. & TAYLOR, A. M. (1959): The zeolite facies, with comments on the interpretation of hydrothermal syntheses. — Geochim. Cosmochim. Acta, **17**, 53—107. [254ff]

COOMBS, D. S. & WHETTEN, J. T. (1961): Composition of analcime from sedimentary and burial metamorphic rocks. — Bull. geol. Soc. Amer., **78**, 269—282. [256]

CORRENS, C. W. (1937): Die Sedimente des äquatorialen Atlantischen Ozeans. Geochemie der Sedimente. — Wiss. Ergebn. dt. Atlant. Expedition Meteor 1925—1927, Bd. III, 3. Teil, 205—245. [182]

— (1939): Die Sedimentgesteine. In: BARTH, T. F. W., CORRENS, C. W. & ESKOLA, P.: Die Entstehung der Gesteine. — Springer, Berlin. [14, 114, 116, 182, 299]

— (1940): Die chemische Verwitterung der Silikate. — Naturwiss., **28**, 369—376. [25]

— (1949): Growth and dissolution of crystals under linear pressure. — Disc. Faraday Soc., **5**, 267—271. [242, 299]

— (1950): Zur Geochemie der Diagenese: I. Das Verhalten von $CaCO_3$ und SiO_2. — Geochim. Cosmochim. Acta, **1**, 49—54. [250, 296]

— (1961): The experimental weathering of silicates. — Clay miner. Bull., **4**, 249—265. [25]

— (1963): Experiments on the decomposition of silicates and discussion of chemical weathering. — Clays and Clay minerals, Vol. 10, Proc. 10th nat. Conf. on Clays and Clay minerals. Pergamon Press, Oxford, 443—459. [25]

CORRENS, C. W. & ENGELHARDT, W. v. (1938): Neue Untersuchungen über die Verwitterung des Kalifeldspates. — Chem. d. Erde, **12**, 1—22. [27ff]

CORRENS, C. W. & STEINBORN, W. (1939): Experimente zur Messung und Erklärung der sogenannten Kristallisationskraft. — Z. Kristallogr. (A), **101**, 117—133. [14f, 242f, 299]

COULTER, H. W. & MIGLIACCIO, R. R. (1966): Effects of the earthquake of March 27, 1964, at Valdez, Alaska, — U. S. geol. Surv. prof. Pap., 542—C. [96]

CROWLEY, A. J. C. (1939): Possible criterion for distinguishing marine and non-marine sediments. — Bull. amer. Assoc. Petrol. Geol., **23**, 1716—1720. [327]

CULKIN, F. (1965): The major constituents of sea water. — In: Chemical oceanography (eds.: RILEY, J. P. & SKIRROW, G.), Vol. I. — Academic Press, London, 121—161. [117, 214]

CURTIS, R. G., EVANS, G., KINSMAN, J. J. & SHEARMAN, D. J. (1963): Association of dolomite and anhydrite in recent sediments of the Persian Gulf. — Nature, **197**, 679—680. [188]

DAPPLES, E. C. (1959): The behaviour of silica in diagenesis. In: IRELAND, H. A. (ed.): Silica in Sediments. — Soc. Econ. Palaeont. Miner., Spec. Publ. Nr. 7, Tulsa, 36—54.
[241, 250, 293]
— (1967): Diagenesis of sandstones. In: LARSEN, G. & CHILINGAR, G. V. (eds.), Diagenesis in Sediments. — Elsevier, Amsterdam, London, New York, 91—125. [247, 250, 293]
DAUBRÉE, A. (1879): Etudes synthétiques de géologie expérimentale. — Paris. [104, 107]
DEFFEYES, K. S. (1959): Zeolites in sedimentary rocks. — J. sediment. Petrol., 29, 602—609.
[254]
DEGENS, E. T. & CHILINGAR, G. V. (1967): Diagenesis of subsurface waters. In: LARSEN, G. & CHILINGAR, G. V. (eds.): Diagenesis in sediments. — Elsevier, Amsterdam, London, New York, 477—501. [208]
DEGENS, E., HUNT, J. M., REUTER, J. H. & REED, W. E. (1964): Data on the distribution of amino acids and oxygen isotopes in petroleum brine waters of various geologic ages. — Sedimentology, 3, 199—225. [238]
DE GROOT, K. (1969): The chemistry of submarine cement formation at Dohat Hussain in the Persian Gulf. — Sedimentology, 12, 63—68. [308]
DE KIMPE, C., GASTUCHE, M. C. & BRINDLEY, G. W. (1961): Ionic coordination in aluminosilicic gels in relation to clay mineral formation. — Amer. Miner., 46, 1370—1381. [45]
— — (1966): Low temperature syntheses of Kaolin minerals. — Amer. Miner., 49, 1—16. [45]
DE SITTER, L. U. (1947): Diagenesis of oil-field brines. — Bull. amer. Assoc. Petrol. Geol., 31, 2030—2040. [215 f, 232]
DEUSER, W. G. (1970): Extreme C^{13}/C^{12} variation in Quaternary dolomites from the continental shelf. — Earth Planet. Sci. Lett., 8, 118—124. [322]
DICKEY, P. A. (1969): Increasing concentration of subsurface brines with depth. — Chem. Geol., 4, 361—370. [217]
DICKINSON, G. (1953): Geological aspects of abnormal reservoir pressures in the Gulf coast region of Louisiana, U. S. A. — Bull. amer. Assoc. Petrol. Geol., 37, 410—432. [283]
DICKSON, F. W., BLOUNT, CH. W. & TUNEL, G. (1963): Use of hydrothermal solution equipment to determine the solubility of anhydrite in water from 100 °C to 275 °C and from 1 bar to 1000 bars pressure. — Amer. J. Sci., 261, 61—78. [187]
DIETRICH, G. & KALLE, K. (1957): Allgemeine Meereskunde. — Gebr. Borntraeger, Berlin, 492 pp. [17, 81 f]
DILLO, H. G. (1960): Sandwanderungen in Tideflüssen. — Mitt. Franzius Inst. Grund- u. Wasserbau, T. H. Hannover, H. 17, 135—253. [70 f]
DIXON, J. B. & JACKSON, M. L. (1962): Properties of intergradient chlorite-expansible layer silicates of soils. — Soil Sci. amer. Proc., 26, 358—362. [43]
DOEGLAS, D. J. (1950): De interpretatie van Korrelgrootteanalysen. — Verh. Nederl. Geol.- Mijnbouwk. Genoot., Deel 15, 247—328. [132]
— (1968): Grain size indices, classification and environment. — Sedimentology, 1, 167—190. [9]
DOTT, R. H. (1963): Dynamics of subaqueous gravity depositional processes. — Bull. amer. Assoc. Petrol. Geol., 47, 104—128. [144]
DROSTE, J. B., BHATTACHARYA, N. & SUNDERMAN, J. A. (1962): Clay mineral alteration in some Indiana soils. — Clays and Clay Minerals, Proc. 9th nat. Conf. on Clays and Clay Minerals. Pergamon Press, Oxford, 329—342. [42 f]
DRYDEN, A. L. & DRYDEN, C. (1946): Comparative rates of weathering of some common heavy minerals. — J. sediment. Petrol., 16, 91—96. [24]
DRYGALSKI, E. v. & MACHATSCHEK, F. (1942): Gletscherkunde. — In: Enzyklopädie der Erdkunde. Wien. [164]
DUNNINGTON, H. v. (1967): Aspects of diagenesis and shape change in stylolitic limestone reservoirs. — Proc. 7th Wld. Petrol. Congr., Mexico 2, Panel, Disc. No. 3, after TRURNIT 1967. [313]
DUNOYER DE SEGONZAC, G. (1969): Les minéraux argilleux dans la diagenèse. Passage au métamorphisme. — Mém. Serv. Carte géol. Als. Lorr., 29, 1—320. [304]
DUNOYER DE SEGONZAC, G., FERRERO, J. & KUBLER, B. (1968): Sur la cristallinité de l'illite dans la diagenèse et l'anchimétamorphose. — Sedimentology, 10, 137—143. [301 f]
DUNOYER DE SEGONZAC, G. & MILLOT, G. (1962): Pyrophyllite du diagenèse gans le Dévonien inférieur du synclinal de Laval (Massif armoricain). — C. r. Acad. Sci. Paris, 225, 3438—3440. [302]

DURUM, W. H., HEIDEL, S. G. & TISON, L. J. (1960): World wide runoff of dissolved solids. — Assoc. internat. d'Hydrol. sci. Assemblée gén. de Helsinki 25.7. — 6.8.1960. Commiss des eaux de surface. Publ. No. 51, Gentbrugge (Belg.), 618—628. [17, 115]

EBHARDT, G. (1968): Experimentelle Untersuchungen zur Kompaktion karbonatischer Sedimente. — Ph. D. Thesis, Würzburg. [304 ff]
ECKART, C. (1952): The propagation of waves from deep to shallow water. In: Gravity waves. — Nat. Bur. Standarts Circ., 521, after INMAN 1963, 165—173. [84]
ECKHARDT, F. J. (1958): Über Chlorite in Sedimenten. — Geol. Jb., 75, 437—474. [300, 302]
EDELMAN, C. H. & DOEGLAS, D. J. (1932): Reliktstrukturen detritischer Pyroxene und Amphibole. — Miner. Petrogr. Mitt., 42, 482—490. [24]
— — (1934): Über Umwandlungserscheinungen an detritischem Staurolith und anderen Mineralien. — Miner. Petrogr. Mitt., 45, 225—234. [24]
EHLMANN, A. J., HULINGS, N. C. & GLOVER, E. D. (1963): Stages of glauconite formation in modern foraminiferal sediments. — J. sediment. Petrol., 33, 87—96. [121]
EHLMANN, A. J. & SAND, L. B. (1959): Occurrence of shales partially altered to pyrophyllite. — Clays and clay minerals, 6th nat. Conf. 1957, 386—391. [120, 303]
EINSELE, G. (1967): Sedimentary processes and physical properties of cores from the Red Sea, Gulf of Aden and off the Nile delta. — In: RICHARDS, A. F. (ed.): Marine Geotechnique. — Univ. of Illinois Press, Urbana, Ill., 154—169. [95]
EINSELE, G. & MOSEBACH, R. (1955): Zur Petrographie, Fossilerhaltung und Entstehung der Gesteine des Posidonienschiefers im Schwäbischen Jura. — N. Jb. Geol. Paläont. Abh., 101, 319—430. [296]
ELLIS, A. J. (1959): The solubility of calcite in carbon dioxide solutions. — Amer. J. Sci., 257, 354—365. [174 f]
EMERY, K. O. (1955): Grain size of marine beach gravels. — J. Geol., 63, 39—49. [139]
— (1968): Relict sediments on the continental shelves of the world. — Bull. amer. Assoc. Petrol. Geol., 52, 445—464. [142]
— (1968): Shallow structures of continental shelves and slopes. — Southeastern Geol. 9, 173—194. [142]
EMERY, K. O. & RITTENBERG, S. C. (1952): Early diagenesis of California basin sediments in relation to origin of oil. — Bull. amer. Assoc. Petrol. Geol., 36, 735—806. [192, 271]
ENGELHARDT, W. v. (1937 a): Mineralogische Beschreibung eines mecklenburgischen Bodenprofils. — Chemie d. Erde, 11, 17—37. [23]
— (1937 b): Über die Schwermineralsande der Ostseeküste zwischen Warnemünde und Darsser Ort und ihre Bildung durch die Brandung. — Z. angew. Miner., 1, 30—59. [140]
— (1940 a): Die Unterscheidung wasser- und windsortierter Sande auf Grund der Korngrößenverteilung ihrer leichten und schweren Gemengteile. — Chemie d. Erde, 12, 445—465. [128, 163]
— (1940 b): Zerfall und Aufbau von Mineralen in norddeutschen Bleicherdewaldböden. — Chemie d. Erde, 13, 1—43. [23, 45]
— (1960): Der Porenraum der Sedimente. — Springer, Berlin, Göttingen, Heidelberg, 207 pp.
[191 ff, 195 ff, 202 ff, 207 f, 216 f, 253, 274, 277]
— (1961): Zum Chemismus der Porenlösung der Sedimente. — Bull. Geol. Inst. Univ. Uppsala, 40, 189—204. [208, 217]
— (1967): Interstitial solutions and diagenesis in sediments. — In: G. LARSEN & G. V. CHILINGAR (eds.): Diagenesis in sediments. — Elsevier, Amsterdam, London, New York, 503—521. [208, 260]
ENGELHARDT, W. v. & GAIDA, K. H. (1963): Concentration changes of pore solutions during the compaction of clay sediments. — J. sediment. Petrol., 33, 919—930. [233 f, 265 f, 268 ff]

ENGELHARDT, W. v. & HAUSSÜHL, S. (1965): Festigkeit und Härte von Kristallen. — Fortschr. Miner., 42, 5—49. [105 ff]
ENGELHARDT, W. v. & PITTER, H. (1951): Über die Zusammenhänge zwischen Porosität, Permeabilität und Korngröße bei Sanden und Sandsteinen. — Heidelb. Beitr. Miner. Petrogr., 2, 477—491. [201]
ERDMANN, E. (1894): Bidrag till kannedomen om rullstenas bildande. — Geol. Fören. Stockh. Förh., 4, 407—417. [104]

FACCA, G. (1951): Three gas bearing fields in the Po-valley (North Italy). — Proc. 3rd World Petrol. Congr., Sect. I, Leiden, 256—265. [283]

FAIRBRIDGE, R. W. (1957): The dolomite question. — In: Regional aspects of carbonate deposition. — Soc. Econ. Paleont. Miner., Spec. Publ. No. 5, 125—178. [319]
FENIAK, M. W. (1944): Grain sizes and shapes of various minerals in igneous rocks. — Amer. Miner., 29, 415—421. [2, 3]
FISCHER, A. G. & GARRISON, R. E. (1967): Carbonate lithification on the sea floor. — J. Geol., 75, 488—496. [309]
FISCHER, K. F. & MEIER, W. M. (1965): Kristallchemie der Zeolithe. — Fortschr. Miner., 42, 50—86. [254]
FISK, H. N. (1955): Sand facies of recent Mississippi delta deposits. — Proc. 4th Wld. Petrol. Congr., Sect. I, Rome, 377—398. [136]
FISK, H. N., MCFARLAN, E., KOLB, C. R. & WILBERT, L. J. (1964): Sedimentary framework of the modern Mississippi delta. — J. sediment. Petrol., 24, 76—99. [132]
FOLK, R. L. & WARD, W. C. (1957): Brazos river bar: a study in the significance of grain size parameters. — J. sediment. Petrol., 27, 3—26. [127]
FOSTER, M. D. (1963): Interpretation of the composition of vermiculites and hydrobiotites. — Proc. 10th National Conference on Clays and Clay Minerals. — Pergamon Press, Oxford, 70—89. [42]
FOTHERGILL, C. A. (1955): The cementation of oil reservoir sands and its origin. — Proc. 4th Wld. Petrol. Congr., Sect. I, Rome, 300—312. [249, 252]
FRIEDL, K. (1956): Die Tiefenwässer der Gösting-Domung. — Erdöl u. Kohle, 9, 505—511. [217. 227 f]
FRIEDMAN, G. M. (1961): Distinction between dune, beach and river sands from their textural characteristics. — J. Sediment. Petrol., 31, 514—529. [162]
— (1962): On sorting, sorting coefficients and the lognormality of the grain size distribution of sandstones. —J. Geol., 70, 737—756. [124]
— (1964): Early diagenesis and lithification in carbonate sediments. — J. sediment. Petrol., 34, 777—813. [178, 182, 309 f, 315, 317]
— (1965): Occurrence and stability relationships of aragonite, high-magnesian calcite and low-magnesian calcite under deep-sea conditions. — Bull. geol. Soc. Amer., 76, 1191—1196. [182]
— (1967): Dynamic processes and statistical parameters compared for size frequency distribution of beach and river sands. — J. sediment. Petrol., 37, 327—354. [124, 133, 139]
— (1968): The fabric of carbonate cement and matrix and its dependence on the salinity of water. — In: G. MÜLLER & G. M. FRIEDMAN (eds.): Recent developments in Carbonate Sedimentology in Central Europe. — Springer, Berlin, 11—20. [310]
FRIEDMAN, G. M., FABRICAND, B. P., IMBIMBO, E. S., BREY, M. E. & SANDERS, J. E. (1968): Chemical changes in interstitial waters from continental shelf sediments. — J. sediment. Petrol., 38, 1313—1319. [238]
FRIEDMAN, G. M. & SANDERS, J. E. (1967): Origin and occurrence of dolostones. — In: G. V. CHILINGAR, H. J. BISSEL & R. W. FAIRBRIDGE (eds.): Carbonate rocks, Part A. — Elsevier, Amsterdam, London, New York,268—348. [319, 323 f]
FRITZ, P. (1966): Zur Genese von Dolomit und zuckerkörnigem Kalk im Weißen Jura der Schwäbischen Alb (Württemberg). — Arb. Geol. Paläont. Inst. T. H. Stuttgart, N. F., Nr. 50, 104 pp. [324]
FRUTH, L. S., ORME, G. R. & DONATH, F. A. (1966): Experimental compaction effects in carbonate sediments. — J. sediment. Petrol., 36, 747—754. [305]
FÜCHTBAUER, H. (1948): Einige Beobachtungen an authigenen Albiten. — Schweiz. Miner. Petr. Mitt., 28, 709—716. [327]
— (1950): Die nicht-karbonatischen Bestandteile des Göttinger Muschelkalkes. — Heidelb. Beitr. Miner. Petr., 2, 235—254. [327 f]
— (1961): Zur Quarzneubildung in Erdöllagerstätten. — Erdöl u. Kohle, 14, 169—173. [239, 247, 249, 262]
— (1967 a): Der Einfluß des Ablagerungsmilieus auf die Sandsteindiagenese im Mittleren Buntsandstein. — Sediment. Geol., 1, 159—179. [252, 254, 261]
— (1967 b): Influence of different types of diagenesis on sandstone porosity. — Proc. 7th Wld. Petrol. Congr., Mexico, Reservoir Geol., 353—369. [250, 254]
— (1970): Sandsteine, Konglomerate und Breccien. — In: H. FÜCHTBAUER & G. MÜLLER: Sedimente und Sedimentgesteine. — Schweizerbart Stuttgart. [239, 249, 258,260, 304]
— (1972): Influence of salinity on carbonate rocks in the Zechstein formation, north western Germany. — Geology of saline deposits. Proc. Hannover Sympos. 1968, Earth Sci., 7, 23—31. [327 f]

FÜCHTBAUER, H. & GOLDSCHMIDT, H. (1959): Die Tonminerale der Zechsteinformation. — Beitr. Miner. Petrogr., 6, 320—345. [299, 303]
— — (1963): Beobachtungen zur Tonmineraldiagenese. — Internat. Clay Conf. Stockholm, Pergamon Press, Vol. I, 99—111. [252, 277, 299, 302, 315]
— — (1964): Aragonitische Lumachellen im bituminösen Wealden des Emslandes. — Beitr. Miner. Petrogr., 10, 184—197. [302]
— — (1965): Beziehungen zwischen Calciumgehalt und Bildungsbedingungen der Dolomite. — Geol. Rdsch., 55, 29—40. [323]
FÜCHTBAUER, H. & MÜLLER, G. (1970): Sedimente und Sedimentgesteine. Sediment-Petrologie, Teil II. — Schweizerbart, Stuttgart. [3, 109, 179, 293, 317, 328]
FÜCHTBAUER, H. & REINECK, H. E. (1963): Porosität und Verdichtung rezenter mariner Sedimente. — Sedimentology, 2, 294—306. [191]
FÜHRBÖTER, A. (1967): Zur Mechanik der Strömungsriffel. — Mitt. Franzius Inst. Grund- u. Wasserbau, T. H. Hannover, H. 29, 1—35. [71]
FYFE, W. S. & BISCHOFF, J. L. (1965): The calcite-aragonite problem. — In: L. C. PRAY & R. C. MURRAY (eds.): Dolomitization and limestone diagenesis: a symposium. — Soc. Econ. Paleont. Miner., Spec. Publ., 13, 3—13. [316]

GADOW, S. (1965): Verwitterung und Mineralneubildung in Bodenprofilen auf Stubensandstein. — Beitr. Miner. Petrogr., 11, 449—469. [23]
GAERTNER, H. R. v. & SCHELLMANN, W. (1965): Rezente Sedimente im Küstenbereich der Halbinsel Kaloum, Guinea. — Miner. Petr. Mitt., 10, 349—367. [122]
GARRELS, R. M. & CHRIST, C. L. (1965): Solutions, minerals, and equilibria. — Harper & Row, New York. [33 ff, 45, 122, 169 f, 176, 178, 184, 253]
GARRELS, R. M., THOMPSON, M. E. & SIEVER, R. (1961): Control of carbonate solubility. — Amer. J. Sci., 259, 24—45. [177]
— — — (1962): A chemical model for sea water at 25 °C and one atmosphere total pressure. — Amer. J. Sci., 260, 57—66. [177]
GARRISON, R. E. & FISCHER, A. G. (1969): Deep-water limestones and radiolarites of the Alpine Jurassic. — In: G. M. FRIEDMAN (ed.): Depositional Environments in Carbonate Rocks: a Symposium. — Soc. Econ. Paleont. Miner., Spec. Publ., 14, 20—55. [309]
GAVISH, E. & FRIEDMAN, G. M. (1969): Progressive diagenesis in Quaternary to the late Tertiary carbonate sediments: sequence and time scale. — J. sediment. Petrol., 39, 980—1006. [310 f]
GEE, H. (1932): Preliminary experiments in precipitation by removal of carbon dioxide under aseptic conditions. — Bull. Scripps Inst. Oceanogr., Techn. Ser., 3, 180. [178]
GEHLEN, K. v. (1960): Die röntgenographische und optische Gefügeanalyse von Erzen, insbesondere mit dem Zählrohrtexturgoniometer. — Beitr. Miner. Petrogr., 1, 340—388. [288]
GERMANN, K. (1966): Ablauf und Ausmaß diagenetischer Veränderungen im Wettersteinkalk (alpine Mitteltrias). — Ph. D. Thesis, München. [307, 310]
GEVIRTZ, J. L. & FRIEDMAN, G. M. (1966): Deep-sea carbonate sediments of the Red Sea and their implications on marine lithification. — J. sediment. Petrol., 36, 143—151. [310]
GILBERT, CH. M. (1949): Cementation of some California Tertiary reservoir sands. — J. Geol., 57, 1—17. [246]
GILBERT, G. K. (1914): Transportation of debris by running water. — U. S. geol. Surv. prof. Pap., 86. [70]
GINSBURG, R. N. (1957): Early diagenesis and lithification of shallow-water carbonate sediments in South Florida. — In: Regional Aspects of Carbonate Deposition. — Soc. Econ. Paleont. Miner., Spec. Publ., 5, 80—100. [304, 310]
GIPSON, M. (1965): Application of the electron microscope to the study of particle orientation and fissility in shale. — J. sediment. Petrol., 35, 408—414. [290 f]
— (1966): A study of the relations of depth, porosity, and clay mineral orientation in Pennsylvanian shales. — J. sediment. Petrol., 36, 888—903. [290 f]
GIRESSE, P. (1965): Observations sur la présence de glauconie actuelle dans les sédiments ferrugineux peu profonds du bassin gabonais. — X. r. Acad. Sci. Paris, 260, 5597—5600. [123]
GJEMS, O. (1960): Some notes on clay minerals in pozdol profiles in Fennoscandia. — Clay Miner. Bull., 4, 208—211. [42]
GLASS, H. D., POTTER, P. E. & SIEVER, R. (1956): Clay mineralogy of some basal Pennsylvanian sandstones, clays and shales. — Bull. amer. Assoc. Petrol. Geol., 40, 750—754. [252]

GLENN, R. C. & NASH, V. E. (1964): Weathering relationships between Gibbsite, Kaolinite, Chlorite and expansible layer silicates in selected soils from the Lower Mississippi Coastal Plain. — Clays and Clay Minerals. Proc. 12th nat. Conf., Pergamon Press, Oxford, 529—548. [43]

GLOVER, J. E. & HOSEMANN, P. (1967): Authigenic high sanidine from Western Australia. — Nature, 214, 262. [251]

GMELIN (1961): Handbuch der Anorganischen Chemie, 8. Aufl., System Nr. 28, Calcium. — Verlag Chemie, Weinheim. [174]

GOLDICH, S. S. (1938): A study in rock weathering. — J. Geol., 46, 17—58. [24]

GOLDSMITH, J. R. & GRAF, D. L. (1958): Structural and compositional variations in some natural dolomites. — J. Geol., 66, 678—693. [323]

GOLDSMITH, J. R., GRAF, D. L. & JOENSUU, O. I. (1955): The occurrence of magnesian calcite in nature. — Geochim. Cosmochim. Acta, 7, 212—230. [179]

GORANSON, R. W. (1940): Physics of stressed solids. — J. Chem. Phys., 8, 323—334. [242]

GORDON, M., TRACEY, J. L. & ELLIS, M. W. (1958): Geology of the Arkansas bauxite region. — U. S. geol. Surv. prof. Pap., 299. [51]

GORHAM, E. (1955): On the acidity and salinity of rain. — Geochim. Cosmochim. Acta, 7, 231—239. [18 f]

— (1957): The influence and importance of daily weather conditions in the supply of chloride, sulfate and other ions to fresh waters from atmospheric precipitation. — Phil. Trans. roy. Soc. London, Ser. B, 241, 147—178. [18 f]

GRAF, D. L., FRIEDMAN, J. & MEENTS, W. F. (1965): The origin of saline formation waters II: Isotopic fractionation by shale micropore systems. — Illinois State geol. Surv. Circ. 393, 31 pp. [234]

GRAF, D. L., MEENTS, W. F., FRIEDMAN, J. & SHIMP, N. F. (1966): The origin of saline formation waters III: Calcium chloride waters. — Illinois State geol. Surv. Circ. 398, 60 pp. [216 f, 219, 222, 224, 238]

GRANT, W. H. (1963): Weathering of Stone Mountain granite. — Clays and Clay Minerals, Proc. 11th nat. Conf., Pergamon Press, Oxford, 65—73. [45]

GREENBERG, S. A. & PRICE, E. W. (1957): The solubility of silica in solutions of electrolytes. — J. Phys. Chem., 61, 1539—1541. [37]

GRIFFIN, G. M. (1962): Regional clay mineral facies — Product of weathering intensity and current distribution in the northeastern Gulf of Mexico. — Bull. geol. Soc. Amer., 73, 737—768. [143]

GRIFFITHS, J. C. (1952): Reaction relation in the finer-grained rocks. — Clay Miner. Bull., 1, 251—256. [41]

GRIGGS, D. T. (1936): The factor of fatigue in rock exfoliation. — J. Geol., 44, 783—796. [12 f]

GRIM, R. E. (1953): Clay mineralogy. — Mc Graw-Hill, New York, London, Toronto, 384 pp. [44, 119, 300]

GRIM, R. E. & JOHNS, W. D. (1954): Clay mineral investigation of sediments in the northern Gulf of Mexico. — Clays and Clay Minerals, Proc. 2nd nat. Conf., Columbia — Nat. Acad. Sci. — Nat. Res. Council, Publ. 327, Washington, 81—103. [119]

GRIMM, W. D. (1957): Stratigraphische und sedimentpetrographische Untersuchungen in der Oberen Süßwassermolasse zwischen Inn und Rott (Niederbayern). — Beih. Geol. Jb., 26, Hannover, 97—200. [24]

GRIPP, K. (1933): Tunnelfährten aus Feuerstein und die Entstehung des Feuersteins. — Mitt. Miner.-Geol. Staatsinst. Hamburg, 14, 23—39. [324 f]

— (1954): Kritik und Beitrag zur Frage der Entstehung der Kreide-Feuersteine. — Geol. Rdsch., 42, 248—262. [324]

GROGAN, R. M. (1945): Shape variation of some Lake Superior beach pebbles. — J. sediment. Petrol., 15, 3—10. [111]

GUILCHER, A. (1963): Continental shelf and slope (Continental margin). — In: M. N. HILL (ed.): The sea, Vol. 3. — Interscience Publ., New York, London, 281—311. [93, 142]

GUTENBERG, B. (1939): Tsunamis and earthquakes. — Bull. seismol. Soc. Amer., 29, 517—526. [96]

HAGLUND, D. S., FRIEDMAN, G. M. & MILLER, D. S. (1969): The effect of fresh water on the redistribution of uranium in carbonate sediments. — J. sediment. Petrol., 39, 1283—1296. [317]

HALLA, F. & RITTER, F. (1935): Eine Methode zur Bestimmung der Änderung der freien Energie bei Reaktionen des Typs $A(S) + B(S) = AB(S)_2$ und die Anwendung auf das Dolomitproblem. — Z. phys. Chem., **175**, 63—82. [182]

HAMILTON, E. L. & MENARD, H. W. (1956): Density and porosity of sea floor surface sediments off San Diego, California. — Bull. amer. Assoc. Petrol. Geol., **40**, 754—761. [191]

HANSHAW, B. B. (1962): Membrane properties of clays. — Ph. D. Thesis, Harvard Univ., Cambridge, Mass. [233]

HARDER, H. (1965): Experimente zur „Ausfällung" der Kieselsäure. — Geochim. Cosmochim. Acta., **29**, 429—442. [118, 121]

HARDER, H. & FLEHMIG, W. (1967): Bildung von Quarz aus verdünnten Lösungen bei niedrigen Temperaturen. — Naturwiss., **54**, 140. [38, 121]

HARDER, H. & MENSCHEL, G. (1967): Quarzbildungen am Meeresboden. — Naturwiss., **54**, 561. [122]

HARDIE, L. A. (1967): The gypsum-anhydrite equilibrium at one atmosphere pressure. — Amer. Miner., **52**, 171—200. [186 f]

HARRISON, J. R. & MURRAY, H. H. (1959): Clay mineral stability and formation during weathering. — Clays and Clay Minerals, Proc. 6th nat. Conf., Pergamon Press, Oxford, 144—153. [43]

HARVEY, H. W. (1955): The chemistry and fertility of sea water. — Cambridge. [181]

HAY, R. L. (1966): Zeolites and zeolitic reactions in sedimentary rocks. — Bull. geol. Soc. Amer. Spec. Pap., **85**. [254]

HEALD, M. T. (1950): Authigenesis in West-Virginia sandstones. — J. Geol., **58**, 624—633. [253]

— (1955): Stylolites in sandstones. — J. Geol., **63**, 101—114. [244 ff]

— (1956 a): Cementation of Simpson and St. Peter sandstones in parts of Oklahoma, Arkansas and Missouri. — J. Geol., **64**, 16—30. [245 f]

— (1956 b): Cementation of Triassic arkoses in Connecticut and Massachusetts. — Bull. geol. Soc. Amer., **67**, 1133—1154. [251]

— (1959): Significance of stylolites in permeable sandstones. — J. sediment. Petrol., **29**, 251—253. [245 f]

HEALD, M. T. & ANDEREGG, R. C. (1960): Differential cementation in the Tuscarora sandstone. — J. sediment. Petrol., **30**, 568—577. [247]

HEDBERG, H. D. (1936): Gravitational compaction of clays and shales. — Amer. J. Sci., **31**, 241—287. [273 f, 277, 282]

HEEZEN, B. C. (1963): Turbidity currents. — In: M. N. HILL (ed.): The Sea, Vol. 3. — Interscience Publ., New York, London, 742—775. [93]

HEEZEN, B. C. & EWING, M. (1952): Turbidity currents and submarine slumps and the Grand Banks earthquake. — Amer. J. Sci., **250**, 849—873. [96 ff]

— — (1955): Orléansville earthquake and turbidity currents. — Bull. amer. Assoc. Petrol. Geol., **39**, 2505—2514. [96 f]

HEEZEN, B. C. & MENARD, H. W. (1963): Topography of the deep-sea floor. — In: M. N. HILL (ed.): The Sea, Vol. 3. — Interscience Publ., New York, London, 232—280. [93]

HEIM, D. (1960): Über die Petrographie und Genese der Tonsteine aus dem Rotliegenden des Saar-Nahe-Gebietes. — Beitr. Miner. Petrogr., **7**, 281—317. [303]

HEIMANN, R. & WILLGALLIS, A. (1969): Kinetik und Morphologie der Auflösung von Quarz. — N. Jb. Miner. Mh., 145—160. [244]

HELING, D. (1963): Zur Petrographie des Stubensandsteins. — Beitr. Miner. Petrogr., **9**, 251—284. [250, 252, 260]

— (1965): Zur Petrographie des Schilfsandsteins. — Beitr. Miner. Petrogr., **11**, 272—296. [251, 253 f, 260]

— (1967): Die Porositäten toniger Keuper- und Jurasedimente Südwestdeutschlands. — Contr. Miner. Petrol., **15**, 224—232. [275, 277, 288]

— (1969): Relationships between initial porosity of Tertiary argillaceous sediments and paleosalinity in the Rheintalgraben (SW-Germany). — J. sediment. Petrol., **39**, 246—254. [193, 276 ff]

— (1970 a): Die Petrographie des Lokisandsteins im Ruhrcarbon (Oberes Westfal C): Ein Beitrag zur Diagenese tief versenkter Sandsteine. — Chemie d. Erde, **28**, 294—330. [250, 253]

— (1970 b): Microfabrics of shales and their rearrangement by compaction. — Sedimentology, **15**, 247—260. [288]

HÉNIN, S. & CAILLÈRE, S. (1961): Vue d'ensemble sur le problème de la synthèse des minéraux phylliteux à basses températures. — Coll. Inst. C. N. R. S. Genése et synthèse des argiles, Paris, 31—43. [44]

HITCHON, B. (1968): Rock volume and pore volume data for plains region of Western Canada sedimentary basin between latitudes 49° and 60° N. — Bull. amer. Assoc. Petrol. Geol., 52, 2318—2323. [210, 212 f]

HITCHON, B. & FRIEDMAN, I. (1969): Geochemistry and origin of formation waters in the western Canada sedimentary basin. — Geochim. Cosmochim. Acta., 33, 1321—1349. [224 f]

HJULSTRÖM, F. (1934/35): Studies of the morphological activity of rivers as illustrated by the river Fyris. — Bull. geol. Inst. Univ. Uppsala, 25, 221—452. [67, 72 f]

HOEFS, J. (1970): Kohlenstoff- und Sauerstoffisotopenuntersuchungen an Karbonatkonkretionen und umgebendem Gestein. — Contr. Miner. Petrol., 27, 66—79. [298]

HOEPPENER, R. (1956): Zum Problem der Bruchbildung, Schieferung und Faltung. — Geol. Rdsch., 45, 247—283. [245]

HOFMANN, U. & HAUSDORF, A. (1945): Über das Sedimentvolumen und die Quellung von Bentonit. — Koll. Z., 110, 1—17. [193]

HOLLAND, H. D., KIRSIPU, T. V., HUEBNER, J. S. & OXBURGH, U. M. (1964): On some aspects of the chemical evolution of cave waters. — J. Geol., 72, 36—67. [182]

HOPPE, G. (1962): Petrogenetisch auswertbare morphologische Erscheinungen an akzessorischen Zirkonen. — N. Jb. Miner. Abh., 98, 35—50. [8]

— (1963): Die Verwendbarkeit morphologischer Erscheinungen an akzessorischen Zirkonen. — Abh. dt. Akad. Wiss., Kl. Bergbau, Hüttenwes., Montangeol., 1, Berlin, 130 pp. [8]

HOPPE, H. J. (1941): Untersuchungen an Palagonit-Tuffen und über ihre Bildungsbedingungen. — Chemie d. Erde, 13, 484—514. [32]

— (1947): Verwitterungsversuche am Olivin. — Ph. D. Thesis, Göttingen. [31 f]

HORN, D. (1965): Diagenese und Porosität des Dogger-beta-Hauptsandsteines in den Ölfeldern Plön-Ost und Preetz. — Erdöl u. Kohle, 18, 249—255. [247, 253]

HORN, M. K. & ADAMS, J. A. S. (1966): Computer-derived geochemical balances and element abundances. — Geochim. Cosmochim. Acta, 30, 279—297. [123]

HOUTZ, R. E. (1962): The 1953 Suva earthquake and tsunami. — Bull. seismol. Soc. Amer., 52, 1—12. [96]

HOWER, J. (1961): Some factors concerning the nature and origin of glauconite. — Amer. Miner., 46, 313—332. [120 f]

HOWER, J. & MOWATT, TH. C. (1966): The mineralogie of illites and mixed layer illite/montmorillonites. — Amer. Miner., 51, 825—854. [41, 43]

HSI, H. R. & CLIFTON, D. F. (1962): Flocculation of selected clays by various electrolytes. — Clays and Clay Minerals, Proc. 9th nat. Conf., Pergamon Press, Oxford, 269—275. [193]

HSU, K. J. (1963): Solubility of dolomite and composition of Florida ground waters. — J. Hydrol., 1, 288—310. [182]

HUMMEL, K. (1922): Die Entstehung eisenreicher Gesteine durch Halmyrolyse. — Geol. Rdsch., 13, 40—81. [114]

HUNT, J. M. (1961): Distribution of hydrocarbons in sedimentary rocks. — Geochim. Cosmochim. Acta, 22, 37—49. [321]

ILLIES, H. (1949): Über die erdgeschichtliche Bedeutung der Konkretionen. — Z. dt. geol. Ges., 101, 95—98. [294, 324 f, 327]

— (1954): Zur Entstehung der Kreidefeuersteine. — Geol. Rdsch., 42, 262—264. [324]

INGLE, J. C. (1966): The movement of beach sand. — Elsevier, New York, 232 pp. [137]

INGRAM, R. L. (1953): Fissility of mudrocks. — Bull. geol. Soc. Amer., 64, 869—878. [291]

INMAN, D. L. (1949): Sorting of sediments in the light of fluid mechanics. — J. sediment. Petrol., 19, 51—71. [131]

— (1952): Measures for describing the size distribution of sediments. — J. sediment. Petrol., 22, 125—145. [124, 126 f]

— (1963): Ocean waves and associated currents. — In: F. P. SHEPARD: Submarine Geology. — Harper & Row, New York, Evaston, London, 49—81. [82, 85, 89]

INMAN, D. L. & BAGNOLD, R. A. (1963): Littoral processes. — In: M. N. HILL (ed.): The Sea, Vol. 3. — Interscience Publ., New York, London, 529—553. [90, 138]

INNERBRITZEN, A. L. (1959): Gravels of Alameda Creek, California. — J. sediment. Petrol., 29, 212—220. [111]

JACKSON, M. L. (1963a): Interlayering of expandable layer silicates in soils by chemical weathering. — Clays and Clay Minerals, Proc. 11th nat. Conf., Pergamon Press, Oxford, 29—46. [42, 43]
— (1963b): Aluminum bonding in soils: A unifying principle in Soil science. — Proc. Soil Sci. Soc. Amer., 27, 1—10. [43]
JAMIESON, J. C. (1953): Phase equilibrium in the system calcite-aragonite. — J. Chem. Phys., 21, 1385—1390. [178]
JASMUND, K. (1955): Die silikatischen Tonminerale. — Verlag Chemie, Weinheim. [39, 120]
JOHNS, W. D. (1963): Die Verteilung von Chlor in rezenten marinen und nichtmarinen Sedimenten. — Fortschr. Geol. Rheinl.-Westf., 10, 215—230. [119]
JOHNS, W. D. & GRIM, R. E. (1958): Clay mineral composition of recent sediments from the Mississippi river delta. — J. sediment. Petrol., 28, 186—199. [119, 143]
JONAS, E. C. (1964): Petrology of the Dry Branch, Georgia, Kaolin deposits. — Clays and Clay Minerals, Proc. 12th nat. Conf., Pergamon Press, Oxford, 199—205. [45]

KAARSBERG, E. A. (1959): Introductory studies of natural and artificial argillaceous aggregates by sound-propagation and X-ray diffraction methods. — J. Geol., 67, 447—472. [289, 291]
KALINSKE, A. A. (1943): Turbulence and the transport of sand and silt by wind. — Ann. New York Acad. Sci., 33, 41—54. [65f, 68]
KAPLAN, I. R., EMERY, K. O. & RITTENBERG, S. C. (1963): The distribution and isotopic abundance of sulphur in recent marine sediments off southern California. — Geochim. Cosmochim. Acta, 27, 297—331. [235]
KÁRMÁN, TH. v. (1930): Mechanische Ähnlichkeit und Turbulenz. — Nachr. Akad. Wiss. Göttingen Math.-phys. Kl., 58—68. [57, 67, 70]
KASTNER, M. & WALDBAUM, D. R. (1968): Authigenic albite from Rhodes. — Amer. Miner., 53, 1579—1602. [327]
KELLER, W. D. (1963): Diagenesis in clay minerals — a review. — Clays and Clay Minerals, Proc. 11th nat. Conf. Pergamon Press, Oxford, 136—157. [118, 300]
— (1964): The origin of high-alumina clay minerals. A review. — Clays and Clay Minerals, Proc. 12th nat. Conf., Pergamon Press, Oxford, 129—151. [43, 51]
KELLER, W. D., BALGORD, W. D. & REESMAN, A. L. (1963): Dissolved products of artificially pulverized silicate minerals and rocks. — J. sediment. Petrol., 33, 191—204. [25f]
KENNEDY, G. C. (1950): A portion of the system silica—water. — Econ. Geol., 45, 629—653. [244]
KENNEDY, J. F. (1964): The formation of sediment ripples in closed rectangular conduits and in the desert. — J. geophys. Res., 69, 1517—1524. [71]
KINSMAN, D. J. J. (1969): Interpretation of Sr^{2+} concentrations in carbonate minerals and rocks. — J. sediment. Petrol., 39, 486—508. [317]
KNOKE, R. (1966): Untersuchungen zur Diagenese an Kalkkonkretionen und umgebenden Tonschiefern. — Contr. Miner. Petrol., 12, 139—167. [296, 298f]
KÖSTER, E. (1964): Granulometrische und morphometrische Meßmethoden. — Enke, Stuttgart. [99]
KOLDEWIJN, B. W. (1958): Sediments of the Paria-Trinidad Shelf. — Rep. Orinoco Shelf exped., Vol. 3. — Thesis Univ. Amsterdam. [123]
KOSSOVSKAYA, A. G., DRITS, V. A. & ALEXANDROVA, V. A. (1965): On trioctahedral micas in sediments. — Internat. Clay Conf. 1963, Vol. 2, Pergamon Press, Oxford, 147—169. [42]
KOZENY, J. (1927): Über die kapillare Leitung des Wassers im Boden. — Sitz.-Ber. österr. Akad. Wiss. II a, 136, 271—306. [200]
KRAMER, J. R. (1959): Correction of some earlier data on calcite and dolomite in sea water. — J. sediment. Petrol., 29, 464—469. [182]
— (1969): Subsurface brines and mineral equilibria. — Chem. Geol., 4, 37—50. [211, 215, 237]
KRAUSKOPF, K. B. (1956): Dissolution and precipitation of silica at low temperatures. — Geochim. Cosmochim. Acta, 10, 1—26. [37f]
— (1959): The geochemistry of silica in sedimentary environments. In: H. A. IRELAND (ed.): Silica in Sediments. — Soc. Econ. Palaeont. Miner., Spec. Publ., 7, 4—19. [37f]
KREJCI-GRAF, K., HECHT, F. & PASLER, W. (1957): Über Ölfeldwässer im Wiener Becken. — Geol. Jb., Hannover, 74, 161—210. [216, 227f]
KROMER, H. (1963): Untersuchungen über den Mineralbestand des Knollenmergel-Keupers in Württemberg. — Ph. D. Thesis, Tübingen. [302]

KRÜGER, G. (1939): Verwitterungsversuche am Leuzit. — Chemie d. Erde, **12**, 236—264. [30]
KRUIT, C. (1955): Sediments of the Rhone delta. — Monton & Co., 'S-Gravenhage, 141 pp. [136 f]
KRUMBEIN, W. C. (1940): Flood gravel of San Gabriel Canyon, California. — Bull. geol. Soc. Amer., **51**, 639—676. [110 f]
— (1941 a): Measurement and geological significance of shape and roundness of sedimentary particles. — J. sediment. Petrol., **11**, 64—72. [103, 111 f]
— (1941 b): The effects of abrasion on the size, shape and roundness of rock fragments. — J. Geol., **49**, 482—520. [104 ff]
— (1942): Flood deposits of Arroyo Seco, Los Angeles Co., California. — Bull. geol. Soc. Amer., **53**, 1355—1402. [109 ff]
KRUMBEIN, W. C. & SLOSS, L. L. (1963): Stratigraphy and Sedimentation. 2nd ed. — Freeman & Co., San Francisco, London, 660 pp. [161]
KRUMBEIN, W. E. (1972): Rôle des microorganismes dans la genèse, la diagenèse et la dégradation des roches en place. — Ecol. Biol. du Sol, **9**, 305—341. [19]
KRYNINE, P. D. (1946): The turmaline group in sediments. — J. Geol., **54**, 635—710. [259]
KRYUKOV, P. A. & KOMAROVA, N. A. (1956): Etudes des solutions de sols, de limons et de roches-mères. — Rapp. 6ième Congr. internat. Sci. du sol. 2ième Commiss. (Chimie du Sol), Moscou, 151—184. [233]
KUBLER, B. (1966): La cristallinité de l'illite et les zones tout à fait supérieures de métamorphose. — Coll. sur les Etages Tecton., Univ. Neuchâtel, Baconière, Neuchâtel, 105—122. [301]
KUENEN, PH. H. (1950): Marine Geology. — Wiley, New York, 551 pp. [142]
— (1951): Properties of turbidity currents of high density. — Soc. Econ. Palaeont. Miner., Spec. Publ., **2**, 14—33. [93, 99]
— (1956): Experimental abrasion of pepples: 2. Rolling by currents. — J. Geol., **64**, 336—368. [104, 106 f]
— (1958): Some experiments on fluviatile rounding. — Kon. Ned. Akad. Wet. Amsterdam, Ser. B, **61**, 1, 47—53. [108]
— (1959): Experimental abrasion: 3. Fluviatile action on sand. — Amer. J. Sci., **257**, 172—190. [108]
— (1960): Experimental abrasion: 4. Eolian action. — J. Geol., **68**, 427—449. [157]
— (1964): Deep sea sands and ancient turbidites. — In: A. H. BOUMA & A. BROUWER (eds.): Turbidites. — Elsevier, Amsterdam, London, New York, 3—33. [93]
— (1967): Emplacement of flysch-type sand beds. — Sedimentology, **9**, 203—243. [93]
KUENEN, PH. H. & HUMBERT, F. L. (1964): Bibliography of turbidity currents and turbidites. — In: A. H. BOUMA & A. BROUWER (eds.): Turbidites. — Elsevier, Amsterdam, London, New York, 222—246. [93]
KUENEN, PH. H. & MENARD, H. W. (1952): Turbidity currents, graded and non-graded deposits. — J. sediment. Petrol., **22**, 83—96. [99, 144]
KUENEN, PH. H. & MIGLIORINI, C. I. (1950): Turbidity currents as a cause of graded bedding. — J. Geol., **58**, 91—127. [93, 99]
KUENEN, PH. H. & PERDOK, W. G. (1962): Experimental abrasion. 5. Frosting and defrosting of quartz grains. — J. Geol., **70**, 648—658. [157]
KULKE, H. (1969): Petrographie und Diagenese des Stubensandsteins (Mittlerer Keuper) aus Tiefbohrungen im Raum Memmingen (Bayern). — Contr. Miner. Petrol., **20**, 135—163. [250, 252 f]
KUZNETSOV, S. I., IVANOV, M. V. & LYALIKOVA, N. N. (1963): Introduction to geological Microbiology (transl. by P. T. BRONEER, ed. C. H. OPPENHEIMER). — Mc Graw Hill, New York, San Francisco, Toronto, London, 252 pp. [234]

LAND, L. S. (1966): Diagenesis of metastable skeletal carbonates. — Thesis, Mar. Sci. Center, Leigh, Univ. Bethlehem, Pa., 141 pp. [310 f]
— (1967): Diagenesis of skeletal carbonates. — J. sediment. Petrol., **37**, 914—930. [310 f]
LANDON, R. E. (1930): Analyses of beach pebble abrasion and transportation. — J. Geol., **38**, 437—446. [109]
LANG, H. B. (1964): Dolomit und zuckerkörniger Kalk im Weißen Jura der mittleren Schwäbischen Alb. — N. Jb. Geol. Paläont. Abh., **120**, 253—299. [323]
LANGHEINRICH, G. (1966): Syndiagenetische Fossildeformation im untersten Lias (Hettangium) von Göttingen. — N. Jb. Geol. Paläont. Mh., 666—680. [292]

LANGMUIR, D. (1964): Stability of carbonates in the system $CaO—MgO—CO_2—H_2O$. — Ph. D. Thesis, Harvard Univ., Cambridge, Mass. [182]

LEITMEIER, H. (1910): Zur Kenntnis der Karbonate. Die Dimorphie des Kohlensauren Kalkes. — N. Jb. Miner. **1910 I**, 49—74. [178]

LEMCKE, K., ENGELHARDT, W. v. & FÜCHTBAUER, H. (1953): Geologische und sedimentpetrographische Untersuchungen im Westteil der ungefalteten Molasse des süddeutschen Alpenvorlandes. — Beih. Geol. Jb., **11**, 110 pp. [24]

LEMCKE, K. & TUNN, W. (1956): Tiefenwasser in der süddeutschen Molasse und in ihrer verkarsteten Malmunterlage. — Bull. Ver. Schweiz. Petrol., Geol., Ing., **23**, 35—56. [226f]

LEOPOLD, L. B. (1953): Downstream change of velocity in rivers. — Amer. J. Sci., **251**, 606—624. [129]

LEOPOLD, L. B. & MADDOCK, T. (1953): The hydraulic geometry of stream channels and some physiographic implications. — U.S. geol. Surv. prof. Pap., **252**. [129f]

LEOPOLD, L. B., WOLMAN, M. G. & MILLER, J. P. (1964): Fluvial processes in geomorphology. — Freeman & Co., San Francisco, London, 522 pp. [129, 133]

LERBEKMO, J. F. & PLATT, R. L. (1962): Promotion of pressure-solution of silica in sandstones. — J. sediment. Petrol., **32**, 514—519. [245]

LINDSTRÖM, M. (1963): Sedimentary folds and the development of limestone in an Early Ordovician sea. — Sedimentology, **2**, 243—275. [310]

LIPPMANN, F. (1954): Über einen Keuperton von Zaisersweiher bei Maulbronn. — Heidelb. Beitr. Miner. Petrogr., **4**, 130—134. [303]

— (1955): Ton, Geoden und Minerale des Barrême von Hoheneggelsen. — Geol. Rdsch., **43**, 475—503. [296ff]

— (1960a): Corrensit. — In: HINTZE-CHUDOBA, Handb. d. Mineralogie, Erg. Bd. II, 688—691. [303]

— (1960b): Versuche zur Aufklärung der Bildungsbedingungen von Calcit und Aragonit. — Fortschr. Miner., **38**, 156—161. [179]

— (1968): Synthesis of $BaMg(CO_3)_2$ (Norsethite) at 20 °C and the formation of dolomite in sediments. — In: G. MÜLLER & G. M. FRIEDMAN (eds.): Recent developments in carbonate petrology in Central Europe. — Springer, Berlin, Heidelberg, New York, 33—37. [321]

— (1970): Functions describing preferred orientation in flat aggregates of flake-like minerals and in other axially symmetric fabrics. — Contr. Miner. Petrol., **25**, 77—94. [288ff]

— (1973): Sedimentary carbonate minerals. — Springer, Berlin, Heidelberg, New York. [185, 315, 319, 321f]

LIU, H. K. (1957): Mechanics of sediment-ripple formation. — J. Hydraul. Div., Proc. amer. Sov. civ. Eng., **83**, Pap. 1197. [70f]

LIVINGSTONE, D. A. (1963): Chemical composition of rivers and lakes. — In: M. FLEISCHER (ed.): Data of Geochemistry. — U.S. geol. Surv. prof. Pap., **440**, ch. G, 1—G 64. [19f, 117]

LÖHNERT, E. (1967): Grundwasserversalzungen im Bereich des Salzstocks von Altona-Langenfelde. — Abh. u. Verh. Naturwiss. Verein Hamburg, **11**, 29—46. [231]

LOHR, J. (1969): Die seismischen Geschwindigkeiten der jüngeren Molasse im ostschweizerischen und deutschen Alpenvorland. — Geophys. Prospect., **17**, 111—125. [292]

LONGUET-ESCARD, J. (1950): Fixation des hydroxides par la montmorillonite. — Trans. 4th internat. Congr. Soil. Sci., **3**, 40. [43]

LONGUET-HIGGINS, M. S. (1953): Mass transport in water waves. — Phil. Trans. roy. Soc. London, Ser. A, **245**, 535—581. [90f]

LOWENSTAM, H. A. (1961): Mineralogy, O^{18}/O^{16} ratios and strontium and magnesium contents of recent and fossil brachiopods and their bearing on the history of the oceans. — J. Geol., **69**, 241—260. [117]

— (1963): Biologic problems relating to the composition and diagenesis of sediments. — In: T. W. DONNELLY (ed.): The Earth sciences, Problems and progress in current research. — Univ. Chicago Press, 137—195. [315]

LUCAS, J. (1962): Remarques sur les minéraux argileux interstratifiés et leur genèse. — Genèse et synthese des argiles. Coll. internat. C.N.R.S., Paris, 177—190. [42]

LUDWIG, G. & VOLLBRECHT, K. (1957): Die allgemeinen Bildungsbedingungen litoraler Schwermineralkonzentrate und ihre Bedeutung für die Auffindung sedimentärer Lagerstätten. — Geologie, **6**, 233—277. [140]

MacCarthy, G. R. (1931): Coastal sands of the eastern United States. — Amer. J. Sci., Ser. 5, **22**, 35—50. [113]
— (1933): The rounding of beach sands. — Amer. J. Sci., **225**, 205—224. [113]
MacEwan, D. M. C., Ruiz Amil, A. & Brown, G. (1961): Interstratified clay minerals. — In: Brown, G. (ed.): The X-ray identification and crystal structures of clay minerals. — Min. Soc. London, 393—445. [42 f]
Mackenzie, F. T. & Garrels, R. M. (1956): Silicates: Reactivity with sea water. — Science, **150**, 57—58. [118]
— — (1966): Chemical mass balance between rivers and oceans. — Amer. J. Sci., **264**, 507—525. [117 f]
Mackenzie, F. T., Garrels, R. M., Bricker, O. P. & Bickley, F. (1967): Silica in sea water: Control by silica minerals. — Science, **155**, 1404—1405. [118]
Manheim, F. T. & Bischof, J. L. (1969): Geochemistry of pore waters from SHELL-OIL Company drill holes on the continental slope of the northern Gulf of Mexico. — Chem. Geol., **4**, 63—82. [231]
Manheim, F. T. & Horn, M. K. (1968): Composition of deeper subsurface waters along the Atlantic continental margin. — Southeastern Geol., **4**, 215—236. [225 f]
Marsal, D. (1967): Statistische Methoden für Erdwissenschaftler. — Schweizerbart, Stuttgart. [124]
Marsal, D. & Philipp, W. (1970): Compaction of sediments. A simple mathematical model for calculating the gravitational porosity-depth equilibrium-curve of shales. — Bull. Geol. Inst. Univ. Uppsala, N. S., **2**, 59—66. [282, 286]
Marshal, P. E. (1927): The wearing of beach gravels. — Trans. Proc. New Zealand Inst., **58**, 507—532. [104]
Marshall, C. E. (1954): Multifunctional ionization as illustrated by the clay minerals. — Clays and Clay Minerals, Proc. 2nd nat. Conf., Columbia. — Nat. Acad. Sci. — Nat. Res. Counc. Washington, Publ. **327**, 364—385. [119]
— (1964): The physical chemistry and mineralogy of soils, Vol. I, Soil materials. — John Wiley, New York, London, Sydney, 388 pp. [267]
Martin-Vivaldi, J. L. & MacEwan, D. M. C. (1951): Triassic chlorites from the Jura and the Catalan coastal range. — Clay Miner. Bull., **3**, 177—183. [43]
Matthes, S. (1950): Vorkommen von Vermikulit in mitteldeutschen Serpentiniten. — N. Jb. Miner. Mh., 29—62. [42]
Mattiat, B. (1969): Eine Methode zur elektronenmikroskopischen Untersuchung des Mikrogefüges in tonigen Sedimenten. — Geol. Jb. Hannover, **88**, 87—111. [291]
McKelvey, J. G. & Milne, I. H. (1962): The flow of salt solution through compacted clay. — Proc. 9th nat. Conf. on Clays and Clay Minerals, Pergamon Press, Oxford, 248—259. [233]
Meade, R. H. (1964): Removal of water and rearrangement of particles during the compaction of clayey sediments — a review. — U. S. geol. Surv. prof. Pap., **497**—B, 1—23. [193, 268, 271, 290 f]
— (1966): Factors influencing the early stages of the compaction of clays and sands — a review. — J. sediment. Petrol., **36**, 1085—1101. [268]
— (1968 a): Compaction of sediments underlying areas of land subsidence in Central California. — U. S. geol. Surv. prof. Pap., **497**—D, 1—39. [272 f, 297 f]
— (1968 b): Relations between suspended matter and salinity in estuaries of the Atlantic seabord, U.S.A. — Internat. Assoc. Sci. Hydrol. Gen. Assembly, Bern, **4**, 96—109. [143]
Meents, W. F., Bell, A. H., Rees, O. W. & Tilbury, W. G. (1956): Illinois oil field brines. — Div. Illinois State Geol. Surv., Urbana. [216 f, 219 ff]
Mehmel, M. (1937): Ab- und Umbau am Biotit. — Chemie d. Erde, **11**, 307—332. [31 f]
Meisl, St. (1970): Petrologische Studien im Grenzbereich Diagenese—Metamorphose. — Abh. Hess. Geol. Landesamt Bodenforsch., H. 57, 93 pp. [247, 250, 252, 254]
Melou, M. & Plusquellec, Y. (1967): Répartition de la pyrophyllite dans quelques niveaux du Brioverien et du Primaire Armoricain. — C.R. Acad. Sci. Paris, **265**—D, 14—16. [303]
Menard, H. W. (1964): Marine Geology of the Pacific. — McGraw-Hill, New York. [93]
Michel, G. (1965): Zur Mineralisation des tiefen Grundwassers in Nordrheinwestfalen, Deutschland. — J. Hydrol., **3**, 73—87. [230]
Michel, G & Rüller, K. H. (1964): Hydrochemische Untersuchungen des Grubenwassers der Zechen der Hüttenwerk Oberhausen A. G. — Bergbau-Arch., **25**, 21—27. [229 f]

MILLER, D. G. & RICHARDS, A. F. (1969): Consolidation and sedimentation-compression studies of a calcareous core, Exuma Sound, Bahamas. — Sedimentology, **12**, 301—316. [304]
MILLIMAN, J. D. (1966): Submarine lithification of carbonate sediments. — Science, **153**, 994—997. [309]
MILLOT, G. (1964): Géologie des argiles. — Masson, Paris. [42, 44 f, 50, 119 f, 300]
MILLOT, G. & CAMEZ, TH. (1963): Genesis of vermiculite and mixed layer vermiculite in the evolution of the soils of France. — Clays and Clay Minerals, Proc. 10th nat. Conf., Pergamon Press, Oxford, 90—95. [42]
MILLOT, G. LUCAS, J. & WEY, R. (1963): Research on evolution of clay minerals and argillaceous and siliceous neoformations. — Clays and Clay Minerals, Proc. 10th nat. Conf., Pergamon Press, Oxford, 399—412. [51]
MITCHELL, W. A. (1955): A review of the mineralogy of Scottish soil clays. — J. Soil. Sci., **6**, 94—98. [42]
— (1963): Mineralogical aspects of soil formation on a granitic till. — Internat. Clay Conf. 1963, Vol. I, Pergamon Press, Oxford, 131—138. [42]
MONAGHAN, P. H. & LYTLE, M. L. (1956): The origin of calcareous ooliths. — J. sediment. Petrol., **26**, 111—118. [178]
MOORE, G. W., ROBERSON, C. E. & NYGREN, H. D. (1962): Electrode determination of the carbon dioxide content of sea water and deep sea sediments. — U. S. geol. Surv. prof. Pap., 450—B, 83—86. [180 f]
MORAVIETZ, F. H. (1958): Die Anlösungserscheinungen in der Juranagelfluh und ihre Bedeutung für die Diagenese. — Ph. D. Thesis, Tübingen. [246, 312 f]
MORGENSTERN, N. R. (1967): Submarine slumping and the initiation of turbidity currents. — In: A. F. RICHARDS (ed.): Marine Geotechnique, Univ. Illinois Press, Urbana, Ill., 189—220. [96]
MORTENSEN, H. (1933): Die Salzsprengung und ihre Bedeutung für die regional-klimatische Gliederung der Wüsten. — Petermanns geogr. Mitt., 130—135. [15]
MÜLLER, A. H. (1956): Die Knollenfeuersteine der Schreibkreide, eine frühdiagenetische Bildung. — Ber. Geol. Ges., **1**, 136—146. [324 ff]
MÜLLER, G. (1961): Die rezenten Sedimente im Golf von Neapel. 2. Mineral-Neu- und -Umbildungen in den rezenten Sedimenten des Golfes von Neapel. Ein Beitrag zur Umwandlung vulkanischer Gläser durch Halmyrolyse. — Beitr. Miner. Petrogr., **8**, 1—20. [124]
— (1963): Zur Kenntnis dioktaedrischer Vierschicht-Phyllosilikate (Sudoitreihe der Sudoit-Chlorit-Gruppe). — Internat. Clay Conf. 1963, Vol. I, Pergamon Press, Oxford, 121—130. [41, 43]
— (1964): Methoden der Sedimentuntersuchung. — Schweizerbart, Stuttgart, [39, 124 f]
— (1966): Die Sedimentbildung im Bodensee. — Naturwiss., **53**, 237—247. [141]
— (1967): Diagenesis in argillaceous sediments. — In: G. LARSEN & G. V. CHILINGAR (eds.): Diagenesis in Sediments. — Elsevier, Amsterdam, London, New York, 127—177. [293]
— (1970): See H. FÜCHTBAUER & G. MÜLLER.
MÜLLER, H. (1958): Die Petrographie der Röt-Muschelkalkgrenzschichten bei Studnitz nördlich Jena. — Chemie d. Erde, **19**, 391—435. [327 f]
MÜLLER, W. (1965): Auflösungsgeschwindigkeit von Kristallen in Wasser. — Ph. D. Thesis, Tübingen. [22]
MURATA, K. J., FRIEDMAN, I. & MADSEN, B. M. (1969): Isotopic composition of diagenetic marine carbonates in marine Miocene formations of California and Oregon. — U. S. geol. Surv. prof. Pap., 614—B. [322]
MURRAY, J. & RENARD, A. F. (1891): Deep-sea deposits. — Challenger Exped. Rep., London. [123]
MUSKAT, M. (1946): The flow of homogeneous fluids through porous media. — Edwards, Ann Arbor, 763 pp. [196]

NERNST, W. & BRUNNER, E. (1904): Theorie der Reaktionsgeschwindigkeit in heterogenen Systemen. — Z. phys. Chem., **47**, 52—55. [21]
NIEMANN, H. (1960): Untersuchungen am Grauen Salzton der Grube „Königshall-Hindenburg", Reyershausen bei Göttingen. — Beitr. Miner. Petrogr., **7**, 137—165. [303]
NIGGLI, P. (1920): Lehrbuch der Mineralogie. — Borntraeger, Berlin. [241]
— (1952): Gesteine und Minerallagerstätten. Bd. 2: Exogene Gesteine und Minerallagerstätten. — Birkhäuser, Basel. [114]
NORDIN, C. F., jr. & DEMPSTER, G. R. (1963): Vertical distribution of velocity and suspended-sediment concentration, Middle Rio Grande, New Mexico. — U. S. geol. Surv. prof. Pap., **462—B**, 1—20. [57 f, 69 f]

NOTA, D. J. G. (1958): Sediments of the western Guiana shelf. — Meded. Landbouwhogeschool Wageningen, Nederland, 58 (2), 1—98. [138, 142]

O'BRIEN, N. R. (1964): Origin of Pennsylvanian underclays in the Illinois basin. — Bull. geol. Soc. Amer., 75, 823—832. [291]

O'BRIEN, N. R. & HARRISON, W. (1969): Fabric of a non-fissile pleistocene clay. — Naturwiss., 56, 135—136. [291]

ODOM, I. E. (1967): Clay fabric and its relation to structural properties in mid-continent Pennsylvanian sediments. — J. sediment. Petrol., 37, 610—623. [290 f]

OSEEN, C. W. (1927): Hydrodynamik. — Akademische Verlagsges., Leipzig. [60 f]

PAPP, A. (1954): Fazies und Gliederung des Sarmats im Wiener Becken. — Mitt. Geol. Ges. Wien, 47, 35—97. [227]

PARASNIS, D. S. (1960): The compaction of sediments and its bearing on some geophysical problems. — Geophys. J. roy. astron. Soc., 3, 1—28. [265 f, 304 ff]

PARK, W. C. & SCHOT, E. H. (1968): Stylolites: their nature and origin. — J. sediment. Petrol., 38, 175—191. [313]

PATTERSON, S. H. (1964): Halloysite underclay and amorphous inorganic matter in Hawaii. — Clays and Clay Minerals, Proc. 12th nat. Conf., Pergamon Press, Oxford, 153—172. [45]

PAWLUCK, S. (1963): Characteristics of 14 Å clay minerals in the B horizons of podzolized soils of Alberta. — Clays and Clay Minerals, Proc. 11th nat. Conf., Pergamon Press, Oxford, 74—82. [43]

PELTO, C. R. (1956): A study of chalcedony. — Amer. J. Sci., 254, 32—50. [38]

PERRY, E. A. & HOWER, J. (1970): Burial diagenesis in Gulf Coast, pelitic sediments. — Clays and Clay Minerals, 18, 165—177. [301]

PETROV, V. P. (1958): Genetic types of white clays in the USSR and laws governing their distribution. — Clay Miner. Bull., 3, 287—296. [45]

PETTIJOHN, F. J. (1941): Persistence of heavy minerals and geologic age. — J. Geol., 49, 610—625. [257]

— (1957): Sedimentary rocks. — Harper & Broth., New York, 1. ed. 1949, 2. ed. 1957. [99, 103, 124, 257 ff]

PETTIJOHN, F. J. & LUNDAHL, A. C. (1943): Shape and roundness of Lake Erie beach sand. — J. sediment. Petrol., 13. 69—78. [113]

PHILIPP, W., DRONG, H. J., FÜCHTBAUER, H., HADDENHORST, H. G. & JANKOWSKY, W. (1963): Zur Geschichte der Migration im Gifhorner Trog. — Erdöl u. Kohle, 16, 456—468. [247 f, 262]

PICHLER, H. (1963): Geologische Untersuchungen im Gebiet zwischen Roßfeld und Markt Schellenberg im Berchtesgadener Land. — Beih. geol. Jb., 48, 129—204. [310]

PLAPP, J. E. & MITCHELL, J. (1960): A hydrodynamic theory of turbidity currents. — J. geophys. Res., 65, 983—992. [99]

PLESSMANN, W. (1966): Lösung, Verformung, Transport und Gefüge (Beiträge zur Gesteinsverformung im Rheinischen Schiefergebirge). — Z. dt. geol. Ges., 115, 650—663. [246]

PLUMLEY, P. E. (1948): Black hills terrace gravels, a study in sediment transport. — J. Geol., 56, 526—577. [109 f]

POLLACK, J. M. (1961): Significance of compositional and textural properties of South Canadian River channel sands, New Mexico, Texas and Oklahoma. — J. sediment. Petrol., 31, 15—37. [112]

PORRENGA, D. H. (1967): Clay mineralogy and geochemistry of recent marine sediments in tropical areas. — Acad. Proefschr., Univ. Amsterdam. [122 f, 142 f]

POSER, H. & HÖVERMANN, J. (1952): Beiträge zur morphometrischen und morphologischen Schotteranalyse. — Abh. Braunschweiger Wiss. Ges., 4, 12—36. [111]

POSNJAK, E. (1940): Deposition of calcium sulfate from sea water. — Amer. J. Sci., 238, 559—568. [186 ff]

POTTER, P. E. & MAST, R. F. (1963): Sedimentary structures, sand shape fabrics and permeability. — J. Geol., 71, 441—471; 548—565. [198]

POTTER, P. E. & SCHEIDEGGER, A. E. (1966): Bed thickness and grain size: graded beds. — Sedimentology, 7, 233—240. [144]

POWERS, M. C. (1953): A new roundness scale for sedimentary particles. — J. sediment. Petrol., **23**, 117—119. [103]
— (1954): Clay diagenesis in the Chesapeake Bay area. — Clays and Clay Minerals, Proc. 2nd nat. Conf., Columbia. — Nat. Acad. Sci. — Nat. Res. Council, Publ. 327, Washington, 68—80. [119]
— (1957): Adjustment of land derived clays to the marine environment. — J. sediment. Petrol., **27**, 355—372. [119]
— (1959): Adjustment of clays to chemical change and the concept of the equivalence level. — Clays and Clay Minerals, Proc. 6th nat. Conf. 1957, 309—326. [300 f]
— (1967): Fluid release mechanism in compacting marine mudrocks and their importance in oil exploration. — Bull. amer. Assoc. Petrol. Geol., **51**, 1240—1254. [300 f]
PRANDTL, L. (1965): Führer durch die Strömungslehre. — 6. Aufl., Vieweg, Braunschweig. [53, 56 f]
PRAY, L. C. (1960): Compaction in calcilutites (Abstr.). — Bull. geol. Soc. Amer., **71**, 1946. [306]
PRICE, W. A. (1963): Physicochemical and environmental factors in clay dune genesis. — J. sediment. Petrol., **33**, 766—778. [162]
PUCHELT, H. (1964): Zur Geochemie des Grubenwassers im Ruhrgebiet. — Z. dt. geol. Ges., **116**, 167—203. [229 f]
PURSER, B. H. (1969): Syn-sedimentary marine lithification of Middle-Jurassic limestones in the Paris Basin. — Sedimentology, **12**, 205—230. [310]
PUSTOWALOFF, L. W. (1955): Über sekundäre Veränderungen der Sedimentgesteine. — Geol. Rdsch., **43**, 535—550. [259]

RANKAMA, K. & SAHAMA, TH. G. (1950): Geochemistry. — Univ. of Chicago Press. [16, 117]
RAYLEIGH, Lord (1943): The ultimate shape of pebbles, natural and artificial. — Proc. roy. Soc. London, Ser. A, **181**, 107—118. [104, 107, 110]
— (1944): Pebbles, natural and artificial. Their shape under various conditions of abrasion. — Proc. roy. Soc. London, Ser. A, **182**, 321—335. [104, 107, 110]
REICHE, P. (1950): A survey of weathering processes and products. — Univ. of Mexico Publ. in Geol., No. 3. [12]
REINECK, H. E. (1963): Sedimentgefüge im Bereich der südlichen Nordsee. — Abh. Senckenberg. Naturforsch. Ges., Frankfurt a. M., **505**, 1—138. [138, 141]
— (1967): Layered sediments of tidal flats, beaches and shelf bottoms of the North Sea. — In: Estuaries. Amer. Assoc. Advanc. of Sci., 191—206. [141]
REULING, H. T. (1934): Der Sitz der Dolomitisierung: Versuch einer neuen Aufwertung der Bohrergebnisse von Funafuti. — Abh. Senckenberg. Naturforsch. Ges., **428**, 44 pp. [319]
REX, R. W. & MARTIN, B. D. (1966): Clay mineral formation in sea water by submarine weathering of K-feldspar. — Clays and Clay Minerals, Proc. 14th nat. Conf., Pergamon Press, Oxford, 235—240. [118]
RICH, C. J. & COOK, M. G. (1963): Formation of dioctahedral vermiculite in Virginia soils. — Clays and Clay Minerals, Proc. 10th nat. Conf., Pergamon Press, Oxford, 96—106. [43]
RIEKE, H. H., CHILINGAR, G. V. & ROBERTSON, J. O. (1964): High pressure (up to 500 000 psi) compaction studies on various clays. — 22nd internat. Geol. Congr., New Delhi, Proc. [233]
RILEY, N. A. (1941): Projection sphericity. — J. sediment. Petrol., **11**, 94—97. [102]
RIMSAITE, J. (1957): Über die Eigenschaften der Glimmer in den Sanden und Sandsteinen. — Beitr. Miner. Petrogr., **6**, 1—15. [24]
RITTENHOUSE, G. (1943): The transportation and deposition of heavy minerals. — Bull. geol. Soc. Amer., **54**, 1725—1780. [128, 163]
RÖSLER, H. J. & LANGE, H. (1965): Geochemische Tabellen. — VEB Deutscher Verlag für Grundstoffindustrie, Leipzig. [117]
ROGERS, J. W., KRUEGER, W. C. & KROG, M. (1963): Sizes of naturally abraded materials. — J. sediment. Petrol., **33**, 628—632. [114]
ROLLER, P. S. & ERVIN, G. (1940): Ionization of silicic acid. — J. amer. Chem. Soc., **62**, 468. [37]
ROSENBERG, P. E. & HOLLAND, H. D. (1964): Calcite-Dolomite-Magnesite stability relations in solutions at elevated temperature. — Science, **145**, 700—701. [183]
ROSENQUIST, I. TH. (1962): The influence of physico-chemical factors upon the mechanical properties of clays. — Clays and Clay Minerals, Proc. 9th nat. Conf., Pergamon Press, Oxford, 12—27. [193]

Rouse, H. (1937): Modern conceptions of the mechanics of fluid turbulence. — Trans. amer. Soc. Civ. Eng., **102**, 534. [67]
Roy, R. & Romo, I. (1957): New data on vermiculite. — J. Geol., **65**, 603—610. [43]
Rubey, W. W. (1931): Lithologic studies of fine-grained upper Cretaceous sedimentary rocks of the Black Hill region. — U. S. geol. Surv. prof. Pap., **165**, 1—54. [291]
— (1951): Geologic history of sea water. — Bull. geol. Soc. Amer., **62**, 1111—1147. [123]
Ruchin, L. B. (1958): Grundzüge der Lithologie. Lehre von den Sedimentgesteinen (translated from Russ. by A. Schüller). — Akademie Verlag, Berlin, 806 pp. [271]
Russell, D. R. (1937): Mineral composition of Mississippi river sands. — Bull. geol. Soc. Amer., **48**, 1307—1348. [112]
Russell, R. D. & Taylor, R. E. (1937): Roundness and shape of Mississippi River sands. — J. Geol., **45**, 225—267. [103, 112]
Ryan, W. B. F. & Heezen, B. C. (1965): Ionian sea submarine canyons and the 1908 Messina turbidity current. — Bull. geol. Soc. Amer., **76**, 915—932. [96]

Sáenz, I. M. de (1963): Autigener Sanidin. — Schweiz. Miner. Petrogr. Mitt., **43**, 485—492. [251]
Sander, B. (1948): Einführung in die Gefügekunde der geologischen Körper. Erster Teil. — Springer, Wien, Innsbruck, 215 pp. [305]
— (1950): Einführung in die Gefügekunde der geologischen Körper. Zweiter Teil. — Springer, Wien, Innsbruck, 409 pp. [305, 318]
Scheffer, F. & Schachtschabel, P. (1966): Lehrbuch der Bodenkunde. 6. Aufl. — F. Enke, Stuttgart. [16, 46 f]
Scheidegger, A. E. (1960): The physics of flow through porous media. — Univ. of Toronto Press., Toronto. [193, 195 f, 203]
— (1961): Theoretical Geomorphology. — Springer, Berlin, Göttingen, Heidelberg. [67, 82]
— (1965): Textural studies of graded bedding. Observation and theory. — Sedimentology, **5**, 289—304. [144]
Scheidegger, A. E. & Potter, P. E. (1967): Bed thickness and grain-size: Cross-bedding. — Sedimentology, **8**, 39—44. [133]
Scherp, A. (1963): Die Petrographie der palaeozoischen Sandsteine in der Bohrung Münsterland 1 und ihre Diagenese in Abhängigkeit von der Teufe. — Forschr. Geol. Rheinl.-Westf., **11**, 251—282. [253]
Scherp, A. & Stadler, G. (1968): Die Pyrophyllit-führenden Tonschiefer des Ordoviziums im Ebbesattel und ihre Genese. — N. Jb. Miner. Abh., **108**, 142—165. [303]
Schiller, L. (1932): Fallversuche mit Kugeln und Scheiben. — In: Handbuch der Experimentalphysik, Bd. 4, 2. Teil. Ed. L. Schiller, Leipzig. [59 f, 63 f]
Schlanger, S. O. (1963): Subsurface geology of Eniwetok Atoll. — U. S. geol. Surv. prof. Pap., 260—BB. [311, 315, 319 f]
Schlenker, B. (1971): Petrographische Untersuchungen am Gips-Keuper und Lettenkeuper von Stuttgart. — Ph. D. Thesis, Tübingen. [303]
Schmidt, W. (1925): Der Massenaustausch in freier Luft und verwandte Erscheinungen. — Hamburg. [65]
Schneiderhöhn, P. (1954): Eine vergleichende Studie über Methoden zur quantitativen Bestimmung von Abrundung und Form an Sandkörnern. — Heidelb. Beitr. Miner. Petrogr., **4**, 172—191. [101, 103]
— (1957): Sedimentpetrographische Untersuchung einiger Terrassenablagerungen des oberen Euphrat. — N. Jb. Miner. Abh., **91**, 58—88. [134]
Schoeller, H. (1955): Géochimie des eaux souterraines. Application aux eaux des gisements de pétrole. — Rev. Inst. Franç. Pétrole **10**, 181, 219, 507, 671, 823. [208]
Schöner, H. (1960): Über die Verteilung und Neubildung der nichtkarbonatischen Mineralkomponenten der Oberkreide aus der Umgebung von Hannover. — Beitr. Miner. Petrogr., **7**, 76—103. [327 f]
Schoklitsch, A. (1933): Über die Verkleinerung der Geschiebe in Flußläufen. — Sitzungsber. Akad. Wiss. Wien, Math.-naturwiss. Kl., Sec. II, **142**, 343—366. [104, 106]
Schubert, C. (1964): Size-frequency distribution of sand-sized grains in an abrasion mill. — Sedimentology, **3**, 288—295. [107 f, 114]
Schuleikin, W. (1959): Theorie der Meereswellen (Translated from Russ. by E. Bruns). — Akademie Verlag, Berlin. [82]
Schumann, H. (1941): Zur Korngestalt der Quarze in Sanden. — Chemie d. Erde, **14**, 131—151. [4, 113]

Seibold, E. (1962): Kalkkonkretionen und karbonatisch gebundenes Magnesium. — Geochim Cosmochim. Acta, **26**, 899—909. [296]
— (1964): Das Meer. — In: R. Brinkmann: Lehrbuch der allgemeinen Geologie. Bd. I. — F. Enke, Stuttgart. [92 f, 138]
Seilacher, A. (1968): Origin and diagenesis of the Oriskany Sandstone (Lower Devonian, Appalachians) as reflected in its shell fossils. — In: G. Müller & G. M. Friedman (eds.): Recent Developments in Carbonate Sedimentology in Central Europe. — Springer, Berlin, Heidelberg, New York, 175—185. [246]
Sharma, G. D. (1965): Formation of silica cement and its replacement by carbonates. — J. sediment. Petrol., **35**, 733—745. [250]
Shaub, B. M. (1939): The origin of stylolites. — J. sediment. Petrol., **9**, 47—61. [313]
— (1949): Do stylolites develop before or after the hardening of the enclosing rock? — J. sediment. Petrol., **19**, 26—36. [313]
Shelton, J. W. (1964): Authigenic kaolinite in sandstone. — J. sediment. Petrol., **34**, 102—111. [252]
Shepard, F. P. (1963 a): Submarine canyons. — In: M. N. Hill (ed.): The sea. Vol. 3. — Interscience Publ., New York, London, 480—506. [93]
— (1963 b): Submarine geology. 2nd Ed. — Harper & Row, New York. [82, 93]
Shinn, E. A. (1969): Submarine lithification of holocene carbonate sediments in the Persian Gulf. — Sedimentology, **12**, 109—144. [308]
Shirley, M. L. & Ragsdale, J. A. (eds.) (1966): Deltas. — Houston Geol. Soc., 251 pp. [136]
Siemes, H. (1966): Röntgenographische Bestimmung der Texturen von unverformten und experimentell verformten Solnhofener Kalkstein. — Proc. 1st Congr. intern. Soc. Rock Mech., Lissabon, 205—215. [318]
Siever, R. (1957): The silica budget in the sedimentary cycle. — Amer. Miner., **42**, 821—841. [37]
— (1959): Petrology and geochemistry of silica cementation in some Pennsylvanian sandstones. — In: H. A. Ireland (ed.): Silica in sediments. — Soc. Econ. Paleont. Miner., Spec. Publ. Nr. 7, 55—79. [245, 247]
— (1962): Silica solubility, 0°—200°C, and the diagenesis of siliceous sediments. — J. Geol., **70**, 127—150. [37 f]
Siever, R., Beck, K. C. & Berner, R. A. (1965): Composition of interstitial waters of modern sediments. — J. Geol., **73**, 39—73. [238]
Skempton, A. W. (1944): Notes on the compressibility of clays. — Quart. J. geol. Soc. London, **100**, 119—135. [264, 267]
Slaughter, M. & Milne, I. H. (1960): The formation of chlorite-like structures from montmorillonite. — Clays and Clay Minerals, Proc. 7th nat. Conf., Pergamon Press, Oxford, 114—124. [43]
Sloss, L. L. & Ferray, D. E. (1948): Microstylolites in sandstones. — J. sediment. Petrol., **18**, 3—13. [244]
Sneed, E. D. & Folk, L. (1958): Pebbles in the lower Colorado River, Texas: a study in particle morphogenesis. — J. Geol., **66**, 114—150. [101, 109 f]
Spotts, J. H. & Silverman, S. R. (1966): Organic dolomite from Point Fermin, California. — Amer. Miner., **51**, 1144—1155. [322]
Stadler, G. (1963): Die Petrographie und Diagenese der oberkarbonischen Tonsteine in der Bohrung Münsterland I. — Fortschr. Geol. Rheinl.-Westf., II, 283—290. [302]
Stehli, F. G. (1956): Shell mineralogy in paleozoic invertebrates. — Science, **123**, 1031—1032. [315]
Stehli, F. G. & Hower, J. (1961): Mineralogy and early diagenesis of carbonate sediments. — J. sediment. Petrol., **31**, 358—371. [310]
Steinfink, H. & Gebhart, J. E. (1962): Compression apparatus for powder X-ray diffractometry. — Rev. Sci. Instr., **33**, 542—544. [301]
Stephen, I. (1963): Bauxitic weathering at Mount Zomba, Nyasaland. — Clay Miner. Bull., **5**, 203—208. [51]
Stephen, I. & MacEwan, D. M. C. (1951): Some chloritic clay minerals of unusual type. — Clay Miner. Bull., **1**, 157—162. [43]
Sternberg, H. (1875): Untersuchungen über Längen- und Querprofile geschiebeführender Flüsse. — Z. Bauwes., **25**, 483—506. [105 f, 109]
Stockdale, P. B. (1922): Stylolites: Their nature and origin. — Indiana Univ. Stud., **9**, 55, 1—97. [313]

— (1926): The stratigraphic significance of solution in rocks. — J. Geol., 34, 399—414.
[245, 313 f]
STOKER, J. J. (1957): Water Waves. — Interscience Publ., New York. [82]
STORER, H. (1959): Costipazione dei sedimenti argillosi nel bacino Padano. — In: Giacimenti Gassiferi dell'Europa Occidentale, Vol. II, Acad. Lincei, Roma. [277]
STRAKHOV, N. M. (1956): Typen der Dolomitgesteine und ihre Entstehung (Russ.). — Akad. Nauk. SSSR, Moskau, Trudy Geol. Inst. [319]
— (1958): Tatsachen und Hypothesen zur Frage der Bildung von Dolomitgesteinen (Russ.). — Izvest. Akad. Nauk. SSSR, Ser. geol., 3—21. [319]
SUDO, T. (1954): Clay mineralogical aspects of the alteration of volcanic glass in Japan. — Clay Miner. Bull., 2, 96—106. [45]
SWARTZENDRUBER, D. (1962): Non-Darcy flow behavior in liquid-saturated porous media. — J. geophys. Res., 67, 5205—5213. [196]
SWINDALE, L. D. & FAN, P. F. (1967): Transformation of gibbsite to chlorite in ocean bottom sediments. — Science, 157, 799—800. [122]
SZADECZKY-KARDOSS, E. (1933): Die Bestimmung des Abrollungsgrades. — Zbl. Miner. Abt. B, 389—401. [103]

TAFT, W. H. (1967): Physical chemistry of formation of carbonates. — In: G. V. CHILINGAR, H. J. BISSELL & R. W. FAIRBRIDGE (eds.): Carbonate rocks. — Elsevier, Amsterdam, 151—167. [316]
TAN, F. C. & HUDSON, J. D. (1971): Carbon and oxygen isotopic relationships of dolomites and coexisting calcites, Great Estuarine Series (Jurassic), Scotland. — Geochim. Cosmochim. Acta, 35, 755—767. [322]
TARR, W. A. (1917): Origin of chert in the Burlington limestone. — Amer. J. Sci., Ser. 4, 44, 409—452. [324]
— (1926): The origin of chert and flint. — Univ. Missouri Stud., 1, No. 2. [324]
TAYLOR, J. C. M. & ILLING, L. V. (1969): Holocene intertidal calcium carbonate cementation, Quartar, Persian Gulf. — Sedimentology, 12, 69—107. [308]
TAYLOR, R. M. & NORRISH, K. (1966): The measurement of orientation distribution and its application to quantitative X-ray diffraction analysis. — Clay Miner., 6, 127—142. [288]
TERTSCH, H. (1949): Die Festigkeitseigenschaften der Kristalle. — Springer, Wien. [106 f]
TERZAGHI, K. (1925): Erdbaumechanik auf bodenphysikalischer Grundlage. — Deuticke, Leipzig, Wien, 399 pp. [263]
TERZAGHI, R. D. (1940): Compaction of lime and mud as a cause of secondary structure. — J. sediment. Petrol., 10, 78—90. [304, 306]
THIEL, G. A. (1940): The relative resistance to abrasion of mineral grains of sand size. — J. sediment. Petrol., 10, 102—124. [108]
THIEM, H. J. (1967): Quantitative Texturanalyse durch Röntgenintensitätsmessungen und ihre Anwendung auf experimentell verdichtete Kaolinit- und Montmorillonit-Tone. — Ph. D. Thesis, Univ. Tübingen. [270, 289]
THOMPSON, A. (1959): Pressure solution and porosity. — In: H. A. IRELAND (ed.): Silica in sediments. — Soc. Econ. Paleont. Miner., Spec. Publ. No. 7, Tulsa, Okla., 92—110. [244 f]
TOPKAYA, M. (1950): Recherches sur les silicates authigènes dans les roches sédimentaires. — Bull. Lab. Univ. Lausanne, 97, 1—132. [327]
TOWE, K. M. (1962): Clay mineral diagenesis as a possible source of silica cement in sedimentary rocks. — J. sediment. Petrol., 32, 26—28. [301]
TRURNIT, P. (1967a): Morphologie und Entstehung von Druck-Lösungserscheinungen während der Diagenese. — Ph. D. Thesis, Heidelberg. [313 f]
— (1967b): Morphologie und Entstehung diagenetischer Druck-Lösungserscheinungen. — Geol. Mitt. Aachen, 1, 173—204. [313 f]
TUNN, W. (1940): Untersuchungen über die Verwitterung des Tremolit. — Chemie d. Erde, 12, 275—303. [31]
TUREKIAN, K. K. (1969): The oceans, streams and atmosphere. — In: K. H. WEDEPOHL (ed.): Handbook of Geochemistry, Vol. I. — Springer, Berlin, Heidelberg, New York, 297—323. [176, 184]
TURNER, F. J. & VERHOOGEN, J. (1960): Igneous and metamorphic rocks. 2nd ed. — Mc Graw Hill, New York, Toronto, London. [257]

UDDEN, J. (1914): Mechanical composition of clastic sediments. — Bull. geol. Soc. Amer., 25, 655—744. [125]

Usdowski, H. E. (1962): Die Entstehung der kalkoolithischen Fazies des norddeutschen Unteren Buntsandsteins. — Beitr. Mineral. Petrol., 8, 141—179. [313]
— (1967): Die Genese von Dolomit in Sedimenten. — Springer, Berlin, Heidelberg, New York, 95 pp. [183, 319]

Valeton, I. (1953): Petrographie des süddeutschen Hauptbuntsandsteines. — Heidelb. Beitr. Miner. Petrogr., 3, 335—379. [259]
— (1955 a): Beziehungen zwischen petrographischer Beschaffenheit, Gestalt und Rundungsgrad einiger Flußgerölle. — Petermanns geogr. Mitt., 14—17. [110f]
— (1955 b): Veränderungen an Zirkon und Turmalin in Buntsandstein und Keuper. — Heidelb. Beitr. Miner. Petrogr., 5, 100—104. [259]
— (1958): Der Glaukonit und seine Begleitminerale aus dem Tertiär von Walsrode. — Mitt. Geol. Staatsinst. Hamburg, 27, 88—131. [120]
van Andel, Tj. H. (1952): Zur Frage der Schwermineralverwitterung in Sedimenten. — Erdöl u. Kohle, 5, 100—104. [24]
— (1959): Reflections on the interpretation of heavy mineral analyses. — J. sediment. Petrol., 29, 153—163. [24]
van Andel, Tj. H. & Postma, H. (1954): Recent sediments in Gulf of Paria. — North Holland Publ. Co., 143 pp. [123, 143]
van Andel, Tj. H., Wiggers, A. J. & Maarleveld, G. (1954): Roundness and shape of marine gravels from Urk (Netherlands), a comparison of several methods of investigation. — J. sediment. Petrol., 24, 100—116. [110f]
van Moort, J. C. (1971): A comparative study of the diagenetic alteration of clay minerals in mesozoic shales from Papua, New Guinea, and in Tertiary shales from Louisiana, U.S.A. — Clays and Clay Miner., 19, 1—20. [300]
van Olphen, H. (1963): Compaction of clay sediments in the range of molecular particle distances. — Clays and Clay Miner., 11, 178—187. [12, 265, 300]
— (1964): Internal mutual flocculation in clay suspensions. — J. Colloid Sci., 19, 313—322. [265]
Veniale, F. & Van der Marel, H. W. (1963): An interstratified saponite-swelling chlorite mineral as a weathering product of lizardite rock from St. Margherita (Pavia Province), Italy. — Beitr. Miner. Petrogr., 9, 198—245. [43]
Vielstich, W. (1953): Der Zusammenhang zwischen Nernst'scher Diffusionsschicht und Prandtl'scher Strömungsgrenzschicht. — Z. Elektrochem., 57, 646—655. [21]
Vinogradov, A. P. & Ronov, A. B. (1957): Evolution of chemical composition of clays of the Russian platform (abstr. by G. V. Chilingar). — Geochim. Cosmochim. Acta, 12, 172—175. [300]

Wadell, H. (1932): Volume, shape and roundness of rock particles. — J. Geol., 40, 443—451. [101 ff]
— (1935): Volume, shape and roundness of quartz particles. — J. Geol., 43, 250—280. [108]
Wagner, G. (1913): Stylolithen und Drucksuturen. — Geol. Paläont. Abh. N. F., 11 (15), 1—30. [313]
Waldschmidt, W. A. (1941): Cementing materials in sandstones and their probable influence on migration and accumulation of oil and gas. — Bull. amer. Assoc. Petrol. Geol., 25, 1839—1879. [247, 254]
Walger, E. (1961): Die Korngrößenverteilung von Einzellagen sandiger Sedimente und ihre genetische Bedeutung. — Geol. Rdsch., 51, 494—507. [133]
— (1966): Untersuchungen zum Vorgang der Transportsonderung von Mineralen am Beispiel von Strandsanden der westlichen Ostsee. — Meyniana, 16, 55—106. [128, 163]
Walker, G. F. (1950): Trioctahedral micas in the soil clays of north-east Scotland. — Miner. Mag., 29, 72—84. [42]
— (1957): On the differentiation of vermiculites and smectites in clays. — Clay Miner. Bull., 3, 154—163. [42]
Walker, T. R. (1957): Frosting of quartz grains by silica replacement. — Bull. geol. Soc. Amer., 68, 267—268. [250]
— (1960): Carbonate replacement of detrital crystalline silicate minerals as a source of authigenic silica in sedimentary rocks. — Bull. geol. Soc. Amer., 71, 145—152. [250]
— (1962): Reversible nature of chert-carbonate replacement in sedimentary rocks. — Bull. geol. Soc. Amer., 73, 237—242. [250 f, 327]

WALLHÄUSER, K. H. & PUCHELT, H. (1966): Sulfatreduzierende Bakterien in Schwefel- und Grubenwässern Deutschlands und Österreichs. — Contr. Miner. Petrol., 13, 12—30. [234]

WALTON, H. F. (1949): Ion exchange equilibria. — In: F. C. NACHOD (ed.): Ion exchange. — Academic Press, New York, 3—28. [237]

WATTENBERG, H. (1933): Kalziumkarbonat- und Kohlensäuregehalt des Meerwassers. — Wiss. Ergebn. dt. Atlant. Exped. Meteor 1925—1928, Bd. VIII, 1. Teil, Berlin. [182]

— (1936): Kohlensäure und Kalziumkarbonat im Meere. — Fortschr. Miner., 20, 168—195. [178]

— (1937): Die Bedeutung anorganischer Faktoren bei der Ablagerung von Kalziumkarbonat im Meere. — Geol. d. Meere u. Binnengew., 1, 237—259. [178, 181 f]

WATTENBERG, H. & TIMMERMANN, E. (1936): Über die Sättigung des Seewassers an $CaCO_3$ und die anorganogene Bildung von Kalksedimenten. — Ann. Hydrogr., 64, 23—31. [178]

WATTS, E. V. (1948): Some aspects of high pressure in the D 7 zone of the Ventura Avenue field. — Trans. Amer. Inst. Miner. Met. Engrs., 174, 191—205. [283]

WEAVER, CH. E. (1958): A discussion on the origin of clay minerals in sedimentary rocks. — Clays and Clay Minerals, Proc. 5th nat. Conf., Nat. Res. Council Publ. No. 566, 159—173. [117]

— (1959): The clay petrology of sediments. — Clays and Clay Minerals, Proc. 6th nat. Conf. 1957, 154—187. [300, 302]

— (1960): Possible use of clay minerals in search for oil. — Bull. amer. Assoc. Petrol. Geol., 44, 1505—1518. [301]

WEDEPOHL, K. H. (1960): Spurenanalytische Untersuchungen an Tiefseetonen aus dem Atlantik. — Geochim. Cosmochim. Acta, 18, 200—231. [123]

— (1967): Geochemie. — Samml. Göschen, Walter de Gruyter, Berlin. [117]

WEEKS, L. G. (1953): Environment and mode of origin and facies relationships of carbonate concretions in shales. — J. sediment. Petrol., 23, 162—173. [296]

WELLER, J. M. (1959): Compaction of sediments. — Bull. amer. Assoc. Petrol. Geol., 43, 273—310. [282, 306]

WENTWORTH, C. K. (1969): A laboratory and field study of cobble abrasion. — J. Geol., 27, 507—522. [104]

— (1922): A method of measuring and plotting the shapes of pebbles. — U. S. geol. Bull., 730, 91—102. [102 f]

WERNER, E. (1961): Zu Verkittungsvorgängen an Psammiten. — Geol. Rdsch., 51, 507—517. [249]

WETZEL, O. (1933): Die in organischer Substanz erhaltenen Mikrofossilien des baltischen Kreidefeuersteins. — Palaeontographica, 77, 141—186; 78, 1—110. [324]

WETZEL, W. (1923): Sedimentpetrographie. — Fortschr. Miner., Kristallogr., Petrogr., 8, 101—198. [114]

— (1957): Selektive Verkieselung. — N. Jb. Geol. Paläont. Abh., 105, 1—10. [324]

WEY, R. & SIFFERT, B. (1961): Réactions de la silice monomoléculaire en solution avec les ions Al^{3+} et Mg^{2+}. — Coll. Inst. C.N.R.S. Genèse et synthèse des argiles, Paris, 11—23. [44 f]

WEYL, P. K. (1959): Pressure solution and the force of crystallisation — a phenomenological theory. — J. geophys. Res., 64, 2001—2025. [244 f]

WEYL, R. (1950): Schwermineralverwitterung und ihr Einfluß auf die Mineralführung klastischer Sedimente. — Erdöl u. Kohle, 3, 209—211. [24]

— (1952): Zur Frage der Schwermineralverwitterung in Sedimenten. — Erdöl u. Kohle, 5, 29—33. [24]

WHITE, D. E. (1957): Magmatic, connate and metamorphic waters. — Bull. geol. Soc. Amer., 68, 1659—1682. [208]

— (1965): Saline waters of sedimentary rocks. — In: A. YOUNG & J. E. GALLEY (eds.): Fluids in subsurface environments — a symposium. — Mem. amer. Assoc. Petrol. Geol., 4, 342—366. [208]

WHITE, W. A. (1961): Colloid phenomena in sedimentation of argillaceous rocks. — J. sediment. Petrol., 31, 560—570. [291]

WHITEHOUSE, U. G., JEFEREY, L. M. & DEBBRECHT, J. D. (1960): Differential setting tendencies of clay minerals in saline waters. — Clays and Clay Minerals, Proc. 7th nat. Conf., Pergamon Press, Oxford, Paris, 1—79. [143]

WICKMAN, F. E. (1954): The total amount of sediments and the composition of the average igneous rock. — Geochim. Cosmochim. Acta, 5, 97—110. [2]

WIESENEDER, H. (1952): Die Verteilung der Schwermineralien im nördlichen Inneralpinen Wiener Becken und ihre geologische Deutung. — Verh. Geol. Bundesanst. Wien, 207—222. [258]

WIESENEDER, H. & MAURER, I. (1958): Ursachen der räumlichen und zeitlichen Änderung des Mineralbestandes der Sedimente des Wiener Beckens. — Eclog. geol. Helv., **51**, 1155—1172. [258]

WILHELMY, H. (1958): Klimamorphologie der Massengesteine. — Georg Westermann, Braunschweig. [9 f, 12]

WILLIAMSON, E. D. (1917): The effect of strain on heterogeneous equilibrium. — Phys. Rev., **10**, 275—283. [242]

WILSON, M. J. (1966): The weathering of biotite in some Aberdeenshire soils. — Miner. Mag., **35**, 1080—1093. [43, 325, 327]

WINKLER, H. G. F. (1967): Die Genese der metamorphen Gesteine. 2. Aufl. — Springer, Berlin, Heidelberg, New York. [257, 303]

WOLF, K. L. (1957): Physik und Chemie der Grenzflächen, Bd. I. — Springer, Berlin, Göttingen, Heidelberg. [11]

WOLFE, M. J. (1968): Lithification of a carbonate mud: senonian chalk in Northern Ireland. — Sediment. Geol., **2**, 263—280. [307, 313]

WYCKOFF, R. D., BOTSET, M., MUSKAT, M. & REED, D. W. (1934): Measurement of permeability of porous media. — Bull. amer. Assoc. Petrol. Geol., **18**, 161—190. [194]

WYLLIE, M. R. (1948): Some electrochemical properties of shales. — Science, **108**, 684—685. [233]

— (1955): Role of clay in well log interpretation. — First Nat. Conf. Clay Technol. 1952. — Calif. Div. Min. Bull., **169**, 282—305. [233]

YATES, D. J. C. (1954): The expansion of porous glass on the adsorption of non-polar gases. — Proc. roy. Soc. London, Ser. A, **224**, 526—544. [11]

ZANKL, H. (1969): Structural and textural evidence of early lithification in fine-grained carbonate rocks. — Sedimentology, **12**, 241—256. [307]

ZIMMERLE, W. (1963): Zur Petrographie und Diagenese des Dogger-beta-Hauptsandsteines im Erdölfeld Plön-Ost. — Erdöl u. Kohle, **16**, 9—16. [250]

ZINGG, TH. (1935): Beiträge zur Schotteranalyse. — Schweiz. Miner. Petr. Mitt., **15**, 39—140. [100 f, 109, 113]

ZOBELL, C. E. (1958): Ecology of sulfate reducing bacteria. — Producers Month., **22**, 12—29. [234]

— (1963): Organic geochemistry of sulfur. — In: I. A. BREGER (ed.): Organic Geochemistry, Ch. 13. — Internat. Ser. Monogr. on Earth Sci., Pergamon Press, Oxford, London, New York, Paris, **16**, 543—578. [234]

ZOEBELEIN, H. K. (1940): Geologische und sedimentpetrographische Untersuchungen im niederbayerischen Tertiär. — N. Jb. Geol. Paläont. Abh., Beil. Bd. 84, Abt. B, 233—302. [24]

ZÜLLIG, TH. (1956): Sedimente als Ausdruck des Zustandes eines Gewässers. — Schweiz. Z. Hydrol., **18**, 5—143. [271]

Subject Index

A-horizon 23, 46, 48 f
Abrasion, pebbles 103 ff, 106 ff, 109 f
—, sand grains 111 ff
Abrasion resistance 106 f
Activity coefficients 168
—, Ions of sea water 177, 186
Adsorption of water 11
Aeolian dust 161
Aeolian sediments, distinguished from aquatic 161 ff
Aggradation 42, 117, 119
Air, specific gravity and viscosity 146
Albite, dissolution in water 28
—, formation in carbonate rocks 327 f
Alkali feldspar, formation in carbonate rocks 347 f
—, formation in sandstones 251
—, formation in shales 301
Allophane 45
Almandine 6
Aluminium, oxides and hydrated oxides 33 f
Amphibole, see hornblende
Analcite, formation in sandstones 254 ff
—, formation on the sea bottom 124
Anatase 6
—, formation in sandstones 259
Andalusite 6
—, weathering 24
Andradite 6
Angle of repose 94, 160
Anhydrite, equilibrium with gypsum 185 f
—, formation from sea water 187 f
—, formation in sandstones 254
—, solubility 186 f
Anion barriers 232
Antigorite 41
Apatite 6
—, dissolution in sandstones 258
—, formation in sandstones 259
—, weathering 24
Aquatic sediments, distinction from aeolian 161 ff
Aragonite, dissolution in carbonate sediments 317, 319 f
—, dissolution in clays 296
—, formation from seawater 178, 308 f
—, occurrence 311 f
—, solubility 178
—, transformation to calcite 311 f

Arfvedsonite 6
Atmosphere, composition 16
Atoll islands 315 f, 319 f, 321 f
Augite, dissolution in water 26

B-horizon 48 f
Bacteria 19
—, sulfate reducing 234, 299, 309, 323
Barchanes 161
Barite, formation in sandstones 254
—, formation on the sea bottom 123
—, in concretions 298
Base exchange 232, 265
Bauxite 50 f
Beachrocks 310
Bed load 70 ff, 76 f, 149 f
Bentonite 44, 95, 233
Berthierine 41
Biotite 42
—, decomposition in water 24 f, 31
—, formation in sandstones 253
—, weathering 42
Black earth soils 49
Boehmite, in soils 50
—, stability 33 f
—, submarine weathering 122
Bottom transport, in air 149 ff, 156 f
—, in water 70 ff, 76 f
Breakers 89, 141
Brookite 6
—, formation in sandstones 259
Brown earth soils 49

C-horizon 23
$CaCO_3$, equilibrium constants 168, 170
—, solubility 181 ff
Calcarenites, compaction 304 f
—, porosity 304
Calcareous concretions 94 f
Calcilutites, compaction 304 f
—, porosity 304
Calcite
—, C^{13}-content 311
—, concretions 295 ff
—, dissolution in sandstones 249
—, equilibrium with dolomite 185, 323
—, formation from sea water 178
—, formation in sandstones 249
—, O^{18}-content 311

—, pressure solution 242, 244, 313
—, recrystallization 317 f
—, replacement by quartz 251
—, replacement of quartz and feldspar 250 f
—, saturation of sea water 179 f
—, solubility 174 f, 176 f
Canyons, submarine 93
Capillary condensation 12 f
Capillary pressure 205 f
Capillary pressure curve 205
Carbonate rocks
—, abundance 1
—, textures 318
Carbonates
—, concretions 295
—, dissolution in sandstones 249 f
—, formation in sandstones 249 f
—, pressure solution 246, 313
—, replacement by quartz 251
—, replacement of quartz, feldspar 250 f
Carbonate sediments
—, allochemical diagenesis 318 f
—, compaction 304 ff
—, isochemical diagenesis 306 f
—, lithification 308
—, mechanical diagenesis 304 f
—, Sr-content 317
—, U-content 317
Chabasite, formation on the sea bottom 254
Chalcedony 38, 325 f
—, replacement by carbonates 251
Chamosite, formation in sandstones 253
—, formation on the sea bottom 123
Chert
—, concretions 325 f
—, early diagenetic formation 325
—, late diagenetic formation 325
—, pressure solution 244 f
Chlorite 40
—, formation in clays 299, 302
—, formation in sandstones 253
—, formation on the sea bottom 122
—, weathering in soils 43
Chrysotile 41
Clay ironstones 298
Clay minerals, formation by weathering 44 ff
—, formation on sea bottom 120 f
—, in sandstones 252 ff
—, in shales 299 ff
—, in soils 48 ff
—, systematics (classification) 39 ff
Clay pebbles 135
Clays, abundance 2
—, chemical diagenesis 293 ff
—, compaction 262 ff, 292
—, density 272
—, mechanical diagenesis 262 ff
—, permeability 268 f
—, porosity 263 ff, 271 ff
—, texture 267 ff, 288 ff

Cleavage planes 11
Climate 48 f
—, arid 12, 16, 47, 51
—, cool 14, 51
—, humid 14, 16, 47, 51
—, tropical 50 f
Clinoptilolite, formation in sandstones 256
Coastal transport 92
Coccolithophorids 182, 309, 316
Compaction
—, carbonate sediments 304 ff
—, clays 262 ff
—, sands 239 ff
Compaction constants 276 f
Compaction equilibrium 280 f
Compaction flow 259 f, 287 f, 293, 316, 322, 327
Compression index 363 ff
Concretions 294
Connate water 209
Contact angle 204
Continental slope 96
Cordierite 6
Corrensite 303
Cronstedite 41
Cross bedding 133, 160
Cumulative curves 126, 135
Current ripples 71, 75
Czernozems 49

Darcy equation 193 f
Decomposition of rocks and minerals 9 ff
Dedolomitization 323 f
Deformation of fossils 292, 306, 326
Degradation 42, 118, 120
Deltas 136 f
Deposition, deltas 136 f
—, ice 164 f
—, lakes and seas 141 ff, 144 f
—, sea coasts 137 ff, 140 f
—, streams 129 ff, 132 ff
—, wind 158 ff
Desert varnish 51
Desorption of water 12 ff
Diagenesis 8, 118, 192
Diaspore 33, 50 f, 122
Diatoms 325
Dickite 40
Diffusion coefficient 243, 294
Dioctahedral minerals 39 ff
Diopside 7
—, dissolution in water 26
Dissolution, in open systems 25 ff
—, incongruent 25
Dissolution rate 21 f
Dissolution shapes 23 f
Distributaries 136
Dolomite 182 ff
—, C^{13}-content 322
—, Ca:Mg ratio 323

—, concretions 294
—, diagenetic dissolution 324
—, dissolution in sandstones 246
—, early diagenetic formation 319 ff
—, equilibrium with calcite 184 f, 322 f
—, formation from sea water 185
—, formation in sandstones 249 f
—, late diagenetic formation 323 f
—, O^{18}-content 324
—, pressure solution 246
—, replacement by quartz and feldspar 250
—, replacement of quartz 249 f
—, solubility 184
Dolomite rocks 318 ff
—, porosity 322
Dolomitization 318 ff
Dunes 158 ff
—, barchan 160 f
—, clay 162
—, longitudinal 158, 160
—, transverse 158, 160

Earthquakes 96 f
Ebb tide 141
Electric double layer 142, 192 f, 232, 265
Electrochemical potential 232
Eluvial horizon 48
Enstatite, dissolution in water 26
Epidote 6
—, dissolution in sandstones 258
—, weathering 24
Erionite, formation on the sea bottom 254
Evaporation 17 f, 47
Evaporite sediments 167 f
Exchange capacity 118 f, 143, 232
Exchange coefficient 56, 65, 67
Extrusive rocks 1 ff

Feldspar 3, 5
—, dissolution in clays 299
—, dissolution in sandstones 250, 252
—, dissolution in water 26 ff, 28 f
—, formation in carbonate rocks 328
—, formation in sandstones 251 f
—, formation in shales 303
—, replacement by carbonate 251
—, weathering 25
Fire clay mineral 41
Flagellates 325
Flattening 100
Flint, see chert
Flocculation 143, 192
Flood sediments 134 f
Flow, fluids in pore space 193 ff
—, gases in pore space 195
—, in argillaceous sandstones 196
—, laminar 53
—, limits 95
—, mixed phases in pore space 204 ff

—, plastic 94
—, turbulent 55 f
Foraminifera 182, 310, 316
Formation waters 208 ff
—, anion content 214 f, 217, 236
—, Atlantic coast 225 f
—, Ca-content 215 f, 222
—, cation content 214 ff
—, composition 210 ff
—, D-content 223 f
—, deep wells, USA 219
—, diagenesis 231 ff
—, dissolved content 219 ff
—, flow 224 ff
—, Illinois basin 217 ff
—, increase in concentration with depth 217 ff, 232 ff
—, Mg-content 214 ff, 236, 316, 323
—, Molasse basin 226 f
—, ^{18}O-content 223 f, 234
—, Ruhr region 229 f
—, salt domes 231
—, sulfate content 234 f
—, Vienna basin 227 f
—, west Canadian basin 210 ff
Form factor 101 ff
Four-layer minerals 40 ff
Fracture surfaces 10
Friction coefficient 95, 240 f
Frost weathering 14

Garnet 6
—, dissolution in sandstones 258
—, weathering 23 f
Gaussian distribution 126 f
Gibbsite 33 f, 45, 122
—, formation in weathering 45
—, in soils 50
Glacial milk 165
Glaciers 164 ff
Glass, weathering 33
Glauconite 41
—, formation on sea bottom 120 f
Glaucophane 6
Globigerina 309 f
Goethite 34
—, formation on sea bottom 122
—, in soils 52
Graded bedding 144
Grain boundaries 10
Grain size distribution 124 ff
Grain size, primary minerals 3
Gravels 139
Grossularite 6
Groundwater 8, 17, 48
Growth pressure 15, 299
Guide minerals 4
Gypsum 185 ff
—, equilibrium with anhydrite 187 f
—, formation from sea water 166 f, 188

—, formation in clays 299
—, solubility 187 f

HAGEN-POISEUILLE equation 199
Halite 22
Halloysite 39
—, decomposition in sea water 118
—, formation by weathering 45
Hardgrounds 310 f
Harmotome, formation on sea bottom 124
Heavy mineral placers 140
Heavy minerals 6
—, dissolution in sandstones 257 f
—, formation in sandstones 259
—, grain size 163
—, weathering 23 f
Hematite 34 ff
—, in soils 50
Hemi-hydrate 186
Heteropermeable rocks 196 f
Heulandite, formation in sandstones 255 f
Hornblende 6
—, dissolution in sandstones 258
—, dissolution in water 26
—, weathering 24
Humus 47 f
Hydraulic radius 55
Hydromorphic soils 50

Igneous rocks 1 ff
Illite 39, 41, 120
—, aggradation in claystones 300
—, aggradation in sandstones 253
—, aggradation in the sea 119 f
—, crystallinity 301
—, decomposition in sea water 120
—, flocculation 143
—, formation by weathering 41 f
—, in soils 49
Illite clay compaction 268
Illuvial horizon 48
Ilmenite 6
Immiscible phase boundaries 207 f
Insolation 12
Intergradation structures 42
Internal surface 200
Intrastratal solution 257
Intrusive rocks 1 ff
Ion filter 232 f
Ionic strength 168 f
Ion mobility 232 f
Iron, oxides and hydrated oxides 34
Iron sulfides 36
—, formation in sandstones 259
—, formation in clays 298 f
Isopermeable rocks 196

K-feldspar
—, dissolution in water 27 ff
—, formation in carbonate rocks 327 f

—, formation in sandstones 251 f
—, weathering 25
K-salts 167
Kaolinite 39, 40
—, decomposition in sea water 122
—, flocculation 143, 192 f
—, formation by weathering 45
—, formation in clays 299
—, formation in sandstones 252 f
—, in soils 50
—, instability at depth 252 f
Kaolinite clay compaction 265
Knudsen flow 195
KOZENY-CARMAN equation 200
Kurtosis 126 f
Kyanite 6
—, dissolution in sandstones 258
—, weathering 24

Labradorite, dissolution in water 26
Lagoons 139
Lakes, deposition 141 f
Landslides, subaquatic 93
—, submarine 97
Laterites 50
Laumontite, formation in sandstones 257
Lepidocrocite 34
Leucite, dissolution in water 30 f
Levees 136
Lichens 19
Limestones 167 ff
—, deep sea 184
—, formation from sea water 178, 181 f
—, saturation in sea water 181
—, shallow sea 192
Lithographic limestone 318
Load pressure 282 f
Loess 49, 161, 165
Log normal distribution 125 f

Magnetite 6, 34 ff
Manganese nodules 123 f
Marcasite
—, concretions in clays 294, 299
—, formation in sandstones 259
Marsh sediments 141
Median 125 ff
Metahalloysite 39
Metamorphic rocks 1 ff
Metamorphism 188
Mg-calcite 178 f
—, formation from sea water 319 f
—, occurrences 315 f
—, transformation to calcite 315 ff
Mg-salts 167 f
Mica (see also biotite, illite, muscovite) 41 f
—, decomposition in water 26, 31
—, deficient 41
—, formation by weathering 42
—, formation in the sea 121

—, weathering 44
Micritization 317
Microcline, dissolution in water 25 ff
Mixed-layer minerals 42 f
Mixing length 56
Modal values 125
Monazite 7
Montmorillonite 40 f
—, alteration in the sea 118 f
—, decomposition in sea water 118
—, flocculation 142 f, 192 f
—, formation by weathering 42 ff
—, formation in the sea 117 f
—, in soils 48 f, 51
—, ion filtration 232 f
—, transformation in clays 300 f
Montmorillonite clay, compaction 264 ff, 268 ff
Moraines 164 f
—, ground 165
—, superficial 164
Muscovite
—, decomposition in water 31
—, degradation in sandstones 253
—, formation in claystones 300
—, formation in sandstones 253 f
—, weathering 24

Nacrite 40
Natrolite, formation on the sea bottom 254 f
Nepheline, dissolution in water 26
NEWTON formula 60 f
Newtonian flow 94
Nontronite 123
Normal distribution 126 f, 128
N/S-Quotient 47

Ocean, currents 81 f
—, deposits 141 fff
—, transport processes 81 ff
Offretite, formation in sandstones 254
Olivine 7
—, dissolution in water 26, 31
—, weathering 24
Orbital paths 82, 86 f
Organic matter in sediments 321
—, decomposition 297, 309, 321 f, 327
Orthite 7
Oscillation flow 90
Oscillation ripples 90
OSEEN formula 60 f
Overburden pressure 282 f
Oxidation potential 19 f, 34 ff

Parabrown earth soils 49
Pebble shapes 100 ff
Pebble size 101 f
Pelagic sediments 123 f
Percentiles 125 f
Permeability 195 f
—, anisotropism 198

—, clays 203
—, limestones 202 f
—, relative 204
—, sands 201 f
—, sandstones 202 f
—, tensors 198 f
pH 20
Phillipsite, formation on sea bottom 124
phi-scale 125 f
Plant roots 15
Plastic sediments 96
Podsols 24, 48
Pore solutions, see formation waters
—, recent sediments 238
Pore space, relative 190
Porosity 190 ff
—, carbonate rocks 304 f
—, carbonate sediments 305 f
—, clays 192
—, dolomite rocks 323
—, recent sediments 191
—, sands 191 f
—, sandstones 202, 240
Precipitation 17 f
Precipitation sequence from sea water 167 f
Pressure solution 242 ff, 244 f
—, carbonate 246
—, feldspar 246
—, quartz 245 f
Pressure solution surface 313 f
Pseudomorphism 241, 252
—, alteration 252
—, replacement 241
Pteropods 182, 309
Pyrite concretions 298
—, formation in sandstones 259
—, pressure solution 245
Pyroclastic sediments 145
Pyrope 6
Pyrophyllite 39
—, formation in shales 303
Pyroxene 7
Pyrrhotite 7

Quartiles 125 f
Quartz 4 f
—, formation in sandstones 244
—, formation on sea bottom 124
—, pressure solution 244
—, replacement by carbonates 250 f
—, replacement of carbonates 251 f
—, solubility 28, 37 f
Quarzine 326

Radiolaria 325
Radiolarites 121
Rainwater, composition 19
Recalcitization 323 f
Red loams 51
Refraction of flow lines 197 f

Rendzina soils 49
Replacement of immiscible phases 206 ff
Residual dissolution layer 29
Resistance coefficient 58 ff
REYNOLDS number 55
Riebeckite 6
RIECKE principle 240, 313
Rip currents 8, 92
Roundness 102 f
—, pebbles 110 ff
—, sand grains 112, 164
Roundness index 102 f
Rutile 7, 24

Salt expansion 14 f
Sand grains, abrasion 141
—, roundness 113 f
—, shape 114
Sands, aeolian and aquatic 161 f
—, beach 138
—, chemical diagenesis 241 ff
—, compaction 240
—, dune 156, 161 f
—, internal surface 201 f
—, mechanical diagenesis 239 ff
—, permeability 202 f
—, porosity 192, 239
—, river 132 ff
Sandstones, abundance 1
—, chemical diagenesis 241 ff
—, mechanical diagenesis 239 ff
—, permeability 202 f
—, porosity 202, 240
Sand storms 146, 158, 160
Saponite 40 f
—, formation by weathering 44
Sea coasts, deposits 137 f
Sea water, composition 117, 167, 176, 214
—, precipitation sequence 167 f
Sediments, chemical 166 ff
—, clastic 52 ff
Sediment volume 267 ff
Selective transport 112
Semi-permeable membranes 232
Septachlorites 41
Septarian concretions 298
Serpentine 39, 41
Settling velocity, in air 147
—, in water 59, 62
Shales, lateral compaction 292
—, porosity and depth 271 ff
—, texture 290 f
Shape of sediment particles 99 ff
Shear 305
Shear stress 54
—, critical, in air 149 f
—, critical, in water 72, 74
—, maximum 54
Shear velocity 56
Shelf sediments 141

Siderite 36
—, concretions 298
Siliceous sponges 325
Silicification 326 f
Sillimanite 7
—, weathering 24
SiO_2, amorphous, dissolution in limestone 325
—, dissolution in sandstones 248
—, precipitation from sea water 121 f
—, solubility 37 f
Skewness 126
Smectite minerals 41
—, formation by weathering 44
Soil atmosphere 16
Soils 46 ff, 49 ff
Soil solutions 19 f
Sorting 127
Specific surface 201
Spessartite 6
Sphericity 104
Spinel 7
Spodumene 7
Spreading pressure 11
Staurolite 7
—, dissolution in sandstones 258
—, weathering 24
STERNBERG law 105, 109
STOKES formula 59 ff
Strand line 92
Streambed sediments 132 f
Streams
—, bed load 70 ff
—, deposition 132 ff
—, exchange 65, 67
—, gradient 129, 133
—, mean velocity 55, 79, 129
—, model 54
—, sediments 132 ff
—, subfluvial weathering 115 f
—, suspension load 79 f, 132
—, transport capacity 77 ff, 131
—, transport power 77, 129 f
—, velocity distribution 57 f
Stream waters, amounts 17
—, composition 20, 117 f
Stylolites 245, 313 f
Subaerial weathering 8 ff
—, subaquatic 114 ff
—, subfluvial 114 ff
—, sublacustrian 114
—, submarine 114, 116 f
Sudoite 41
—, formation by weathering 43
—, formation in claystones 302
—, formation in sandstones 253
—, formation in the ocean 122
—, in soils 49
Sulfate reduction 234 f, 299, 309, 322
Surface energy 10 f
Surface tension 204 ff

Suspension currents 98 ff, 144
Suspension load, in air 146 ff, 156
—, in water 58 ff, 76, 132
Suspension transport, in air 146 ff
—, in water 58 ff
Swelling pressure 12, 15

Talc 39
Terra rossa 51
Texture analysis 288 ff
Thixotropy 94
Three layer minerals 39 ff
Tides 92, 141
Titanite (sphene) 7
—, pressure solution 246
—, weathering 24
Topaz 7
Tosudite 253
Tourmaline 7
—, formation in carbonate rocks 328
—, formation in sandstones 295
—, pressure solution 246
—, weathering 24
Transformation of layer silicates 41 ff, 118 ff, 121 f, 253 f, 299 ff
Transport capacity, of streams 77 ff, 132
—, of winds 155 ff
Transport power 75 f
Transport processes, in air 145 ff
—, in water 76
Trapped water 205
Tremolite 6
—, dissolution in water 31
Trioctahedral minerals 39 ff
Turbidites 144
Turbulence 55 ff, 146 ff
Two-layer minerals 39 ff

Undertow 89

Varves 142
Vermiculite 40 f
—, formation by weathering 43
—, formation on sea bottom 122
—, in soils 49
Vesuvianite 8
Viscosity
—, of air 146
—, of water 54

—, turbulent 56
Volcanic rocks 1 ff

Wairakite
—, authigenesis in sandstones 255 f
—, formation by weathering 49 ff
—, formation in sandstones 253 f
—, in soils 49 ff
—, on sea bottom 121 ff
—, transformation to micas in clays 300 ff
Wave action 82 ff
Wave amplitudes 86 ff
Wave flow 91
Wave length 84
Wave period 85 ff
Wave reflection 91
Wave refraction 91 f
Wave transport 82 ff
—, mechanical effects 99
—, rolling 70 f, 130 f
—, saltation 74 ff, 128 f, 151 ff
Wave velocity 83, 85 ff
Weathering resistance 22 ff
Wetting 204 ff
Whewellite 298
Wind
—, bed load 149 ff
—, deposition 158 ff
—, mechanical effects 157
—, suspended load 146 ff
—, transport capacity 155 ff
—, velocity distribution 150 ff, 153 f
Windkanter 157
Wind ripples 155
Wind velocities 146 ff, 150 ff
—, critical 150
Wollastonite 8

Xenotime 8

Zeolites, formation in sandstones 254 ff
—, formation in shales 303
—, formation on sea bottom 124, 254 ff
Zircon 8
—, formation in sandstones 259
—, pressure solution 246
—, weathering 24
Zoisite 6